Green Energy and Technology

Climate change, environmental impact and the limited natural resources urge scientific research and novel technical solutions. The monograph series Green Energy and Technology serves as a publishing platform for scientific and technological approaches to “green”—i.e. environmentally friendly and sustainable—technologies. While a focus lies on energy and power supply, it also covers "green" solutions in industrial engineering and engineering design. Green Energy and Technology addresses researchers, advanced students, technical consultants as well as decision makers in industries and politics. Hence, the level of presentation spans from instructional to highly technical.

Indexed in Scopus.

Indexed in Ei Compendex.

Lin Chen
Editor

Clean Energy Research

Now and in the Future

Springer

Editor
Lin Chen
Institute of Engineering Thermophysics Chinese Academy of Sciences & University of Chinese Academy of Sciences, China
Beijing, China

ISSN 1865-3529 ISSN 1865-3537 (electronic)
Green Energy and Technology
ISBN 978-3-031-99311-4 ISBN 978-3-031-99312-1 (eBook)
https://doi.org/10.1007/978-3-031-99312-1

This Springer imprint is published by the registered company Springer Nature Switzerland AG
The registered company address is: Gewerbestrasse 11, 6330 Cham, Switzerland

Preface

With the increasing global sustainable development challenges and the climate emergency, the development and deployment of clean energy technologies have become one most critical priorities of mankind. The 10th International Conference on Advances in Clean Energy Research (ICACER 2025) was successfully held in Nice, France, from April 7 to 9, 2025. As the Program Committee Chair of the ICACER 2025 conference, I proposed and led fruitful discussions and interactions among international experts on clean energy topics. The event was co-sponsored by Polytech Engineering School (University of Lille) and supported by Universitetet i Agder (Norway) and Dunărea de Jos University of Galati (Romania), among other esteemed institutions.

Based on the high-quality discussions before, during, and after the ICACER 225 conference, we recommended authors to extend and submit their results in the form of chapters that contribute to the common goals of this clean energy category. All submitted works underwent a rigorous peer-review process, evaluated by experts from the Scientific Committee, Editorial Board, and external reviewers, each selected based on their subject-matter expertise. The electronic review system ensured efficiency, transparency, and fairness, assessing submissions based on originality, significance, and clarity. The selected papers, along with late-breaking contributions, have been included in this book, forming a cutting-edge technical book, which presents the latest advancements in clean energy research.

This book spans eight critical research tracks, including: Advanced solar energy systems and case studies; hydrogen production, storage, and material challenges; wind energy technology and offshore potential; carbon capture, storage, and resource utilization; bioenergy and sustainable fuel innovations; energy policy, economics, and societal impact; energy system optimization and thermal management; sustainable industrialization and urban energy challenges; and hydrogen policy adoption and transportation strategies. We hope this compilation fosters collaboration among researchers, scientists, engineers, and graduate students, offering valuable insights into emerging trends and technological advancements in clean energy.

As the Program Committee Chair, I would like to thank all the experts involved in this event and the staffs who helped organize the conference and book series, as well as the support from the Chinese Academy of Sciences Project for Young Scientists in Basic Research (YSBR-043, Prof. Lin Chen). We also gratefully acknowledge the kind help from the Springer Editorial team, including Pradheepa Vijay, Michael McCable, Brain Halm, and others. Without them, this book would not have reached our readers in time. Sincerely, we hope this book serves as a valuable resource for researchers and practitioners in related fields, supporting the development and application of innovative clean energy technologies worldwide.

With Warmest Regards,

Beijing, China Lin Chen

Contents

Chapter 1
Photovoltaic Thermal Management Using Phase Change Material

Ahmad Al Miaari, Hafiz Muhammad Ali, and Qamar Abbas

Abstract Photovoltaics is an important technology for producing clean electricity. Photovoltaics can convert solar irradiance into electrical current; however, a portion of the solar irradiance is converted into heat, which raises the panel temperature and reduces its performance. To address this issue, proper cooling should be considered. This study investigates the cooling effect of phase change material on the temperature of a 5-watt photovoltaic panel. Results showed that the temperature of the photovoltaic panel can be reduced by a maximum of 8.6 °C while eliminating temperature fluctuations and maintaining temperature uniformity across the panel.

Keywords Photovoltaics · Phase change material · Thermal management

1.1 Introduction

Phase change materials (PCMs) are gaining popularity as a new passive cooling approach [1]. Recently, they have been applied in various thermal management applications [2], including photovoltaic (PV) temperature control. PV systems play a crucial role in power generation [3], and effective thermal management can significantly enhance their performance [4]. Al-Miaari et al. investigated the effect of suitable and unsuitable types of PCMs on the temperature of a 20-watt PV panel. The results showed that using a PCM with an appropriate melting point could reduce the PV temperature by 10 K [5]. Another study explores a sustainable approach to enhancing the thermal management of PV panels by utilizing shape-stabilized phase change material (SSPCM). A novel SSPCM was developed using oil ash as a cost-effective and environmentally friendly carrier for RT-42 paraffin

A. Al Miaari (✉) · H. M. Ali · Q. Abbas
Mechanical Engineering Department, King Fahd University of Petroleum and Minerals, Dhahran, Saudi Arabia

L. Chen (ed.), *Clean Energy Research*, Green Energy and Technology,
https://doi.org/10.1007/978-3-031-99312-1_1

wax, offering improved stability and thermal conductivity. Experimental results demonstrated that the SSPCM effectively reduced the PV panel's temperature by **6.1 °C**, leading to a 9.02% increase in efficiency compared to an uncooled panel [6]. Another study utilized a eutectic phase change material (PCM) with a high latent heat capacity and an optimal melting point, composed of calcium chloride hexahydrate and ferric chloride hexahydrate in a 70:30 molar mass ratio, to regulate the temperature of the PV panel. Results showed that the power output increased by 9.7% [7]. The PCM has demonstrated great capability in thermal management [8], and due to the importance of PV panels in many applications, such as building [9] and power generation [10]. This study will investigate the effect of using RT-42 PCM on the temperature of a 5-watt PV panel in the climate of Saudi Arabia.

1.2 Experimental Setup

For this experiment, paraffin PCM called RT-42 is used to cool a 5-watt PV panel. RT-42 is an organic paraffin wax produced by Rubitherm, known for its high latent heat capacity, making it an effective thermal energy storage material. The latent heat, melting point, and thermal conductivity of RT-42 were measured using Differential Scanning Calorimetry (DSC) on an STA 449 F3 Jupiter machine and a FOX 50 heat flow meter, respectively, and are summarized in Table 1.1.

A small aluminum container measuring 10 × 10 × 3 cm, filled with RT-42, is attached to the rear of the PV panel to absorb the excess heat generated by the PV and regulate its temperature, as shown in Fig. 1.1. To enhance heat transfer between the PV and the aluminum container, a thermal paste with a conductivity of 2 W/m·K is applied.

Two 5-watt monocrystalline photovoltaic panels, one cooled and the other uncooled as a reference, were used for this experiment. Both panels were mounted on a wooden base oriented due south at an inclination angle of 52°. The experiment was conducted in Saudi Arabia on May 23, 2024. The temperature of the two PV panels was measured and recorded every second using a PROVA 830. The final experimental setup is shown in Fig. 1.2.

Table 1.1 Paraffin RT-42 properties

Physical properties	Paraffin RT-42
Melting temperature (°C)	**38.3**
Latent heat (J/g)	**178**
Thermal conductivity (W/m.k)	**0.2**

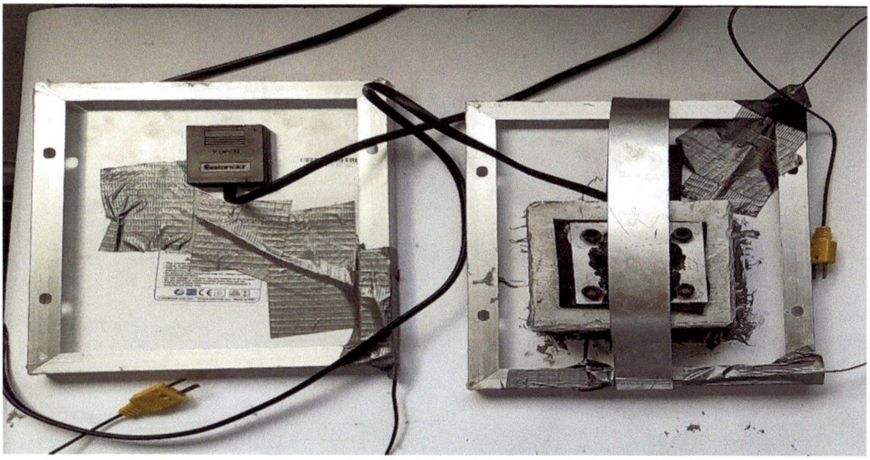

Fig. 1.1 The cooled panel using PCM container on the right and the reference PV panel on the left

Fig. 1.2 The experimental setup

1.3 Results and Discussion

Four K-type temperature sensors are used in this experiment. The first sensor measures the ambient temperature, the second is attached to the rear side of the standalone PV panel, and the last two sensors are placed under the PCM aluminum container. One of these two sensors is used to compare the temperature with the standalone PV, while the other is used to calculate the temperature uniformity beneath the container. The temperature is recorded every second from 4:58 AM to 3:55 PM on a sunny day, with wind speeds ranging from 13 to 31 m/s. The recorded temperatures are presented in Fig. 1.3.

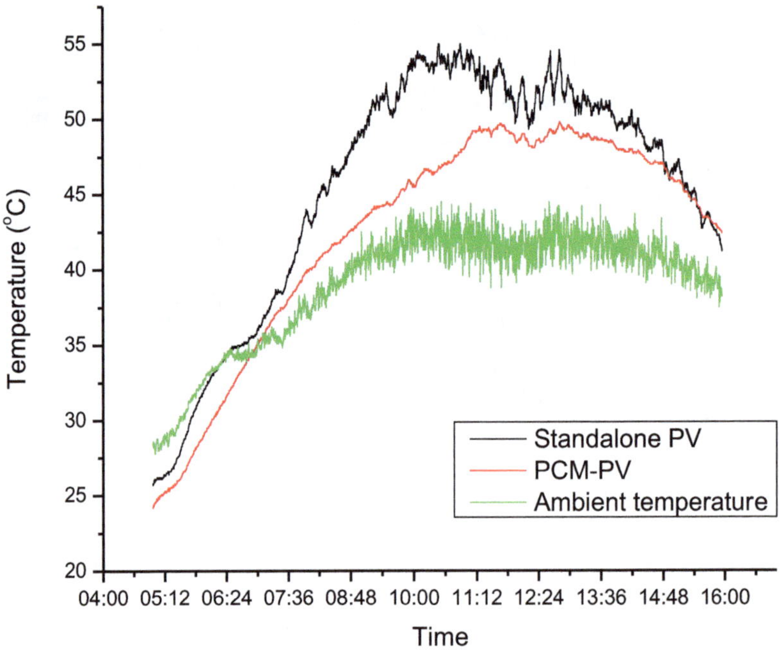

Fig. 1.3 The temperature profiles for the cooled and uncooled PV panels

From Fig. 1.3, it can be observed that the ambient temperature fluctuates over time, increasing from 28.4 °C at 4:58 AM to a maximum of 44.5 °C at 10:32 AM. The average ambient temperature throughout the day is 39 °C. On the other hand, at the beginning of the experiment, it is noticed that the PCM-PV temperature is lower than that of the standalone PV due to the cooling effect of the PCM, with a temperature difference of 1.6 °C. As the sun rises, the temperature of the standalone PV starts to increase, reaching its maximum of 55 °C at 10:30 AM, which is almost the highest peak of the ambient temperature. Meanwhile, the temperature of the PCM-PV at this time is 46.8 °C, which is 8.2 °C lower than that of the standalone PV. The maximum temperature difference between the two panels is 8.6 °C, recorded at 10:03 AM. It is clear from the figure that the cooling effect of the PCM lasted until 3:30 PM, totaling around 10.5 h. It is also noticed that the PCM eliminates the fluctuation of the panel temperature throughout the day compared to the standalone PV, where its temperature is seen to highly fluctuate, which can affect its life time.

It is reported in the literature that the temperature distribution in the PV panels varies, which creates nonuniform temperature profile on the PV panel, affecting its power generation and lifetime. To investigate the temperature uniformity of the PV cooled with the PCM, the other temperature sensor readings are compared with the first one. The coefficient of variation is calculated and plotted in Fig. 1.4.

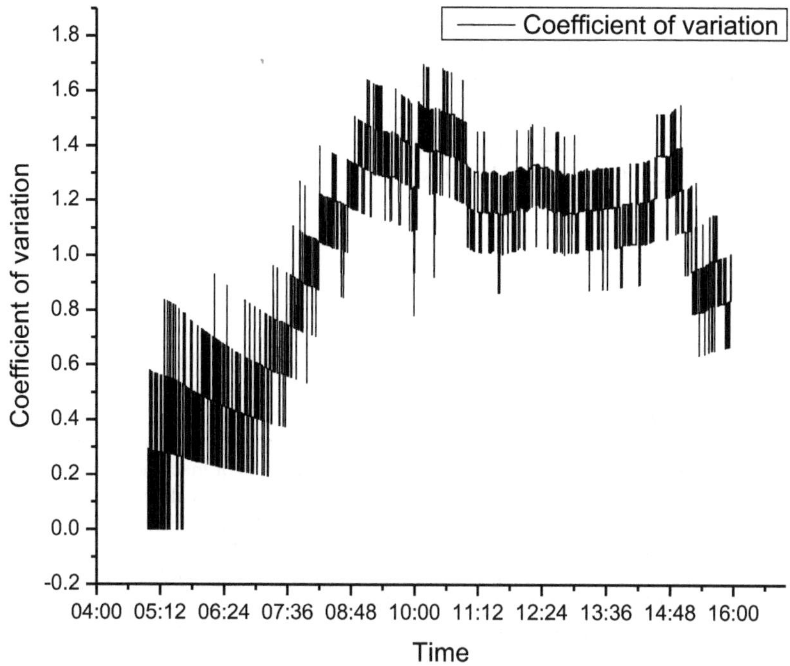

Fig. 1.4 The temperature profiles for the cooled and uncooled PV panels

Figure 1.4 shows the coefficient of variation between the two temperature records under the aluminum container. A coefficient closer to zero indicates a more uniform temperature. According to Fig. 1.4, the coefficient fluctuates between a minimum of 0 and a maximum of 1.69. It is observed that the maximum coefficient occurs during the highest ambient temperature, recorded at approximately 10:30 AM. However, any coefficient below 4% is considered highly uniform. In conclusion, the PCM reduces the temperature of the PV and minimizes fluctuations, maintaining the temperature uniformity of the panel.

1.4 Conclusion

Photovoltaic (PV) technology is a crucial method for generating electrical energy; however, its performance declines as temperature increases. In this study, a phase change material (PCM), specifically paraffin RT-42, enclosed in an aluminum container, is used to cool the PV panel from the rear side. The advantage of using small aluminum containers is that the PCM can be easily accessed and replaced, allowing adaptation to various weather conditions throughout the year. Additionally, the aluminum container itself can be replaced with a new one when the PCM is fully melted during the day. The temperature of the cooled PV panel was recorded and compared to that of a standalone PV panel. Results showed that using PCM reduced the PV temperature by a maximum of 8.6 °C and maintained temperature uniformity, with a maximum coefficient of variation of 1.69. The PCM inside the aluminum container demonstrated significant cooling benefits. Further improvements in cooling efficiency and greater temperature reduction can be achieved by enhancing the thermal conductivity of the PCM, such as by incorporating nanoparticles.

References

1. Al Miaari A, Ali HM (2023) Batteries temperature prediction and thermal management using machine learning: an overview. Energy Rep 10:2277–2305. https://doi.org/10.1016/j.egyr.2023.08.043
2. Al Miaari A, H. M. Ali (2025) "Recent advances on nano-enhanced phase change materials (NEPCMs) for photovoltaic thermal management and role of machine learning: A review of fundamentals, preparation, characterization, and thermo-physical properties," J. Energy Storage, vol. 124, p. 116544, Jul. 2025, https://doi.org/10.1016/j.est.2025.116544
3. Hatamleh Z (2024) Techno-economic comparison between PV and wind to produce green hydrogen in Jordan. pp. 36–43. doi: https://doi.org/10.21741/9781644903216-5
4. Al Miaari A, El Khatib A, Ali HM (2023) Design and thermal performance of an innovative greenhouse. Sustain Energy Technol Assess 57:103285. https://doi.org/10.1016/j.seta.2023.103285
5. Al Miaari A, Ali HM (2023) Technical method in passive cooling for photovoltaic panels using phase change material. Case Stud Therm Eng 49:103283. https://doi.org/10.1016/j.csite.2023.103283
6. Al Miaari A, Mohaisen K, Al-Ahmed A, Ali HM (2025) Experimental investigation on thermal management and performance enhancement of photovoltaic panel cooled by a sustainable shape stabilized phase change material. Case Stud Therm Eng:105763. https://doi.org/10.1016/j.csite.2025.105763
7. Karthick A, Ramanan P, Ghosh A, Stalin B, Vignesh Kumar R, Baranilingesan I (2020) Performance enhancement of copper indium diselenide photovoltaic module using inorganic phase change material. Asia-Pac J Chem Eng 15:5. https://doi.org/10.1002/apj.2480
8. Al Miaari A, Ali HM (2024) A recent review on thermal management of photovoltaic panels using phase change material based on thermal conductivity enhancers for sustainable buildings. J Energy Storage 103:113936. https://doi.org/10.1016/j.est.2024.113936
9. Hatamleh Z, Al Miaari A, El Khatib A (2024) A performance analysis of building integrated PV systems in Jordanian residential buildings. In: 2024 6th international symposium on advanced electrical and communication technologies (ISAECT), pp 1–4. https://doi.org/10.1109/ISAECT64333.2024.10799799
10. El Khatib A, Al Rashidi H, Al Miaari A, Hatamleh Z (2024) Thermal performance evaluation techniques for thin-film semi-transparent PV glazing using experimental testing. In: 2024 6th international symposium on advanced electrical and communication technologies (ISAECT), pp 1–6. https://doi.org/10.1109/ISAECT64333.2024.10799601

Chapter 2
Assessment of the Possibility of Locating Photovoltaic Panels in the Area Around Kindergartens: A Case Study

Anna Justyna Werner-Juszczuk, Dorota Anna Krawczyk, Piotr Rynkowski, Agata Ołtarzewska, Bożena Babiarz, Tomasz Cholewa, Lech Lichołai, Dorota Gawryluk, and Robert Stachniewicz

Abstract This paper focuses on the possibility of replacing fossil fuels with renewable energy sources. The study was conducted for 107 Polish kindergartens, located in Bialystok, in north-eastern Poland, Europe. So far only 10% of kindergartens use PV energy. Those with no PV systems were categorized into 3 types, taking into account the usefulness of the area of the kindergarten's surroundings for localization of the PV. It was found that 23% of plots enable the location of PV panels without restriction, 32% kindergartens could be supplied from a PV system with limited power. However in the case of 45% plots, it is impossible to install a PV system. In total, it is estimated that it is possible to install about 810 kWp of PV on the kindergarten plots, which will result in a total energy output of 824 MWh per year. For kindergartens with no possibility of location PV on the ground (no own plot, small and fully developed plot) the best solution is to look for possibilities to install PV on the roof or to look into solutions to increase the share of RESs in the city plant that supplies education buildings via municipal power grid.

Keywords PV · Energy self-sufficiency · RES · Energy transformation · Kindergarten plot

---

A. J. Werner-Juszczuk (✉) · D. A. Krawczyk · P. Rynkowski · A. Ołtarzewska · D. Gawryluk · R. Stachniewicz
Bialystok University of Technology, Bialystok, Poland
e-mail: a.juszczuk@pb.edu.pl

B. Babiarz · L. Lichołai
Rzeszow University of Technology, Rzeszow, Poland

T. Cholewa
Lublin University of Technology, Lublin, Poland

L. Chen (ed.), *Clean Energy Research*, Green Energy and Technology,
https://doi.org/10.1007/978-3-031-99312-1_2

2.1 Introduction

Retrofitting the building stock is recognized as a key element of global decarbonization strategies [1] and is crucial to reducing the environmental impact of old buildings [2]. In Poland, the process of modernizing buildings has been going on for 30 years and has included, apart from residential buildings, educational facilities like kindergartens [3, 4].

Renovation in kindergartens mainly included increasing the thermal insulation of building partitions, replacing windows and doors with those with a low heat transfer coefficient, and changes in the heating installation and domestic hot water preparation, i.e., replacing radiators, using automatic regulation, insulating pipes and storage tanks [5]. However, in order to ensure zero-emission and energy self-sufficiency of buildings, it is necessary to introduce changes in the sources of heat and electricity, consisting in the use of renewable energy sources (RES) [6].

Among RES, photovoltaic systems play a key role [7] and their use becomes a necessity to ensure the energy self-sufficiency of the buildings [8]. This is related to providing energy for lighting and the operation of the building's technical equipment: mechanical ventilation, office equipment, and food preparation devices. Electricity is also necessary for the operation of heat pumps, which play a significant role in the energy transformation from fossil fuels to renewable energy [9–11]. This is especially important for buildings located in cold climates like Poland, where energy is spent more on heating than cooling [12].

The potential of using PV systems strictly depends on the location of the building. According to data from the Polish Energy Management Agency, the average solar radiation in Poland is approximately 950–1050 kWh/(m^2a). Based on the data, it can be concluded that in Poland, the sunlight conditions are average; therefore, in order to fully cover the electricity demand by PV panels, it is necessary to use large panel surfaces or energy storage facilities [13, 14]. Additionally, in kindergartens, there is a discrepancy between the demand for electricity and its production from solar energy. The highest production from PV in Poland occurs in the months from May to August, while the highest demand for heat and electricity in kindergartens occurs from October to April [15, 16], when there is relatively little sunshine duration. This means that in order to ensure the effective use of solar energy, it is necessary to search for the largest possible area for the PV installation, in order to cover the demand in the transitional and winter periods. The most favourable location of PV panels in urban space is on building roofs [17, 18]. However, in conditions of insufficient roof area, it may sometimes be necessary to locate the PV installation also on the ground in order to cover the building's demand for electricity.

The article assesses the potential of locating a PV installation on a building plot of kindergartens located in the selected city in north-eastern Poland, Bialystok, in a cold climate. The current structure of the power supply to kindergarten buildings is presented. The total energy output potential from PV systems located on the kindergarten's plot in the analysed city was estimated.

2.2 Methodology

The study focused on 107 kindergartens (public and non-public) in Bialystok, north-eastern Poland, for which the structure of the electricity source was analyzed. However, only 75 preschools are independent, separate units, while the rest cannot make their own decisions regarding energy management (e.g., using parts of schools or ground floors of multi-family buildings). For this number of kindergartens, an analysis of the possibility of locating the PV panels on the ground was performed. All analyzed kindergartens were working in 2023–2024.

Data necessary to analyze the area of plots and buildings was collected from local detailed information in Geographic Information Systems [19]. Data on the structure of the electricity supply was obtained from the service [20]. Solar potential in the analysed kindergartens was assessed based on data from the Photovoltaic Geographical Information System [21]. The procedure of the research is presented in Fig. 2.1.

The plots were categorized into three categories. The basis for the division was the size of the non-built area of the kindergarten plot and the development status of the plot.

The shape of the available area, tree cover, existing infrastructure (terraces, parking lots, access roads) and the presence of tall buildings in the vicinity, which cause significant shading of the plot, were taken into account. Each plot was analysed separately. Taking all the above factors into account, kindergartens were assigned to three categories:

- Category 0—no possibility of locating PV installations,
- Category 1—limited possibilities of locating PV (available plot area for PV in the range 40–100 m^2),
- Category 2—very good possibilities of locating PV on the ground (available plot area for PV more than 100 m^2).

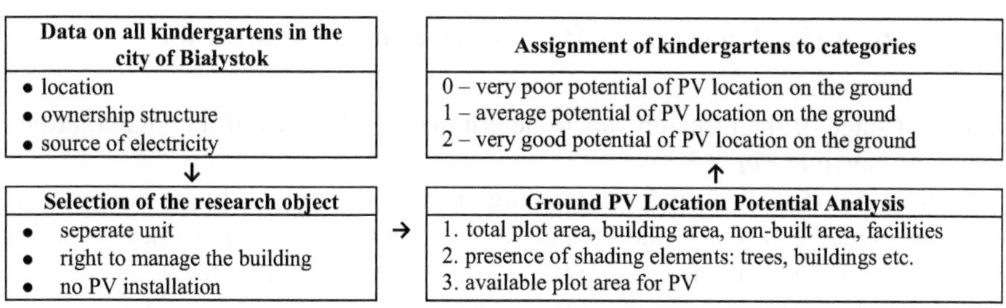

Fig. 2.1 Research scheme

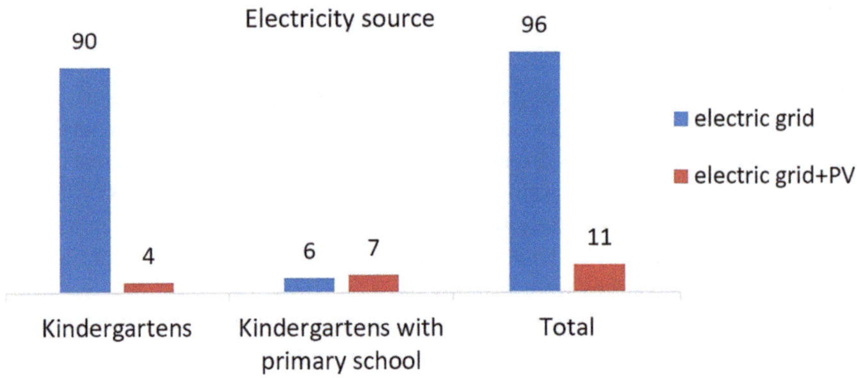

Fig. 2.2 The structure of the electricity supply to kindergartens located in Bialystok

2.3 Results and Discussion

Structure of an Electricity Source

In terms of the source of electricity, all the analysed kindergartens are powered from the municipal power grid (Fig. 2.2), which is fed by the City Heat and Power (CHP) plant. In 10% of kindergartens, additionally, PV panels are located on the roofs of the buildings. It should be noted that in 11 units with a PV installation, 7 are kindergartens, which are located in buildings shared with a primary school. In the case of kindergartens constituting a separate unit, only 4 kindergartens have a PV installation, which is 4%. As can be observed, the use of PV in kindergartens in the analysed city is currently at a very low level, and in the few facilities where PV is installed, it only serves as an additional source of energy.

Potential of the Location of PV on the Kindergarten Plot

The next stage of the research was to determine the plot area of the kindergartens, which is shown in Fig. 2.3.

A majority (58%) of kindergartens are situated on a plot with an area of over 3000 m^2, which can be considered a large plot. About 22% of kindergartens have a medium-sized plot ranging from 1000 to 3000 m^2. On the other hand, the smallest share among the analysed facilities, that is 20%, is constituted by kindergartens with a very small plot of less than 1000 m^2.

In order to determine the available plot area, the building area of each facility was next analysed (see Fig. 2.4).

Only 3 out of 71 kindergartens (4%) have a built-up area of over 50%. Such a small number of kindergartens results from the fact that only facilities that constitute a separate unit and are not located in buildings managed by other entities, e.g.,

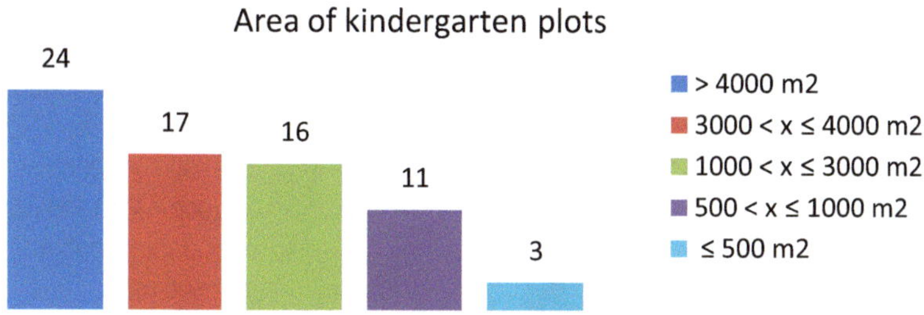

Fig. 2.3 Area of kindergarten plots

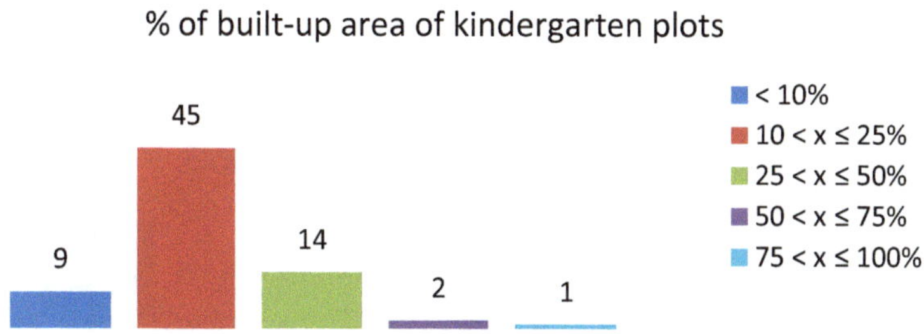

Fig. 2.4 Percentage of built-up area of kindergarten plots

on the ground floors of multi-family buildings, are analysed. 20% of kindergartens have a built-up area of 25–50%. The largest share, on the other hand, is held by kindergartens with a built-up area of 10–25%, as much as 63%. In 13% of kindergartens, almost the entire plot around the building is undeveloped.

Based on the plot size and development area available, the undeveloped area was determined (see Fig. 2.5).

The large available plot should correlate with the large potential for locating additional RES systems, such as PV installations. It should be remembered, however, that the kindergarten area is specific in terms of development. It serves primarily as a recreational function for children (playground) and its presence is required to ensure children's well-being and proper socio-physical development [22]. The installation of photovoltaic panels on the kindergarten plot should primarily take into account the safety of children and the need to ensure accessibility to the playground. For this reason, the undeveloped area of the plot should be large enough to fulfill both functions around the kindergarten.

Nearly half of kindergartens (47%) have an undeveloped plot of 3000 m^2 with the greatest potential for locating PV panels at ground level. 21% of kindergartens have a plot of land ranging from 1500 to 3000 m^2; thus, the potential for PV installation is lower and depends strictly on the location of the building on the plot, especially in the context of the existing infrastructure—entrances, access to the main road, parking lot, etc. Thirty-two percent of kindergartens have an undeveloped area of less than 1500 m^2, which means very poor possibilities for locating PV panels and receiving a proper amount of solar energy. However, in order to finally determine

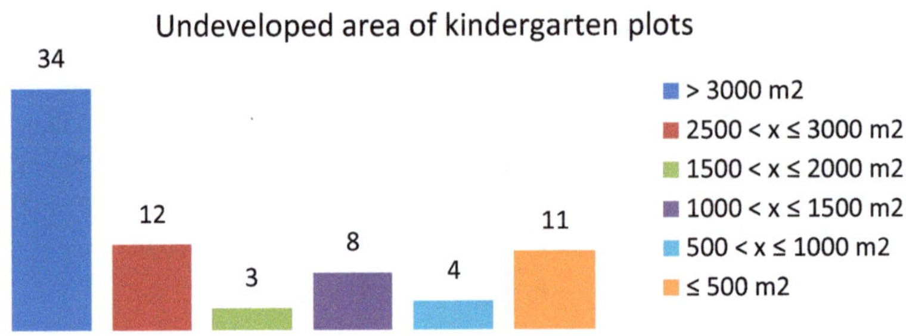

Fig. 2.5 Undeveloped area of kindergarten plots in m^2

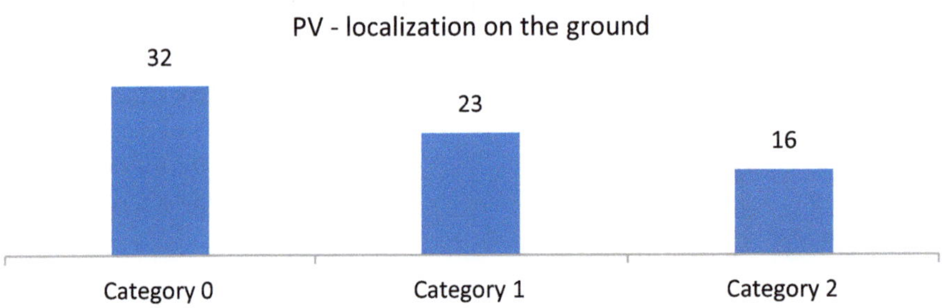

Fig. 2.6 Categories of kindergartens based on their usefulness for the development of PV panels on plots (ground level)

the possibility of locating the PV installation on the ground, the development of the plot of each kindergarten was analyzed. In the case of kindergarten plots, they are often covered with trees, which reduces the possibility of locating PV on the ground. Very often, there are low plantings along the fence (shrubs), the purpose of which is to separate the playground from unauthorized persons. Moreover, there are various types of playground equipment in kindergartens, which also reduces the possibilities of installing the PV system. Sometimes, despite a large plot, the location of the building on the plot is unfavorable, which results in narrow green belts around the building, which also reduces the PV potential.

Taking the above into account, the available area for the installation of PV panels was determined, which was the basis for the assignment of kindergartens to categories. The results are shown in Fig. 2.6. Most kindergartens (45%) were classified as category 0; i.e., there is no potential for locating PV panels on the ground. 32% of kindergartens were assigned to category 1, with medium potential for locating panels on the ground. 23% of kindergartens have good conditions for PV location on the ground (Category 2), and what is important, only 16 of 34 kindergartens with an undeveloped plot of 3000 m^2 were qualified for this group. That indicates that a large un-built area of plot does not correspond with a large potential for PV installation on the ground.

Selected examples of kindergartens that belong to all three categories are presented in Fig. 2.7.

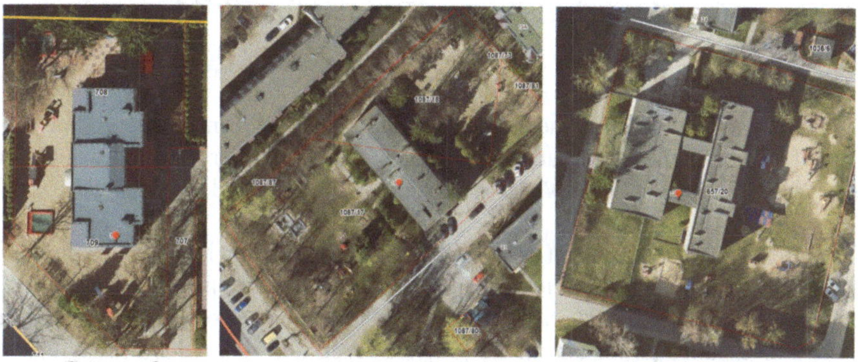

Fig. 2.7 Examples of kindergartens belonging to three categories [18]

A kindergarten of category 0 (see Fig. 2.7) has a small plot with an area of 850 m^2 and a building area of 177 m^2 (21%). The building is located centrally on a trapezoidal plot. In the only accessible places, there are numerous playground elements: sandboxes, swings, and ladders. Additionally, there are numerous plantings in the form of trees and shrubs on the plot, which means that there is no available space for the PV installation.

The kindergarten classified in category 1 (see Fig. 2.7) has a large plot of land with an area of 5136 m^2 and a building area of 474 m^2 (9%). The kindergarten has a large area of land in the front and back of the building, which is only partially developed by playground equipment. However, the possibility of locating PV is limited due to the numerous plantings of trees and shrubs around the building and along the fence. Especially on the north-eastern side of the plot, it is visible that despite the available undeveloped square, the efficiency of the PV installation would be negligible due to the large shade from the trees and the building.

The last kindergarten presented in Fig. 2.5 belongs to category 2, representing very good conditions for locating PV panels. The facility has a large plot with an area of 5366 m^2 and a building area of 1030 m^2 (19%). This kindergarten is located closer to one of the plot boundaries, which means that the southern part of the plot is not shaded. The plantings are located in the eastern-southern part of the plot and on the northern side. Most of the playground equipment is located in the eastern part of the plot. In the southwestern part of the plot, there is a large undeveloped and unshaded area, which has very good conditions for locating a PV installation.

Analysis of the Electrical Energy Potential from PV Located in Kindergarten's Plots

The next step was to determine the possible energy output from a photovoltaic system, if located in the kindergarten plot. This is related to solar radiation, which for Bialystok is equal to 1267 kWh/m^2 per year (see Fig. 2.8) and the efficiency of the PV system.

In Bialystok the optimal tilt angle is 38° and the orientation is south [23]. For this PV orientation, assuming the total energy loss of the PV system at the level of 14% (losses in cables, power inverters, dirt, etc.), it is possible to obtain 1017 kWh per 1 kWp of installed power during the year. However, this production is not uniform and its peak falls in the months from April to September, then we have overproduction of electrical energy in relation to our needs. Production in these months constitutes about 72% of the total production during the year.

In order to estimate the potential for installing PV systems in the kindergartens in Bialystok, the possible area for installing PV panels at ground level was estimated in each of the kindergartens assigned to categories 1 and 2. Despite the fact that there are more kindergartens of category 1, the total area on which it is possible to install PV panels was about 1300 m^2. In the case of kindergartens classified in category 2, it is a total of about 2300 m^2.

To determine the installed power, the technical data of PV panels offered by three leading global companies were used [24–26]. Based on their offer, the range of yields from 1 kWp from 1 m^2 of PV panels was determined (Table 2.1).

The installed power related to 1 m^2 of PV surface ranged from 213 to 235 Wp/m^2. For further analysis, an average value of 225 Wp/m^2 was assumed, which means that a total of 810 kWp of PV can be installed on the premises of kindergartens. This allowed for estimating the energy output from PV panels installed in kindergartens' plots in the city of Bialystok, which is equal to 824 MWh per year (see Fig. 2.9).

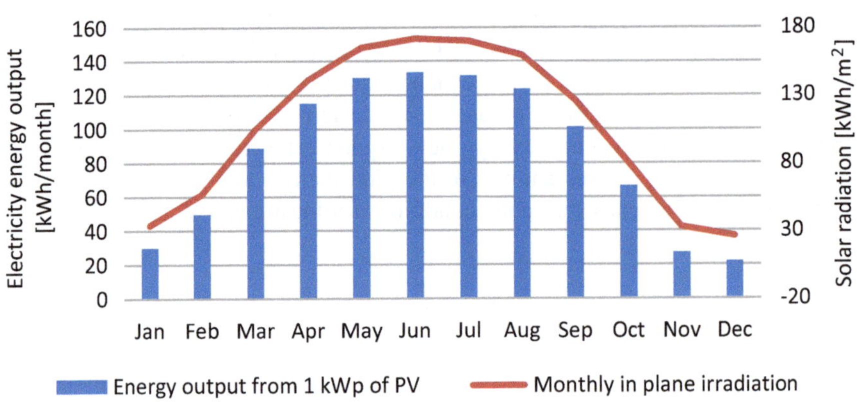

Fig. 2.8 Monthly in-plane radiation and energy output of 1 kWp PV orientated south and tilt angle 38° (based on [21])

Table 2.1 Parameters of PV panels

Producer	PV type	Dimensions [mm]	Brutto area of 1 panel [m²]	P [Wp]	Efficiency [%]	Power [Wp/m²]
JinkoSolar [24]	Tiger Neo Bifacial 54R HC	1762 × 1134	2.00	450	22.52	225
	Tiger Neo Mono 66 HC	2382 × 1134	2.70	635	23.51	235
Longi [25]	LR7-72HGD-610 M	2382 × 1134	2.70	610	22.60	226
	LR5-72HGD-570 M	2278 × 1134	2.58	570	22.60	221
Trinasolar [26]	TSM-NEG9RC.27425	1762 × 1134	2.00	425	21.30	213
	TSM-NE19R 630 W	2382 × 1134	2.70	630	23.30	233

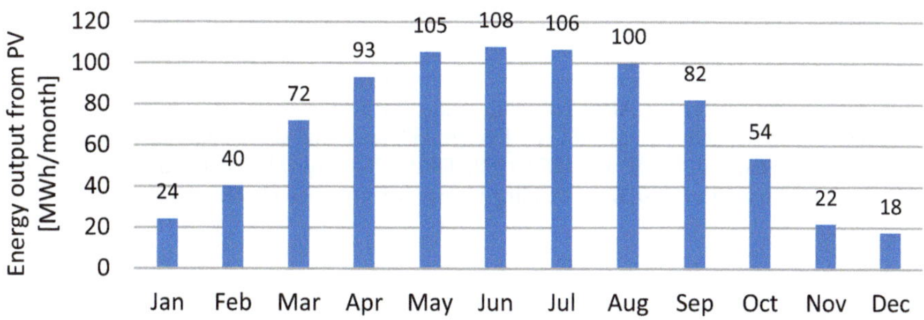

Fig. 2.9 Estimated energy output of PV systems located on the kindergartens plots in Bialystok

68% of this can be produced by installations located in kindergartens classified as category 2.

2.4 Conclusions

In terms of the source of electricity, 90% of kindergartens in analysed city draw energy only from the municipal power grid, and 10% have a mixed system: the power grid and a PV installation located on the roof of the building. This means that there is a need to develop a global strategy to increase the share of renewable energy sources, including PV systems, in educational buildings in order to meet the upcoming requirements for zero-emission buildings.

The analysis showed that locating the PV panels on the ground is not the best solution in urbanized areas due to the presence of neighbouring buildings and existing trees that may cause shading. Only 23% of analyzed kindergartens have conditions to install PV panels that could supply their electricity system on the plot, while in the case of 32% such investments are very restricted and could cover only a small part of energy needs. In case of 45% of preschools, the installation of PV panels on parcels is impossible.

The estimated potential of energy output from PV panels located on the ground of kindergartens in the analysed city is 824 MWh/year, which indicates that the location of panels on the ground can, to some extent, cover the demand for electricity of buildings. However, due to the unevenness of solar radiation in Poland throughout the year, additional solutions should be sought in the form of electricity storage. This system, combined with the energy generation system in photovoltaic modules, has the potential to effectively reduce the demand for electricity from the national power system. Another way to reduce the consumption of energy from fossil fuels is to change the scope of central CHP plants to increase the share of RES in electricity and heat generation, as it is the main source of electricity in Poland.

The next step is to expand the scope of the research, including energy storage systems and analysis of the demand for electricity by individual kindergartens throughout the year and to conduct an economic analysis.

Acknowledgements The research leading to these results has received funding from the commissioned task entitled "VIA CARPATIA Universities of Technology Network named after the President of the Republic of Poland Lech Kaczynski", under the special purpose grant from the Minister of Science, contract no. MEiN/2022/DPI/2575, MEiN/2022/DPI/2577, MEiN/2022/DPI/2578, action entitled "ISKRA—building inter-university research team.

References

1. Walker L, Hischier I, Schlueter A (2022) Does context matter? Robust building retrofit decision-making for decarbonization across Europe. Build Environ 226:109666
2. Solanki SK, Rastogi R, Paul VK (2022) Cost analysis of functional retrofitting measures in buildings. J Inst Eng (India) Ser A 103:725–732
3. Życzyńska A, Suchorab Z, Kočí J, Černý R (2020) Energy effects of retrofitting the educational facilities located in South-Eastern Poland. Energies 13(10):2449
4. Krawczyk DA (2014) Theoretical and real effect of the school's thermal modernization—a case study. Energ Buildings 81:30–37
5. Sadowska B, Piotrowska-Woroniak JB, Woroniak G, Sarosiek W (2022) Energy and economic efficiency of the thermomodernization of an educational building and reduction of pollutant emissions—a case study. Energies 15(8):2886
6. Liu Y, Xue S, Guo X, Zhang B, Sun X, Zhang Q, Wang Y, Dong Y (2023) Towards the goal of zero-carbon building retrofitting with variant application degrees of low-carbon technologies: mitigation potential and cost-benefit analysis for a kindergarten in Beijing. J Clean Prod 393:136316
7. Walch A, Rüdisüli M (2023) Strategic PV expansion and its impact on regional electricity self-sufficiency: case study of Switzerland. Appl Energy 346:121262
8. Senkus P, Glabiszewski W, Wysokińska-Senkus A, Cyfert S, Batko R (2021) The potential of ecological distributed energy generation systems, situation, and perspective for Poland. Energies 14(23):7966

9. Pater S (2023) Increasing energy self-consumption in residential photovoltaic systems with heat pumps in Poland. Energies 16(10):4003
10. Marchwiński J, Kurtz-Orecka K (2020) Influence of photovoltaic installation on energy performance of a nursery building in Warsaw (central European conditions). J Build Eng 32:101630
11. Hassan Q, Algburi S, Sameen AZ, Salman HM, Jaszczur M (2023) Implications of strategic photovoltaic deployment on regional electricity self-sufficiency by 2050: a case study in Iraq. Renew Energy Focus 46:338–355
12. Felius LC, Dessen F, Hrynyszyn BD (2020) Retrofitting towards energy-efficient homes in European cold climates: a review. Energ Effic 13:101–125
13. Knutel B, Pierzyńska A, Dębowski M, Bukowski P, Dyjakon A (2020) Assessment of energy storage from photovoltaic installations in Poland using batteries or hydrogen. Energies 13(15):4023
14. Dzikuć M, Piwowar A, Dzikuć M (2022) The importance and potential of photovoltaics in the context of low-carbon development in Poland. Energy Storage Saving 1(3):162–165
15. Igliński B, Piechota G, Kiełkowska U, Kujawski W, Pietrzak MB, Skrzatek M (2023) The assessment of solar photovoltaic in Poland: the photovoltaics potential, perspectives and development. Clean Techn Environ Policy 25:281–298
16. Jurasz J, Mikulik J (2016) Investigating theoretical PV energy generation patterns with their relation to the power load curve in Poland. Int J Photoenergy 3789840
17. Hołuj A, Ilba M, Lityński P, Majewski K, Semczuk M, Serafin P (2021) Photovoltaic solar energy from urban sprawl: potential for Poland. Energies 14(24):8576
18. Mokhtara C, Negrou B, Settou N, Bouferrouk A, Yao Y (2021) Optimal design of grid-connected rooftop PV systems: an overview and a new approach with application to educational buildings in arid climates. Sustain Energy Technol Assess 47:101468
19. The Geographic Information Systems of the city of Białystok, https://gisbialystok.pl/, Accessed 3 Jan 2025
20. Poland's National Geoportal website, https://mapy.geoportal.gov.pl/imap/Imgp_2.html, last accessed 2025/01/03
21. JRC. Photovoltaic Geographical Information System (PVGIS)—European Commission, https://re.jrc.ec.europa.eu/pvg_tools/en/, Accessed 10 Feb 2025
22. Krawczyk DA, Kłopotowski M, Gawryluk D et al (2024) Evaluation of accessibility of kindergarten playgrounds and outdoor green areas in Polish cities. Urban Des Int
23. Kusznier J, Wojtkowski W (2019) Impact of climatic conditions and solar exposure on the aging of PV panels. In: 15th selected issues of electrical engineering and electronics (WZEE), Zakopane, Poland, pp 1–6
24. Jinko Solar company website, https://jinkosolar.eu/, last Accessed 14 Feb 2025
25. Longi company website, https://www.longi.com/eu/, Accessed 14 Feb 2025
26. Trinasolar company website, https://www.trinasolar.com/us, Accessed 14 Feb 2025

Chapter 3
Solar Energy–Driven Combined Cooling and Cold Thermal Energy Storage System for a District Cooling System

K. Sarath Babu and Zhibin Yu

Abstract Current cooling technologies primarily rely on electricity generated from fossil fuels, contributing to global warming. Consequently, there is an urgent need to transition towards sustainable cooling technologies. In response to this challenge, this study proposes and analyzes a combined cooling and cold thermal energy storage system utilizing a solid-gas thermochemical sorption process. The aim is to meet the cooling demand of a building using solar energy, eliminating the use of fossil fuel-based electricity. Solar energy available during the daytime is used to produce cooling and store surplus cold thermal energy. The stored cold thermal energy is utilized during nighttime when solar energy is not available. Cooling is produced by evaporation of liquid refrigerant while the cold energy storage is achieved by storing surplus liquid refrigerant (ammonia) for later use. A steady-state thermodynamic model is developed to study the performance of the proposed system. To demonstrate the operating strategy of the proposed system, the cooling demand of a building is considered as a case study. The average COP of the proposed system is estimated. An average COP of 0.42 is estimated for the proposed system at a heat source temperature of 100 °C and an evaporating temperature of 0 °C. The analysis of the present case can be effectively used to provide the cooling load of a building without dependency on electricity.

Keywords Cooling · Cold Thermal Energy Storage · Ammonia · Chemisorption · Adsorption

K. S. Babu · Z. Yu (✉)
University of Liverpool, Liverpool, UK
e-mail: Zhibin.Yu@liverpool.ac.uk

L. Chen (ed.), *Clean Energy Research*, Green Energy and Technology,
https://doi.org/10.1007/978-3-031-99312-1_3

3.1 Introduction

Refrigeration plays a pivotal role in modern industries. The growing demand for cooling systems is driven by rapid industrialization and population growth. However, this increasing demand has led to significant environmental challenges, including ozone layer depletion and global warming [1]. As a result, there is an urgent need to develop renewable energy-based refrigeration systems to reduce dependency on fossil fuels [2]. Among renewable energy sources, solar and wind power stand out as viable alternatives. However, these sources are intermittent and unstable; for instance, solar energy is only available during the daytime but is both clean and uniformly distributed [3]. Due to the intermittent nature of solar energy, continuous cooling throughout the day cannot be achieved. Therefore, reliance on cold energy storage technology is necessary.

The primary methods for cold energy storage are sensible cold energy storage and latent cold energy storage (PCM). Sensible cold energy storage relies on chilled water [4], but it has significant drawbacks, including the need for large storage volumes, low storage density, and cold loss during extended storage periods. On the other hand, phase change materials (PCMs), which utilize liquid-solid phase change, offer higher energy density [5]. However, PCMs face challenges such as phase segregation, low thermal conductivity, and limited storage density [6].

Chemisorption cold thermal energy storage emerges as a promising alternative. This method utilizes solar energy to produce gaseous refrigerants, which are then condensed and stored, leveraging the liquid-to-gas phase change to achieve higher storage density and greater flexibility [7]. For example, Gang Li et al. investigated an adsorption-based cold energy storage system using zeolite/water vapor as a sorption pair. Their system demonstrated an energy storage density of 400 kJ/kg with a storage efficiency of 44%. However, this pair operates at sub-ambient pressures and suffers from low storage capacity, limiting its practicality [8]. Anish Modi et al. studied solar photovoltaic (PV) refrigerators where PV panels and battery stacks store excess solar energy as electrical energy for use during non-sunshine hours. Despite its advantages, this system is economically unfeasible due to the high costs of solar PV and lithium-ion batteries, along with the environmental impact associated with battery manufacturing [9]. To address the fluctuating nature of solar irradiation, Gao et al. [10] proposed a compression-assisted chemisorption cycle for refrigeration and cold energy storage. This system achieved a coefficient of performance (COP) of 5.5 at a solar hot water temperature of 90 °C. However, the integration of compressors and solar PV increases costs and results in a bulky system. Zhang et al. developed a solar photovoltaic-driven refrigeration system coupled with chemisorption-based cold thermal energy storage. In this design, solar PV powers a vapor compression cycle during the day for refrigeration and cold energy storage, while the stored cold energy is released at night. This system is particularly beneficial for regions with an insufficient electricity supply [11].

In summary, a combined solar energy-driven cooling and cold thermal energy storage system offers significant potential to replace fossil fuel-based systems and address fluctuating cooling demands, even during periods when solar energy is unavailable. However, research on chemisorption cooling systems integrated with cold thermal energy storage remains limited. Moreover, there is a notable gap in the literature regarding practical case studies of such systems for large-scale applications.

This study investigates a chemisorption-based combined cooling and cold thermal energy storage system powered by solar thermal energy. Unlike previous research that primarily focuses on converting solar energy into electrical energy for cooling, this approach directly utilizes solar thermal energy, eliminating the need for expensive batteries. Additionally, the proposed system is evaluated through a case study for a residential building, demonstrating its feasibility and effectiveness in real-world scenarios.

3.2 Cycle Description

The solar thermal energy-driven chemisorption system for combined cooling and cold thermal energy storage is illustrated in Fig. 3.1. The proposed system comprises two sorption beds, a condenser, an evaporator, a storage tank, and an expansion valve. Ammonia is used as a refrigerant, and calcium chloride is used as a sorbent. The system operates in two distinct modes, which are described as follows:

Mode 1: Cooling and cold energy storage mode. As illustrated in Fig. 3.1a, during the daytime, solar energy is utilized to heat Sorption Bed 1, resulting in the release of ammonia vapor. This vapor is subsequently cooled and condensed in the condenser. A portion of the condensed liquid ammonia is throttled by the expansion valve to meet the cooling demand at the evaporator, while the surplus liquid refrigerant is stored in the storage tank to meet cooling demands during nighttime, where solar radiation is not available. The refrigerant from the expansion valve passes through the evaporator, where it evaporates, producing a cooling effect. The refrigerant vapor from the evaporator is then adsorbed by Sorption

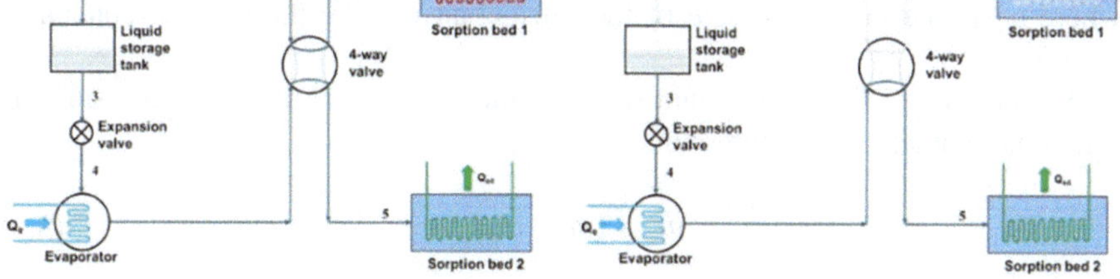

Fig. 3.1 Schematic of the proposed system during (**a**) cooling and cold energy storage mode and (**b**) cold energy release mode

Bed 2. The heat of reaction from the sorption bed is dissipated by cooling water supplied at ambient temperature.

Mode 2: Cold energy release mode. As illustrated in Fig. 3.1b, during nighttime, solar energy is unavailable, preventing Sorption Bed 1 from releasing ammonia vapor. The liquid ammonia stored in the storage tank during the daytime is now utilized. This liquid ammonia is throttled by an expansion valve. The ammonia from the expansion valve passes through the evaporator, where it evaporates to meet the cooling demand at nighttime. The refrigerant vapor from the evaporator is subsequently adsorbed by Sorption Bed 2. The heat of reaction from Sorption Bed 2 is dissipated by cooling water supplied at ambient temperature. On the following day, this process will repeat, starting with heating Sorption Bed 2.

The advantage of the proposed system lies in the utilization of ammonia as both the refrigerant and medium for cold thermal energy storage. Ammonia possesses a higher latent heat of evaporation, specifically 1262 kJ/kg at 0 °C, which is approximately three times greater than the phase change enthalpy of PCMs. Consequently, higher storage density can be achieved with ammonia than with PCMs.

3.3 Thermodynamic Model

The mathematical expression for the cooling load at the evaporator is given by

$$Q_e = \int_0^{\Delta t} \dot{m}_e \left(h_5 - h_4\right) dt \tag{3.1}$$

where Δt represents the time period in seconds. $\dot{m}_e$ denotes the mass flow rate of the refrigerant passing through the evaporator in kg/s. The cooling demand is specified for a building, and consequently, the mass flow rate of the refrigerant is determined based on this specified cooling demand.

The condenser heat load is calculated using the following mathematical expression.

$$Q_c = \int_0^{\Delta t} \dot{m}_c \left(h_1 - h_2\right) dt \tag{3.2}$$

where $\dot{m}_c$ is the mass flow rate of the refrigerant passing through the condenser in kg/s.

The amount of liquid ammonia stored in the storage tank is estimated using the following mathematical expression.

$$m_{st} = \left(\dot{m}_c - \dot{m}_e\right) \Delta t \tag{3.3}$$

The amount of heat required at the Sorption Bed 1 for ammonia desorption is determined using the following mathematical expression.

$$Q_{de} = \int_0^{\Delta t} \frac{\dot{m}_c \cdot \Delta H}{M_{\mathrm{NH_3}}} dt \tag{3.4}$$

Here ΔH is the reaction enthalpy kJ/mol and $M_{\mathrm{NH_3}}$ is the molecular weight of ammonia gas.

The amount of heat released at the Sorption Bed 2 during ammonia adsorption is determined using the following mathematical expression.

$$Q_{ad} = \int_0^{\Delta t} \left[\frac{\dot{m}_e \Delta H}{M_{\mathrm{NH_3}}} - \dot{m}_e c_{p,\mathrm{NH_3}} \left(T_{amb} - T_a \right) \right] dt \tag{3.5}$$

Here, $c_{p,\mathrm{NH_3}}$ is the specific heat of ammonia gas in kJ/kg. K and T_a is the equilibrium adsorption temperature of the salt corresponding to the evaporator pressure.

The coefficient of performance (COP) of the proposed cycle is given by the following expression.

$$\mathrm{COP} = \frac{Q_e}{Q_{des}} \tag{3.6}$$

3.4 Results and Discussion

Operating Conditions

A case study of the proposed cycle for a large residential building has been conducted to demonstrate its working principle and benefits. The cooling demand for maintaining thermal comfort in the residential building, as depicted in Fig. 3.2, is analyzed [12]. The building in question is a 40-floor structure with a floor area of 324 m^2, located in Makkah. Due to Makkah's arid and hot climate, particularly during summer daytime, the cooling demand is significantly high. Furthermore, Saudi Arabia benefits from abundant solar energy and predominantly clear skies throughout most of the year, presenting a potential opportunity for solar-based cooling solutions. The cooling demand profile for the summer season is presented. Figure 3.2 illustrates the variation in cooling demand and ambient temperature over a 24-h period. The temperature and cooling demand data presented in Fig. 3.2 represent the average values over a three-year period (2010, 2011, and 2012), specifically recorded on the 21st day of each month. The temperature data were obtained from meteorological station records for Makkah, Saudi Arabia. The building's cooling demand was calculated using the Cooling Load Temperature Difference (CLTD) method, which is a widely used approach for estimating cooling requirements in buildings. It is observed that the cooling demand exhibits a proportional relationship with ambient temperature fluctuations. The cooling demand is at its minimum

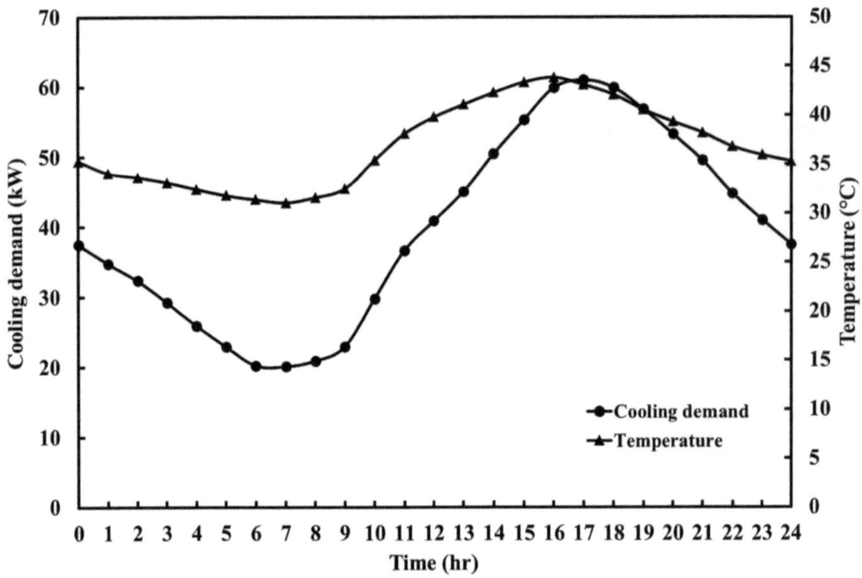

Fig. 3.2 Cooling demand of a residential building [12]

during the early morning hours, specifically at 06:00, corresponding to the lower ambient temperatures. As the day progresses, the cooling demand increases, reaching a peak of 61 kWh at 17:00, coinciding with the highest ambient temperature. Subsequently, the cooling demand decreases during the evening and nighttime hours. The slight lag in cooling demand relative to the peak ambient temperature may be attributed to the thermal inertia of the building.

The operational study of the proposed system is conducted under conditions analogous to those of the building. The evaporator temperature is maintained at 0 °C, while the condenser temperature varies in accordance with the ambient temperature. For effective heat transfer, the condensing temperature is assumed to be 10 K higher than the ambient temperature. The adsorption and desorption of ammonia on $CaCl_2$ salt occur as described by Eq. (3.7). The forward reaction represents adsorption, whereas the reverse reaction denotes desorption.

$$CaCl_2.4NH_3 + 4NH_3 \leftrightarrow CaCl_2.8NH_3 + 4\Delta H \quad (3.7)$$

The equilibrium pressure and temperature are related by the following equation.

$$\ln\left(\frac{P_{eq}}{P_o}\right) = \frac{-\Delta H}{RT} + \frac{\Delta S}{R} \quad (3.8)$$

The thermodynamic and thermophysical properties of salt and ammonia is given in Table 3.1.

Table 3.1 Thermophysical properties of salt and ammonia gas [13, 14]

Parameter	Units	Value
Evaporator temperature (T_e)	°C	0
Condensing temperature (T_c)	°C	$T_{amb} + 10$
Adsorption enthalpy (ΔH_a)	kJ/mol	37.7
Desorption enthalpy (ΔH_d)	kJ/mol	47.2
Specific heat of ammonia (C_p)	kJ/kg.K	2.08
Molecular weight of ammonia (M_{NH_3})	kg/mol	0.017

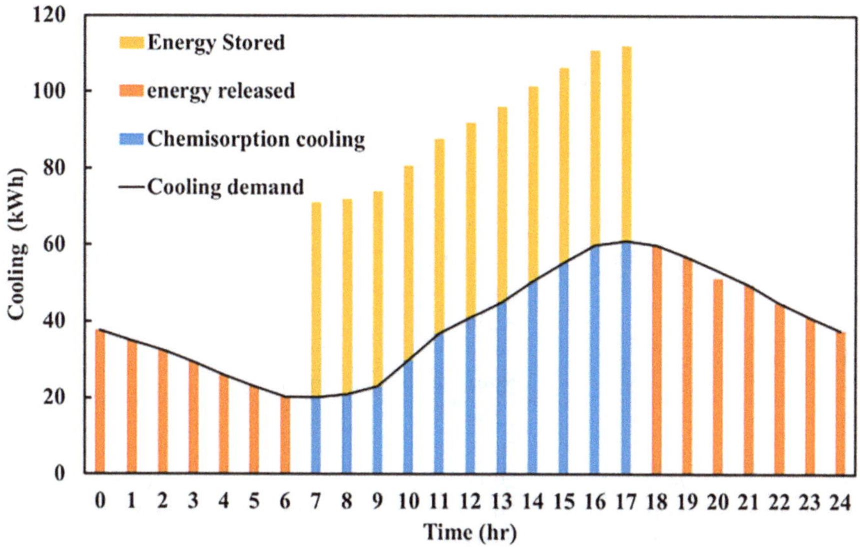

Fig. 3.3 Operating strategy of the proposed cycle

Operational Strategy

Figure 3.3 illustrates an effective cooling demand management strategy that leverages solar energy for cooling and cold thermal energy storage. The total cooling load of the building under consideration is calculated to be 950 kWh, with the daytime cooling demand (07:00 to 17:00 hours) accounting for 443.3 kWh, while the evening and nighttime cooling demand amounts to 506.7 kWh.

The system provides cooling between 07:00 and 17:00, while simultaneously storing cold thermal energy. During this period, 506.7 kWh of cold thermal energy is stored. To achieve this storage, 1.6 tons of liquid ammonia is produced and stored in the liquid storage tank within the same timeframe. The stored cold thermal energy is utilized during two intervals: midnight to 06:00 and 18:00 to midnight. This design enables the storage of cold thermal energy during midday, when solar energy is readily available (07:00 to 17:00). The chemisorption cooling process is synchronized with the building's cooling demand, and cold thermal energy is stored by generating surplus liquid refrigerant. By integrating chemisorption cooling with cold thermal energy storage, the system effectively manages cooling demand throughout the day, even during periods when solar energy is unavailable. Furthermore, the integration of the system with renewable energy sources reduces dependence on conventional fossil fuel-based energy systems, thereby addressing variable cooling demands more sustainably.

The heat source temperature of 100 °C is assumed to be provided by a solar thermal collector. The condenser heat, evaporator load, and adsorption-desorption heat have been calculated and are presented in Table 3.2. The average COP of the proposed cycle is determined to be 0.42 for the building's specified cooling demand. This COP is comparable to other chemisorption-based cooling cycles. For instance, the COP of a chemisorption chiller has been reported as 0.33 [15], while that of a chemisorption air conditioner is 0.46 [16].

Table 3.2 Heat transfer at the system components

Time (h)	Q_c (kWh)	Q_e (kWh)	Q_{ad} (kWh)	Q_{de} (kWh)
1	–	34.83	66.21	–
2	–	32.40	61.56	–
3	–	29.29	55.63	–
4	–	25.91	49.19	–
5	–	22.94	43.52	–
6	–	20.25	38.39	–
7	74.36	20.11	39.16	167.70
8	75.16	20.92	40.85	169.84
9	77.21	22.94	45.01	175.11
10	84.22	29.83	59.44	193.32
11	91.24	36.71	74.25	211.93
12	95.51	40.90	83.49	223.50
13	99.85	45.08	92.70	235.04
14	105.54	50.49	104.50	249.84
15	110.67	55.34	115.27	263.31
16	115.66	59.94	125.12	275.70
17	117.02	61.02	126.87	278.01
18	–	59.94	114.81	–
29	–	56.83	108.70	–
20	–	51.30	98.00	–
21	–	49.54	94.55	–
22	–	44.81	85.41	–
23	–	41.03	78.14	–
24	–	37.52	71.42	–

3.5 Conclusions

This study investigates the performance of a solar energy-driven combined cooling and cold thermal energy storage cycle. The cooling process is facilitated through chemisorption, while the cold thermal energy storage is achieved by directly storing the liquid refrigerant ammonia. A case study was conducted to address the cooling demands of a large residential building. The operational strategy corresponding to a specific cooling load is also examined. The theoretical estimation of the COP for the proposed cycle is calculated to be 0.42, which aligns with values reported in the existing literature. The proposed system is more suitable for refrigerated warehouses, commercial buildings, and district cooling plants for emission-free cooling.

Acknowledgments The authors would like to acknowledge EPSRC in the UK for financial support through the grants EP/X025322/1, EP/T022701/1, EP/V042033/1, EP/V030515/1, and EP/W027593/2, and the Royal Society for their financial support (Ref: IF\R1\231053).

References

1. McMullan JT (2002) Refrigeration and the environment—issues and strategies for the future. Int J Refrig 25(1):89–99
2. Inayat A, Raza M (2019) District cooling system via renewable energy sources: a review. Renew Sust Energ Rev 107:360–373
3. Ge TS, Wang RZ, Xu ZY, Pan QW, Du S, Chen XM et al (2018) Solar heating and cooling: present and future development. Renew Energy 126:1126–1140
4. Saito A (2002) Recent advances in research on cold thermal energy storage. Int J Refrig 25(2):177–189
5. Li SF, Liu ZH, Wang XJ (2019) A comprehensive review on positive cold energy storage technologies and applications in air conditioning with phase change materials. Appl Energy 255:113667
6. Yan C, Shi W, Li X, Zhao Y (2016) Optimal design and application of a compound cold storage system combining seasonal ice storage and chilled water storage. Appl Energy 171:1–11
7. Geilfuß K, Dawoud B (2020) Analytical investigation of a zeolite-NaY-water adsorption heat and cold storage and its integration into a steam power process. Energy 195:116977
8. Li G, Hwang Y, Radermacher R (2014) Experimental investigation on energy and exergy performance of adsorption cold storage for space cooling application. Int J Refrig 44:23–35
9. Modi A, Chaudhuri A, Vijay B, Mathur J (2009) Performance analysis of a solar photovoltaic operated domestic refrigerator. Appl Energy 86(12):2583–2591
10. Gao P, Hu H, Jin S, Wang S, Chen Y, Wu W et al (2022) Solar-driven compression-assisted desorption chemisorption refrigeration/cold energy storage system. Energy Convers Manag 258:115474
11. Zhang W, Fu S, Gao P, Wu W, Pan Q, Wang L (2024) Solar photovoltaic refrigeration system coupled with a flexible, cost-effective and high-energy-density chemisorption cold energy storage module. Energy 132163
12. Mohamed MM, Almarshadi MH (2017) Simulation of district cooling plant and efficient energy air cooled condensers (part I). J Electron Cool Therm Control 7(3):45–62
13. An G, Li Y, Wang L, Gao J (2020) Wide applicability of analogical models coupled with hysteresis effect for halide/ammonia working pairs. Chem Eng J 394:125020. [cacl2 properties]

14. National Institute of Standards and Technology (NIST) (n.d.) Ammonia, NH_3—molecular weight. NIST Chemistry WebBook. Retrieved from https://webbook.nist.gov/chemistry/
15. Oliveira RG, Generoso DJ (2016) Influence of the operational conditions on the performance of a chemisorption chiller driven by hot water between 65 C and 80 C. Appl Energy 162:257–265
16. Oliveira RG, Wang RZ, Kiplagat JK, Wang CY (2009) Novel composite sorbent for resorption systems and for chemisorption air conditioners driven by low generation temperature. Renew Energy 34(12):2757–2764

Chapter 4
Actual Performance of PV Micro-Installations in the Same Location: Case Study for Single Family Houses

Anna Justyna Werner-Juszczuk

Abstract PV is an important element of the energy transformation of the residential construction sector. This paper presents a comparison of the actual performance of two PV micro-installations (up to 10 kWp) located on neighbouring single-family buildings with a different orientation: south (roof tilt angle 42°) and east-west (30°). The buildings are located in a cold climate area, in north-eastern Poland. The analysis was based on actual measurements of electricity production and electricity costs incurred according to invoices for the period 2022–2024. Despite the same climatic conditions and lack of shading, the south-facing PV installation achieved 16% higher electricity production per 1 kWp than the one facing east-west. This PV installation also provides 27% higher monthly savings per 1 kWp, resulting from the installation of the PV. In both cases, the installation of the PV system contributed to a drastic reduction in electricity costs at the level of 93–94%. The SPBT for the PV installation with the south orientation is lower that east-west system: 6.3 versus 7.6 years. In order to ensure energy self-sufficiency of the buildings in both cases it is necessary to use electricity storage facilities.

Keywords Economic Analysis · PV Orientation · PV Panel Tilt · Optimal PV Location

4.1 Preface

Literature Review

The use of photovoltaic panel systems (PV) in the residential construction sector is becoming a necessity to ensure energy self-sufficiency of buildings, but their efficiency strictly depends on the climatic conditions of the location [1]. Since these are

A. J. Werner-Juszczuk (✉)
Bialystok University of Technology, Bialystok, Poland
e-mail: a.juszczuk@pb.edu.pl

L. Chen (ed.), *Clean Energy Research*, Green Energy and Technology,
https://doi.org/10.1007/978-3-031-99312-1_4

constant for a given building, various methods are sought to improve the electricity production from PV. In the case of micro-installations (i.e. power up to 10 kWp), one of the methods is to increase the surface area of the panels or to use the optimal orientation of the panels while maintaining the optimal tilt angle [2]. Research in this area has been conducted for different climatic conditions around the world [3] including low-latitude regions around the equator [4], the Middle East, Africa, the Asia Pacific, America and Europe [5, 6] using various optimization techniques [7]. For the climatic conditions of Poland, similar to the area of Central and Eastern Europe, the optimal tilt angle of PV panels is 30–45° [8, 9] and the best orientation is south [8, 10]. The east-west orientation of PV panels is less advantageous, but as proven in Germany and Austria, using steeper tilts for this orientation can increase the PV yield during morning and afternoon hours [11].

Considering that within the next few decades all residential buildings will have to be zero-energy, regardless of the climatic conditions of their location, it is important to analyse the improvement of the yield from PV panels, especially in climates where conditions are not conducive to the efficient use of solar energy, an example of which is Poland.

There are many examples of models for simulating electricity production by PV systems in the literature, such as the general method verified with statistical indicators presented in [12], the models of Skartveit and Olseth, Perez and Reindl [13], models for PV locations with no meteorological data [14] or isotropic and anisotropic models [15]. However, there are not enough analyses carried out for real objects located in a cold climate, such as the climate of north-eastern Poland. This paper fills this gap by presenting an analysis of the actual electricity production of two photovoltaic systems installed on the roofs of neighbouring single-family houses located in Bialystok, in north-eastern Poland, with different panel orientations (south and east-west) and different tilt angles (42° and 30°). The measured parameters allowed for comparing the efficiency of the PV systems in real operating conditions and for conducting an economic analysis.

4.2 Objects and Methods

Objects of the Analysis

The analysis covered PV systems located on the roofs of two single-family buildings located in Bialystok, in north-eastern Poland (see Fig. 4.1). The average sunshine duration for Bialystok is 1755 h, the highest value occurs from May to August (251–259 h), and the lowest from November to February (29–38 h) [16]. Monthly irradiance on horizontal in years from 2013 to 2023 varied from 14 to 19 kWh/month in winter to 141–173 kWh/month in summer, and in 84–163 kWh/month in spring and 21–91 kWh/month in autumn [17].

The buildings are located in an open space, 100 m from each other, without any elements shading the roof (no other buildings or tall trees).

Building A is a single-storey building with a usable attic and a gable roof with an angle of inclination of 42° oriented south-north. The PV installation consists of 16 panels of a total area of 30.5 m^2, located on the southern slope of the roof (see Fig. 4.1).

Building B is a single-storey building with a non-usable attic, a gable roof with an angle of inclination of 30°, oriented east-west. Twelve panels are located on the eastern slope, 12 panels on the western slope, with a total area of 44 m^2 (see Fig. 4.1).

Both buildings are connected to the city grid and do not have electricity storage facilities. Building A was equipped with a PV installation in December 2021, and building B in March 2022. The characteristics of both PV installations are presented in Table 4.1. The installed power in building B is 1.6 times greater than in building A.

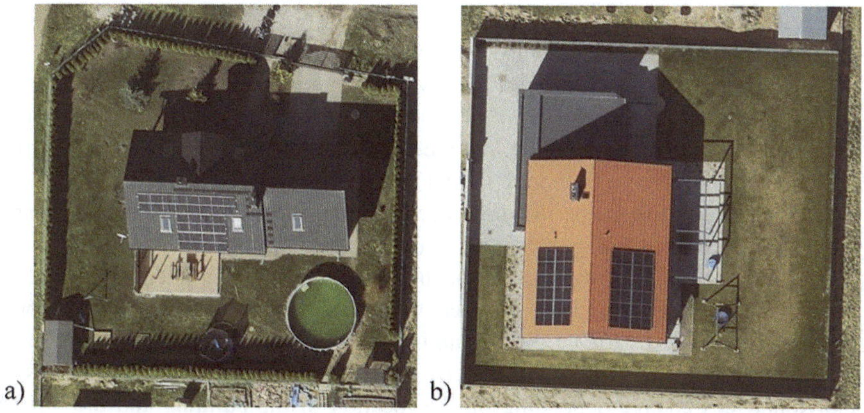

Fig. 4.1 Views of the analysed: (**a**) building A, and (**b**) building B with PV systems (source [18])

Table 4.1 Characteristics of the analysed photovoltaic systems

	Building A	Building B
Roof slope	42°	30°
Type of panels	Q-CELLS G9 390 Wp	JKM405N-6RL3 405 Wp
Cell type	Monocrystalline	N-type Monocrystalline
Module efficiency STC (%)	≥20.6%	21.22
Degradation of each year	0.5%	0.4%
Number of panels	16	24
Surface area of 1 panel	1030 × 1840 mm	981x1855 mm
Inverter	FRONIUS 5.0–3-M	SUN2000-8KTL-M1
Installed power	6.24 kWp	9.27 kWp

Source of Data

The analysis was based on actual data from both buildings. Data on the amount of electricity generated by the PV system was obtained from the *Solar.web* application (building A) and *Fusion Solar* (building B). This data is presented for the period from April 2022 to December 2024.

The actual amount of electricity drawn from the city grid and delivered to the grid by buildings on a monthly basis was obtained from the electricity supply invoices of the Polish Energy Group PGE. Due to the method of invoicing, this data was obtained for the period from May 2022 to August 2024.

Electricity Billing System

The analysed PV installations are settled in the system (net metering), which is applicable for installations installed by March 31, 2022 and which will be applicable until 2037 for installations meeting this condition. This system means that surplus energy is transferred through the PV installation to the grid, and then, for a period of 12 months from the moment of production, the prosumer, i.e. the owner of the PV installation, can collect 80% of the energy previously transferred to the grid. So there is no exchange of money in this system, only the exchange of kWh. This system allows, during the period of reduced energy production, i.e. the autumn and winter, collecting the energy produced in the summer without incurring any costs.

The electricity billing system for the analysed buildings, supplied by the Polish Energy Group PGE, includes a number of fixed fees, which are determined for a monthly period, and variable fees, established for 1 kWh of consumed electricity. The method of calculating electricity costs is presented in Eq. (4.1). The values of individual fees change depending on the location of the building and the billing period (Table 4.2), as well as the applied electricity sales tariff [19, 20].

$$C = \left(C_E + C_q + C_{vnet} + C_{RES} + C_{co}\right) \cdot E + \left(C_{cnet} + C_t + C_P + C_{sub}\right) \cdot m \tag{4.1}$$

where C—cost of electricity [PLN/billing period], C_E—fee for active electricity [PLN/kWh], C_q—quality fee [PLN/kWh], C_{vnet}—variable network fee [PLN/kWh], C_{RES}—renewable energy sources fee [PLN/kWh], C_{co}—cogeneration fee [PLN/kWh], E—the amount of active electricity used [kWh], C_{cnet}—fixed network fee [PLN/month], C_t—transition fee [PLN/month], C_P—power fee [PLN/month], C_{sub}—subscription fee [PLN/month], m—number of months for which the electricity charge is calculated [−].

The fees, presented in Table 4.2, were converted from PLN to Euro, taking into account the exchange rate of 4.21 PLN = 1 €, and include taxes (5% in 2022; 23% in 2023 and 2024).

Table 4.2 Values of electricity fees in analysed buildings [source: own elaboration on invoices]

Fee		Unit	May–Jun 2022	Jul–Dec 2022	Jan–Dec 2023	Jan–Jun 2024
C_E	Active electricity	€/kWh	0.1020	0.1020	0.1210	0.1210
C_q	Quality	€/kWh	0.0024	0.0024	0.0028	0.0028
C_{vnet}	Variable network	€/kWh	0.0554	0.0054	0.0649	0.0649
C_{RES}	Res	€/kWh	0.0002	0.0002	0.0000	0.0000
C_{co}	Cogeneration	€/kWh	0.0010	0.0010	0.0014	0.0018
C_{cnet}	Fixed network	€/month	1.6361	1.6361	1.9166	1.9166
C_t	Transition	€/month	0.0823	0.0823	0.0964	0.0964
C_P	Power	€/month	2.3594	3.3046	3.9004	4.3532
C_{sub}	Subscription	€/month	0.1871	0.1871	0.2191	0.2191

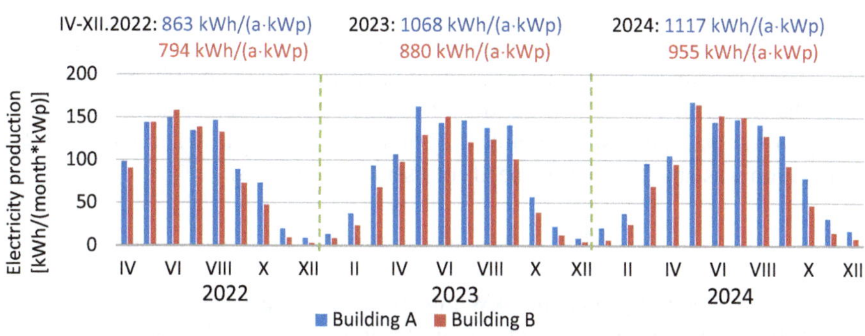

Fig. 4.2 Electricity production by PV systems in building A and B per 1 kWp of installed power and total PV production in each analysed period

4.3 Results

Electricity Production

The PV installation on building B produced more electricity than the system on building A, respectively by 43% in the period April–December 2022, 28% in 2023 and by 33% in 2024. However, comparing the produced energy in relation to 1 kWp of installed power (see Fig. 4.2), the PV installation on building A produced 9% more electricity in the period April–December 2022 and by as much as 21% and 17% in 2023 and 2024 (on average by 16%).

Analyzing the production in individual months, it can be concluded that electricity production in building A was significantly higher than in building B in the winter-autumn period (October–February), from 47% to 218%, depending on the year. Such large differences are partly due to snow cover or frost that could appear on the surface of the PV panels during these months, which, according to [21], can reduce the annual power generation by up to 15%. Observations by users show that snow and frost disappeared from the surface of the panels more quickly on building A than on building B.

In late summer and spring, the differences between the electricity production by the PV system in building A and B are smaller. They amount to 9–11% for April and August and 22–39% for September. The smallest differences between the production of electricity in relation to 1 kWp of installed capacity are for May (0–2%). July 2023 brought higher production by 21% in building A. Building B had higher production than building A (2–5%) only in June and July 2022 and 2024 and in June 2023.

The results clearly indicate that for most of the year the efficiency of the PV installation on building A, oriented to the south, is higher than on building B, with PV panels oriented east-west.

Next, based on invoices, the amount of energy delivered to the grid by PV systems was determined, as well as the amount of energy that the buildings had to take from the city grid. This data allowed us to determine the monthly demand for electricity of the buildings (see Formula 4.2) and to determine to what extent the PV installation was able to meet the needs of users (see Fig. 4.3).

$$E_{dem} = E_{PV} + E_{from\_grid} - E_{to\_grid} \quad (4.2)$$

where E_{dem}—electricity demand [kWh/month], E_{PV}—electricity produced by PV system [kWh/month], E_{from_grid}—electricity taken from grid [kWh/month], E_{to_grid}—electricity transferred to grid [kWh/month].

The demand for electricity of building A from May 2022 to August 2024 was 13,051 kWh (monthly average 466 kWh), and building B: 17,542 kWh (monthly average 627 kWh). The demand for electricity of building B is therefore 1.3 times greater than for building A.

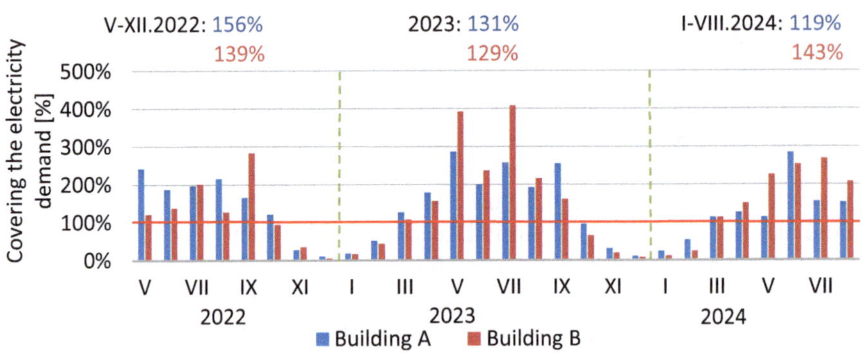

Fig. 4.3 The degree (%) of coverage of the electricity demand by the PV system in the analysed buildings

In Fig. 4.3, the red line runs at the value of 100%, which means that the amount of electricity generated by the PV system on a monthly basis is equal to the electricity demand of the building.

On average, throughout the analysed period, the PV installation in building A covered 133% of the building's electricity demand, and 137% in building B. These values are at a similar level. The monthly settlement shows that the PV installation in both buildings is not able to cover the demand in the months from November to February. However, in September, May, or July, the electricity production surpluses in building B are much higher than in building A, ranging from 107% to 151%.

Economic Analysis

The most important aspect from the point of view of PV system users is the economic aspect. In the analysed period, most of the fees paid by users were fees for fixed parts such as network, transition, power or subscription fees. Partially, despite overproduction (see Fig. 4.3), users incurred costs related to active energy. This results from the discrepancy between electricity production and demand for it during the day and the month.

Based on the prices specified in the invoices of the individual components of the electricity cost in formula (4.1), calculations of the electricity costs were made if the buildings did not have PV installations (see Table 4.3).

The PV installation on both buildings allowed for a reduction in electricity costs in the analysed period by 93% in building A and 94% in building B, despite the lack of energy storage. Currently, users incur a monthly cost of electricity of around 7 €.

Monthly savings resulting from the installation of the PV system in building B are 22% higher than in building A. However, when related to 1 kWp of installed power, the monthly savings for building A are 27% higher than in building B.

The simple payback period (SPBP) of the PV system in building A is shorter by 16.5 months than the SPBT for building B. The SPBT related to 1 kWp of installed power for building A is 1 year, and for building B: 0.8 year. This is partially influenced by the 4% higher cost of 1 kWp of installed power in building A compared to the cost in building B. In both cases, the SPBT of the PV system is shorter than the assumed minimum lifespan of the PV panels (30 years).

The results clearly indicate that the south orientation of PV installation allows for a reduction in the required panel area (30.5 m^2 versus 44 m^2 for east-west orientation) as well as the investment cost of the installation (the costs of the entire installation in building A were 33% lower than in building B). At the same time, both PV installations cover the demand for electricity in the buildings to a similar extent.

Table 4.3 Results of the economic analysis of PV systems

	Building A	Building B
Cost of electricity with PVPhotovoltaic (PV) system		
May 2022–August 2024	183.01 €	201.98 €
Average monthly	6.54 €	7.21 €
Cost of electricity without PVPhotovoltaic (PV) system		
Total: May 2022–August 2024	2648.40 €	3215.57 €
Fee components: C_E	1656.50 €	2016.81 €
C_q	38.15 €	46.44 €
C_{vnet}	766.67 €	960.61 €
C_{RES}	0.58 €	0.73 €
C_{co}	20.44 €	24.92 €
C_{cnet}	51.42 €	51.42 €
C_t	2.59 €	2.59 €
C_P	106.18 €	106.18 €
C_{sub}	5.88 €	5.88 €
Average monthly	94.59 €	114.84 €
Monthly savings from installing a PV system	88.05 €	107.63 €
Monthly savings from installing a PV system per 1 kWp	14.11 €	11.07 €
Total cost of PV system	6612.83 €	9857.48 €
Total cost of PV system per 1 kWp	1059.75 €	1014.14 €
SPBT	6.3 year	7.6 year
SPBT per 1 kWp	1 year	0.8 year

4.4 Conclusions

The location of the PV installation on a south-facing roof with a 42° inclination angle is more advantageous than on east-west roof slopes with a 30° inclination angle for the cold climate of north-eastern Poland. This is evidenced by the higher electricity yield (on average by 16%) and 27% higher monthly savings resulting from the installation of the PV system. The simple payback period (SPBT) of the south-oriented PV installation is also shorter than that of the east-west-oriented installation: 6.3 versus 7.6 years.

In both cases, the installation of the PV system contributed to a drastic reduction in electricity costs at the level of 93–94%. Despite the electricity overproduction by both PV systems at the level of 133% in building A and 137% in building B, the billing system with the electricity supplier caused users to bear the costs of active energy drawn from the grid. The reason for this is the low production of electricity, especially in the autumn-winter period (November–February). This indicates the validity of using electricity storage facilities to ensure the self-sufficiency of buildings.

The presented analysis of the actual electricity PV production, investment outlays and costs of purchasing electricity clearly indicates the justification for prudent design of buildings in terms of orientation and the tilt angle. Favourable orientation to the south allows the use of a smaller surface of PV panels with a simultaneous higher yield, and also reduces the investment costs of the system.

Acknowledgements This work was supported by the Bialystok University of Technology and financed by the Ministry of Science and Higher Education of the Republic of Poland (grant number WZ/WB-IIL/2/2023).

References

1. Olczak P, Olek M, Matuszewska D, Dyczko A, Mania T (2021) Monofacial and bifacial micro PV installation as element of energy transition—the case of Poland. Energies 14:499
2. Olczak P, Komorowska A (2021) An adjustable mounting rack or an additional PV panel? Cost and environmental analysis of a photovoltaic installation on a household: a case study in Poland. Sustain Energy Technol Assess 47:101496
3. Jacobson MZ, Jadhav V (2017) World estimates of PV optimal tilt angles and ratios of sunlight incident upon tilted and tracked PV panels relative to horizontal panels. Sol Energy 169:55–66
4. Mukisa N, Zamora R (2022) Optimal tilt angle for solar photovoltaic modules on pitched rooftops: a case of low latitude equatorial region. Sustain Energy Technol Assess 50:101821
5. Rehman T, Qaisrani MA, Shafiq MB, Baba YF, Aslfattahi N, Shahsavar A, Cheema TA, Park CW (2025) Global perspectives on advancing photovoltaic system performance—a state-of-the-art review. Renew Sust Energ Rev 207:114889
6. Chen XM, Li Y, Zhao BY, Wang RZ (2022) Are the optimum angles of photovoltaic systems so important? Renew Sust Energ Rev 124:109791
7. Yadav AK, Chandel SS (2013) Tilt angle optimization to maximize incident solar radiation: a review. Renew Sust Energ Rev 23:503–513
8. Olczak P (2022) Energy productivity of microinverter photovoltaic microinstallation: comparison of simulation and measured results—Poland case study. Energies 15:7582
9. Bukowski M, Majewski J, Sobolewska A (2021) Macroeconomic electric energy production efficiency of photovoltaic panels in single-family homes in Poland. Energies 14:126
10. Ciuman P, Kaczmarczyk J, Winnicka-Jasłowska D (2025) Investigation of energy-efficient solutions for a single-family house based on the 4E idea in Poland. Energies 18:449
11. Hartner M, Ortner A, Hiesl A, Haas R (2015) East to west—the optimal tilt angle and orientation of photovoltaic panels from an electricity system perspective. Appl Energy 160:94–107
12. Bakirci K (2012) General models for optimum tilt angles of solar panels: Turkey case study. Renew Sust Energ Rev 16(8):6149–6159
13. Danandeh MA, Mousavi SMG (2018) Solar irradiance estimation models and optimum tilt angle approaches: a comparative study. Renew Sust Energ Rev 92:319–330
14. Chinchilla M, Santos-Martín D, Carpintero-Rentería M, Lemon S (2021) Worldwide annual optimum tilt angle model for solar collectors and photovoltaic systems in the absence of site meteorological data. Appl Energy 281:116056
15. Tamoor M, Bhatti AR, Farhan MA, Sherefa A (2024) Optimizing tilt angle of PV modules for different locations using isotropic and anisotropic models to maximize power output. Sci Rep 14:30197
16. Institute of Meteorology and Water Management Homepage, https://klimat.imgw.pl/pl/climate-normals/USL, Accessed 30 Jan 2025
17. JRC. Photovoltaic Geographical Information System (PVGIS)—European Commission, https://re.jrc.ec.europa.eu/pvg_tools/en/, Accessed 30 Jan 2025
18. Geoportal Information System Homepage, https://gisbialystok.pl/, Accessed 28 Jan 2025
19. Sakowicz RA, Werner-Juszczuk AJ (2022) Operating costs of air and ground source heat pumps. Instal 7–8:33–37
20. Polish Energy Group Distribution Homepage, https://pgedystrybucja.pl/, Accessed 30 Jan 2025
21. Ruan T, Wang F, Topel M, Laumert B, Wang W (2024) A new optimal PV installation angle model in high-latitude cold regions based on historical weather big data. Appl Energy 359:122690

Chapter 5
Numerical Analysis of Jet-Stabilized Hydrogen Combustion in Co-Flow Configuration

Sheher Bano, Johannes Mohs, Manfred Wirsum, and Ekachai Juntasaro

Abstract Gas turbines are commonly used in energy sectors, primarily relying on fossil fuels, which produce significant NOx and CO_2 emissions, contributing to environmental pollution. Hydrogen has emerged as a clean alternative. However, its high reactivity compared to natural gas poses challenges such as self-ignition and flashback in premixed combustion chambers. To address this, jet stabilization mechanisms have been proposed. This study numerically analyzes injector designs for jet stabilization in hydrogen combustion. Using a 2-D structured mesh with SST k-ω turbulence and species transport models, we evaluate how fuel and air inlet diameters and velocities influence flame stabilization and NOx emissions. The findings offer guidance for designing injectors that achieve stable combustion and reduced emissions, supporting hydrogen as a clean energy source for gas turbines.

Keywords Jet stabilized combustion · Hydrogen combustion · NOx emissions

S. Bano (✉)
The Sirindhorn International Thai German Graduate School of Engineering, KMUTNB, Bangkok, Thailand

Institute of Power Plant Technology, Steam and Gas Turbines, RWTH Aachen University, Aachen, Germany

J. Mohs · M. Wirsum
Institute of Power Plant Technology, Steam and Gas Turbines, RWTH Aachen University, Aachen, Germany

E. Juntasaro
The Sirindhorn International Thai German Graduate School of Engineering, KMUTNB, Bangkok, Thailand

L. Chen (ed.), *Clean Energy Research*, Green Energy and Technology,
https://doi.org/10.1007/978-3-031-99312-1_5

5.1 Introduction

Gas turbines are widely used in power generation, industry, and aviation due to their reliability, efficiency, and multi-load capability [1]. However, conventional turbines burn fossil fuels, emitting carbon dioxide (CO_2), nitrogen oxides (NO_x), and other greenhouse gases, contributing to air pollution and global warming. Hydrogen offers a clean alternative fuel for gas turbines, eliminating CO_2 emissions [2, 3]. However, its high reactivity, low ignition energy, short delay time, and fast flame speed increase flashback risk compared to natural gas [4]. Stable, low-emission combustion requires careful injector design.

Hence, considerable efforts have been made in recent years to obtain stable premixed hydrogen flames in a combustion chamber using either swirl stabilization, staged combustion, micro-flames, jet stabilization, etc.

Swirl-stabilized flames are the industry standard for both diffusion and premixed flames, with extensive literature available. In contrast, jet stabilization has only recently gained attention due to its promise for H_2 combustion. This work investigates the general geometric influences on jet stabilization. Jet stabilization is effective for hydrogen combustion due to its higher reactivity. It avoids the low velocity recirculation zone, which leads to flashback. So, swirl stabilization is effective for low-reactivity fuels, where low flame speed reduces the risk of upstream flame propagation [5].

For jet-stabilized flames, instead of a swirling flow, a high-velocity jet of fuel and air is used to create a recirculation zone to stabilize the flame, as shown in Fig. 5.1. A design proposed by the German Aerospace Center [6–8] employs multiple coaxial fuel-air injectors arranged circumferentially and offset from the symmetry axis. The general concept is shown in Fig. 5.1. Fuel and air inlets are offset from the symmetrical axis and arranged coaxially around the circumference. Air and fuel mix in the tube before being injected into the combustion chamber. The high injection velocity creates a central recirculation zone that stabilizes combustion. This mixing concept is taken as a reference, and an altered design is used in this study.

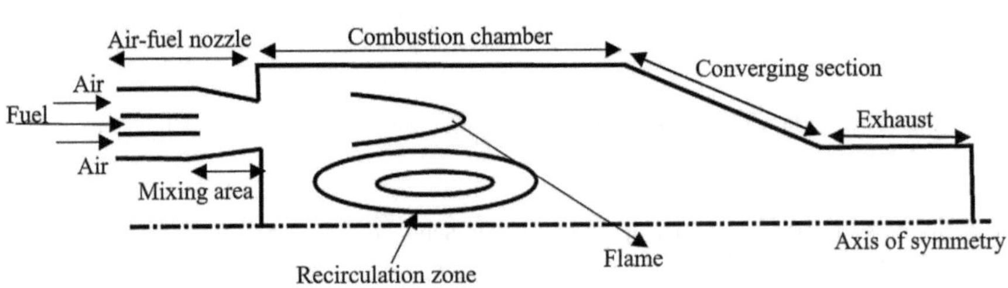

Fig. 5.1 Schematic of Felix Grimm concept

5.2 Proposed Design

The proposed design uses a toroidal recirculation zone per injector at the mixing tube outlet, replacing the large central recirculation zone formed by off-axis injectors, as shown in Fig. 5.2. Instead of a circumferential arrangement, the injectors are configured vertically and horizontally in a multi-nozzle concept as employed by Micromix (MMX), Doosan cluster combustor, etc. [2, 9]. The proposed injector concept is shown in Fig. 5.2.

A high-velocity fuel jet is injected into a converging co-flowing air jet, entering the mixing area. A velocity gradient between air and fuel in the mixing area creates a turbulent shear layer, enhancing mixing. The fuel-air mixture is then directed into the combustion zone in a free jet arrangement. At the mixing zone exit, a second shear layer forms between the fuel-air mixture and the hot combustion gases. This creates a toroidal recirculation zone around the fuel-air jet. Downstream, a flame front forms where the fuel-air jet breaks up, and the recirculated hot exhaust gases ignite the mixture.

This research numerically analyzes jet-stabilized hydrogen combustion to investigate flame stabilization and emission trends, aiming to advance jet-stabilized mechanisms for industrial gas turbines. Table 5.1 presents the performance data of the gas turbine used in the study [10].

5.3 Applied Design Boundary Condition

This analysis determines appropriate dimensional and boundary condition parameters by deriving relationships between them, focusing on maintaining a constant equivalence ratio while varying other parameters. An injector equivalence ratio of 0.43 is used, with 73% of the total air allocated for combustion and the remainder for cooling. The equivalence ratio φ is given as

$$\left(\frac{\dot{m}_f}{\dot{m}_a}\right)_{act} = \left(\frac{(\rho AV)_f}{(\rho AV)_a}\right)_{act} = \varphi \cdot \left(\frac{\dot{m}_f}{\dot{m}_a}\right)_{st} \tag{5.1}$$

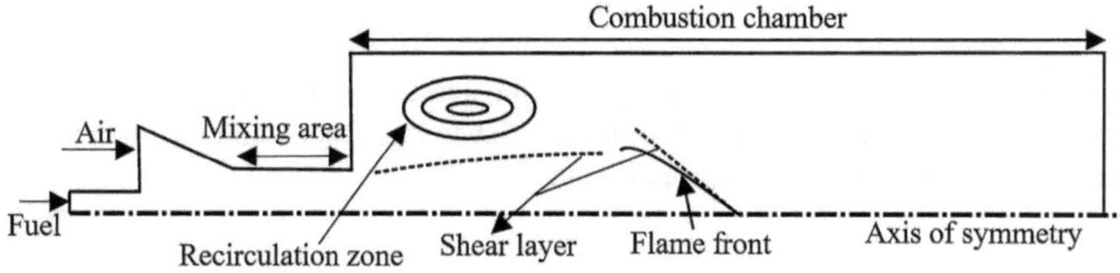

Fig. 5.2 Schematic of the proposed design of jet stabilization mechanism

Table 5.1 Performance data of gas turbine Siemens [10]

Performance parameters	Value
Pressure (bar)	22
Power (MW)	60
Temperature of air (K)	778.9
Mass flow rate of air (kg/s)	133.7
Mass flow rate of fuel (kg/s)	1.216
Temperature of fuel (K)	300
Overall equivalence ratio	0.3

The mass of fuel and air depends on density, inlet-velocity, and area. With constant temperature, density remains unchanged, so velocity and area are varied for the parametric study. The study examines three cases:

In Case 1, baseline dimensional parameters are selected to meet design and performance requirements. Fuel velocity is then systematically increased until flame blow-off occurs. Since the equivalence ratio is constant, changes in fuel velocity correspond to changes in air velocity.

$$\frac{V_f}{V_a} = \varphi \cdot \left(\frac{\dot{m}_f}{\dot{m}_a} \right)_{st} \cdot \left(\frac{(\rho A)_a}{(\rho A)_f} \right)_{const} \tag{5.2}$$

In Case 2, the air inlet dimension and velocity are adjusted while keeping the fuel inlet dimension and velocity constant until air velocity becomes significantly low.

$$(AV)_a = \frac{1}{\varphi \cdot \left(\frac{\dot{m}_f}{\dot{m}_a} \right)_{st}} \cdot \left(\frac{(\rho AV)_f}{\rho_a} \right)_{const} \tag{5.3}$$

In Case 3, the fuel dimensions and boundary parameters are varied until reaching the point when air velocity exceeds the velocity of the fuel.

$$(AV)_f = \varphi \cdot \left(\frac{\dot{m}_f}{\dot{m}_a} \right)_{st} \cdot \left(\frac{(\rho AV)_a}{\rho_f} \right)_{const} \tag{5.4}$$

In this analysis, the length of the injector, the length of the mixing area, the diameter of the mixing area, the length of the combustion chamber, and the diameter of the combustion chamber are kept constant.

5.4 Numerical Study

Numerical simulation on the jet stabilization combustion mechanism is performed using ANSYS Fluent 2024 R1.

Geometry

To reduce computational cost, this study employs a 2D axisymmetric geometry, as initial 3D analysis showed close quantitative similarity to 2D results. Table 5.2 lists the baseline dimensional parameters of the analyzed injector.

Boundary Conditions and Simulation Model

Both the fuel and air inlets are modelled as velocity inlets, with jet velocities determined from the parametric study. A pressure outlet is applied, and operating pressure, fuel temperature, and air temperature are specified in Table 5.1. The domain is shown in Fig. 5.3. The SST k-ω turbulence model is used for its accuracy in modelling high shear rates. Combustion is modelled using a species transport approach with the eddy dissipation concept and Conaire et al. [11] H_2, while the Thermal NO_x model is used for modelling NO_x emissions [12].

Mesh Independence Study

Two mesh types were considered: structured and unstructured. The structured mesh uses quadrilateral elements sized at 0.2 mm with a wall model for the near-wall region. The unstructured mesh also uses quadrilateral elements, with a 0.5 mm base size, an inflation layer for the boundary layer, and a central refinement zone with

Table 5.2 Baseline design dimensional parameters

Parameters	Value
Diameter of fuel inlet (mm)	5
Diameter of oxidizer inlet (mm)	22
Length of injector (mm)	12
Diameter of mixing area (mm)	15
Length of mixing area (mm)	12
Diameter of combustion chamber (mm)	60
Length of combustion chamber (mm)	300

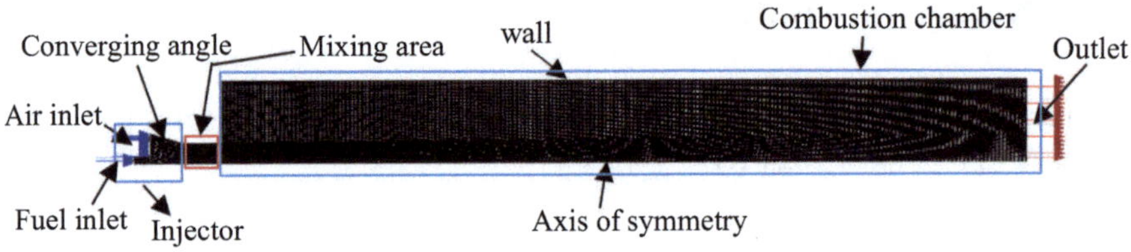

Fig. 5.3 Mesh and boundary conditions of the axisymmetric domain

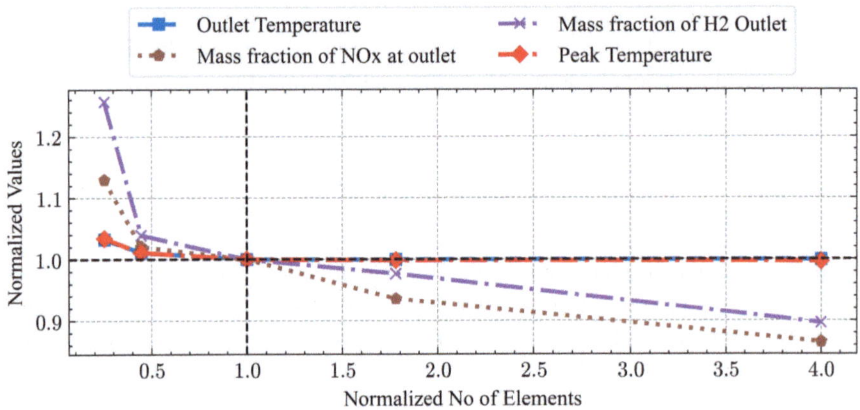

Fig. 5.4 Mesh independence study

0.25 mm elements at the axis of symmetry. Both meshes yielded similar results, but the structured mesh was chosen for its higher accuracy and computational efficiency.

A mesh independence study was performed on the structured mesh, with numerical convergence set to 10^{-6} for all equations. Monitored parameters include outlet temperature, peak temperature, mass fraction of NO_x, and H_2 at the outlet. After analyzing the graph shown in Fig. 5.4, a mesh with 0.983 million elements and a 0.1 mm element size was selected, balancing accuracy and computational efficiency.

Numerical Results

This section presents simulation results, focusing on the temperature distribution and the flow field for the baseline design parameters. Additionally, parametric study results are provided as normalized data relative to the baseline for comparative analysis.

Flame stabilization is visualized in the flow field as shown in Fig. 5.5, illustrating the mechanism described in Sect. 2.

The temperature distribution of the jet stabilization mechanism is shown in Fig. 5.6. The maximum temperature occurs at the jet break-up zone, creating a

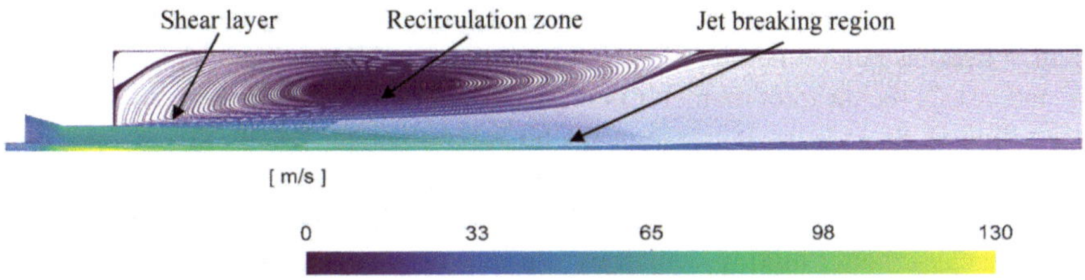

Fig. 5.5 Flow field using velocity magnitude

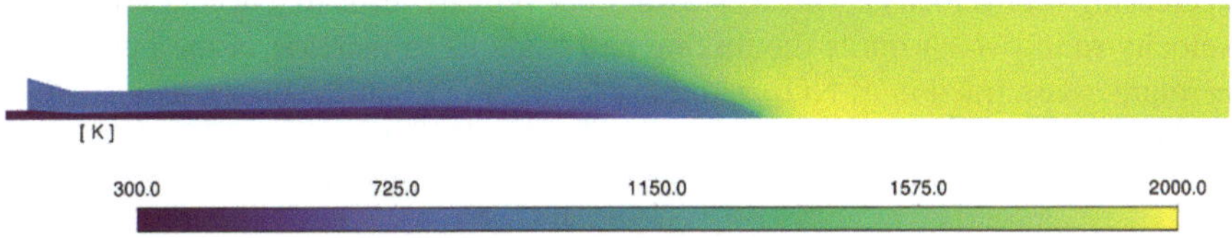

Fig. 5.6 Static temperature distribution contour

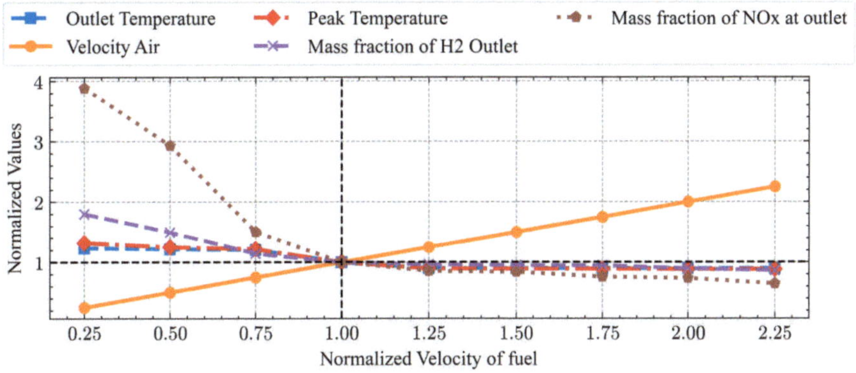

Fig. 5.7 Normalized parametric study graph of Case 1

flame front. The maximum temperature at the flame front is 1973.19 K, which is 1.029 times higher than the exit temperature of the domain, indicating partially premixed combustion.

The parametric study evaluates the effects of boundary and dimensional parameters on jet-stabilized combustion. Figure 5.7 presents the results for Case 1 (increasing fuel velocity), normalized to the baseline design simulation.

As seen in Fig. 5.7, with the increase in the velocity of the fuel jet, all other parameters besides the velocity of the air are decreasing. High velocity enhances turbulent mixing and leads to shorter residence time in the combustion chamber. Due to better mixing, the mass fraction of H_2 at the outlet, the outlet temperature, and peak temperature decreases. Since NO_x formation is highly temperature dependent, a decrease in temperature leads to a decrease in NO_x at the outlet.

In Case 2, the air inlet diameter is increased while the inlet velocity is decreased according to Sect. 3. An increase in the air inlet diameter results in a steeper converging angle, causing an increased constriction in the fuel jet in the mixing tube, and further increasing its velocity. Both the slower air velocity and the higher fuel velocity cause better mixing, which in turn improves combustion. This reduces the mass fraction of H_2 at the outlet and lowers the peak combustion temperature, which further reduces the mass fraction of NO_x at the outlet, as shown in Fig. 5.8.

A similar effect can be observed in Case 3 (increasing fuel inlet diameter, decreasing fuel velocity). The decreasing fuel velocity leads to a decreased fuel-air velocity ratio, which limits the mixing and leads to an increase in the outlet temperature, mass fraction of NO_x, and mass fraction of H_2 at the outlet, as visualized in Fig. 5.9.

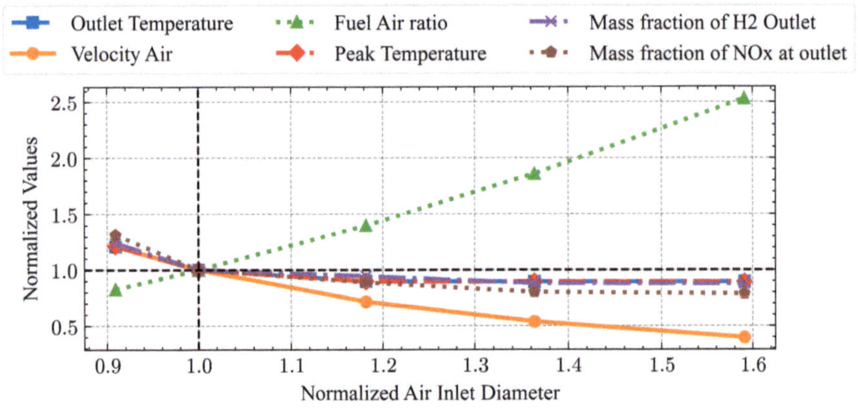

Fig. 5.8 Normalized parametric study graph of Case 2

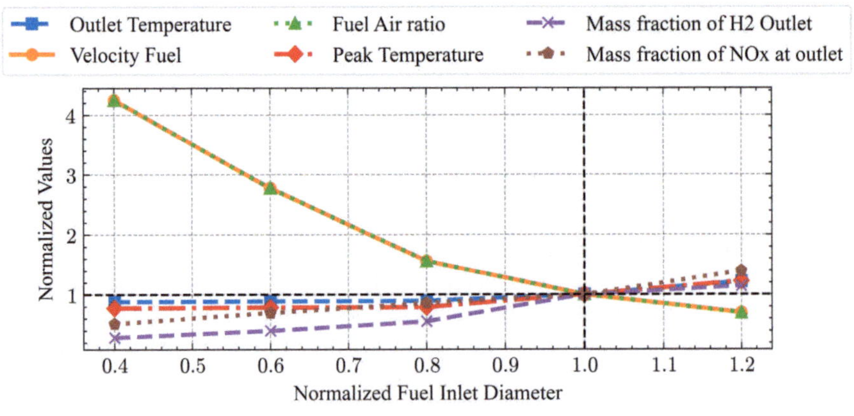

Fig. 5.9 Normalized parametric study graph of Case 3

The parametric analysis identifies two key factors influencing the combustion process: the fuel-air velocity ratio and the overall velocity of the mixed jet entering the combustion zone. A higher fuel-air velocity ratio improves premixing through increased shear between the fuel and air flows. An increased overall velocity enhances recirculation in the combustion zone, promoting mixing until flame blow-off occurs. However, the mixing within the mixing tube is insufficient, leading to an air layer along the tube wall. While this reduces the risk of boundary layer flashback, a more uniform premixed mixture could improve flame stability and lower NO_x emissions. Additionally, the effect of dilution air introduced through dilution holes in a practical combustion chamber is not modeled. This could further decrease NO_x emissions. Future research should analyze 3D models of alternative mixing tube geometries, such as those incorporating turbulence generators, using CFD.

5.5 Conclusion

This research numerically analyses hydrogen combustion in a jet-stabilized combustor, focusing on achieving stable flames with low emissions. The study highlights the impact of design parameters on flame stabilization, with a parametric analysis performed on a 2D structured mesh. The results show that excessive changes in design parameters adversely affect stability. Three design modifications were analyzed: increasing fuel velocity enhances mixing but reduces combustion due to shorter residence time; increasing air diameter reduces air velocity, lowering flame temperature and NO_x mass fraction; increasing fuel diameter raises both flame temperature and NO_x mass fraction. These findings underscore the need for a balanced design to maintain equilibrium. This study contributes to the development of precise jet-stabilized injector designs for clean, stable hydrogen combustion, laying the foundation for future research. Experimental validation will follow to address flashback and further assess the effects of design parameters, ensuring safe and stable hydrogen combustion in gas turbines.

References

1. Mane S (2023) Advancements in gas turbine engine technology: a conceptual aspect. Int J Enhanc Res Sci Technol Eng 12(7):37–41
2. H. H.-W. Funke, N. Beckmann, J. Keinz, and A. Horikawa, "30 Years of Dry-Low-NOx Micromix Combustor Research for Hydrogen-Rich Fuels—An Overview of Past and Present Activities," J. Eng. Gas Turbines Power, vol. 143, no. 071002, Mar. 2021, https://doi.org/10.1115/1.4049764
3. A. Haj Ayed, K. Kusterer, H. H.-W. Funke, J. Keinz, C. Striegan, and D. Bohn, "Experimental and numerical investigations of the dry-low-NOx hydrogen micromix combustion chamber of an industrial gas turbine," Propuls. Power Res., vol. 4, no. 3, pp. 123–131, Sept. 2015, https://doi.org/10.1016/j.jppr.2015.07.005

4. T. Tan, W. Fan, and R. Zhang, "Study on hydrogen staged combustion for gas turbine," Fuel, vol. 349, p. 128700, Oct. 2023, https://doi.org/10.1016/j.fuel.2023.128700
5. G. C. Krieger, A. P. V. de Campos, F. L. Sacomano Filho, and R. C. de Souza, "A swirler stabilized combustion chamber for a micro-gas turbine fuelled with natural gas," J. Braz. Soc. Mech. Sci. Eng., vol. 34, pp. 441–449, Dec. 2012, https://doi.org/10.1590/S1678-58782012000400004
6. F. Grimm, "Low-Order Reactor Network Based Prediction of Pollutant Emissions Applied to FLOX Combustion," Energies, vol. 15, p. 1740, 2022
7. F. Grimm, T. Lingstädt, P. Kutne, and M. Aigner, "Numerical and Experimental Study of a Jet-and-Recirculation Stabilized Low Calorific Combustor for a Hybrid Power Plant," Energies, vol. 14, no. 3, Art. no. 3, Jan. 2021, https://doi.org/10.3390/en14030537
8. S. Hasemann, H. Seliger, P. Kutne, and M. Aigner, "Experimental and Numerical Design Study for a Small Scale Jet-Stabilized Micro Gas Turbine Combustor," presented at the ASME Turbo Expo 2018: Turbomachinery Technical Conference and Exposition, American Society of Mechanical Engineers Digital Collection, Aug. 2018. https://doi.org/10.1115/GT2018-75050
9. E.-S. Cho, H. Jeong, J. Hwang, and M. Kim, "A Novel 100% Hydrogen Gas Turbine Combustor Development for Industrial Use," presented at the ASME Turbo Expo 2022: Turbomachinery Technical Conference and Exposition, American Society of Mechanical Engineers Digital Collection, Oct. 2022. https://doi.org/10.1115/GT2022-80619
10. C. Goßrau, N. H. Petersen, T. Sieker, and M. Wirsum, "Environmental Impact Analysis of H_2-fired Gas Turbines with SCR in a Flexible Operation Scenario," presented at the 55th Kraftwerkstechnisches Kolloquium (KWTK 2023), Dresden, Germany, Oct. 10–11, 2023. [Online]. Available: https://www.researchgate.net/publication/374631157_Environmental_Impact_Analysis_of_H2-fired_Gas_Turbines_with_SCR_in_a_Flexible_Operation_Scenario
11. M. Ó Conaire, H. J. Curran, J. M. Simmie, W. J. Pitz, and C. K. Westbrook, "A comprehensive modeling study of hydrogen oxidation," Int. J. Chem. Kinet., vol. 36, no. 11, pp. 603–622, 2004, https://doi.org/10.1002/kin.20036
12. I. ANSYS, ANSYS Fluent Theory Guide, Canonsburg, PA, USA: ANSYS, Inc., 2024

Chapter 6
Performance Analysis of Biomass Gasification with Biogas Co-Feeding for Hydrogen Production

Dang Saebea and Yaneeporn Patcharavorachot

Abstract To improve the hydrogen yield in biomass gasification, the addition of biogas for co-feeding with biomass in gasification was proposed in this work. The gasification model developed in Aspen Plus software was validated with experimental data. The performance of biomass gasification with biogas co-feeding using steam as a gasifying agent was investigated. The effect of the steam-to-fuel ratio on the gasification of mixed biomass and biogas was studied. The results show that the simulation results of biomass gasification were consistent with experimental data. Biomass gasification with biogas co-feeding can raise the amount of hydrogen and carbon monoxide in gas products by 22.12% and 18.44%, respectively. Moreover, the increase in the steam-to-fuel ratio enhances hydrogen in syngas. However, the system efficiency decreases with increasing steam-to-fuel ratio.

Keywords Steam gasification · Biogas · Biomass · Hydrogen production · Process simulation

---

D. Saebea (✉)
Department of Chemical Engineering, Faculty of Engineering, Interdisciplinary Centre for Decarbonisation & Green Energy, Burapha University, Chonburi, Thailand

Department of Chemical Engineering, Faculty of Engineering, Biochemical and Environmental Living Labs for Sustainable Development Sustainability (BELLS), Burapha University, Chonburi, Thailand
e-mail: dangs@eng.buu.ac.th

Y. Patcharavorachot
Department of Chemical Engineering, School of Engineering, King Mongkut's Institute of Technology Ladkrabang, Bangkok, Thailand

L. Chen (ed.), *Clean Energy Research*, Green Energy and Technology,
https://doi.org/10.1007/978-3-031-99312-1_6

6.1 Introduction

The increasing global demand for sustainable and renewable energy solutions has sparked a rise in interest in developing alternative energy sources. Hydrogen, recognized for its high energy content and clean combustion, has emerged as a promising candidate for future energy systems. Hydrogen can be used to generate electricity via fuel cells. Additionally, it can be used to produce higher-value chemicals such as ammonia, methanol, and dimethyl ether.

Biomass gasification is the most effective method for producing hydrogen. This thermochemical process transforms solid biomass into a combustible gas mixture called syngas, primarily composed of hydrogen (H_2), carbon monoxide (CO), and carbon dioxide (CO_2). Biomass gasification presents several advantages over conventional fossil fuel-based energy production, including the potential for reducing greenhouse gas emissions, utilizing locally available resources, and using a wide variety of feedstock [1].

The composition and yield of hydrogen in syngas depend on various factors, such as the type of biomass used, types of gasifying agents, and gasification conditions [2]. The gasifying agent type has a great impact on the quality of syngas. The typical gasifying agents used in biomass gasification are air, steam, and oxygen. Hydrogen production from biomass gasification using various agents, i.e., air, oxygen, and steam, was compared by Shayan et al. [3]. Their results indicated that biomass gasification using a steam agent obtained the highest hydrogen production, and the calorific value of the producer gas can reach higher than 11 MJ/Nm^3. Moreover, steam gasification had the highest exergy efficiency.

The hydrogen content produced from the steam biomass gasification process is still lower than that obtained from the steam reforming process of natural gas, as a conventional process for hydrogen production. Nakyai et al. [4] reported that methane co-feeding can improve hydrogen yield in biomass gasification. Biomass gasification with methane co-feeding achieved energy efficiency and exergy efficiency of 87.14% and 71.80%, which is higher than that of biomass gasification without methane co-feeding.

Although methane co-feeding in biomass gasification can improve the system efficiency, feeding natural gas or methane from fossils results in the production system still relying on fossil fuels and still having the problem of increasing the amount of carbon dioxide gas. Therefore, to make the hydrogen production system sustainable and environmentally friendly, biogas co-feeding instead of methane is an interesting way to increase hydrogen production. Because biogas can be produced from biomass or waste through anaerobic digestion, the main component of biogas is 60–80% methane [5, 6]. Moreover, another main component of biogas is carbon dioxide, which can be used as a gasifying agent. There is still a limit to investigating the impact of biogas co-feeding in biomass gasification. Therefore, this work aims to analyze hydrogen production from the biomass gasification process with biogas co-feeding using a model simulated in Aspen Plus.

6.2 Process Description

The biomass gasification process was simulated in Aspen Plus based on the mass balance, energy balance, and chemical reactions. Assumptions for the simulation of biomass gasification are as follows:

- Steady state condition.
- Isothermal condition.
- Char is composed of carbon and ash.
- Toluene, naphthalene, and benzene were considered as tar.
- H_2S, NH_3, and HCl were considered as inorganic contaminants.

In biomass gasification, the syngas as product gas consists of hydrogen, carbon monoxide, carbon dioxide, methane, and other compounds, which are slightly polar hydrocarbon components. The Peng-Robinson equation of state with Boston-Mathias modifications was used for thermodynamic characteristics. Biomass is a non-conventional solid, while syngas products from biomass gasification are conventional components. MIXNCPSD stream class was specified. The sugar bagasse as biomass was considered. The ultimate and proximate analysis of sugar bagasse is included in Table 6.1 [7].

The biomass gasification process is divided into four zones: drying and decomposition, pyrolysis, and gasification. Figure 6.1 shows the simulation of biomass gasification in this work. In the first zone, drying and decomposition occur simultaneously. The decomposition step converts non-conventional components into conventional components, i.e., ASH, C, H, N, Cl, S, O, and moisture. A RYield block was used for the DECOMP block to represent the decomposition step, and a calculator module was used to calculate the yield components based on the ultimate analysis of sugar bagasse. For the pyrolysis step, the PYRO block uses the RGibbs block, which is a reactor based on the Gibbs free energy minimization method. This step is the conversion of the ELEMENTS stream to CO, CO_2, CH_4, NH_3, H_2S, HCl, and tar. The contaminating compounds, i.e., HCl, NH_3, and H_2S, were separated from the PYRO1 stream by the GAS-SEP block [8]. Then, the PYRO2 stream from the GAS-SEP block is introduced into the GASF block, which was simulated for the gasification step by using the RGibbs block. Steam as a gasifying agent is introduced to the GASF block. The SYNGAS1 stream from the GASF block is mixed with the CONTAM1 stream by the MIXER block. Ash is separated from the

Table 6.1 Proximate and ultimate analyses of sugarcane bagasse

Proximate analysis (wt%)		Ultimate analysis (dry and free of ashes wt %)		Heating value (MJ/kg)	
Fixed carbon	3.36	C	44.6	Lower	17.4
Volatile matter	88.12	H	5.9	Higher	18.7
Ash	6.52	O	42.1		
Moisture	2	N	0.2		
		Cl	0.3		

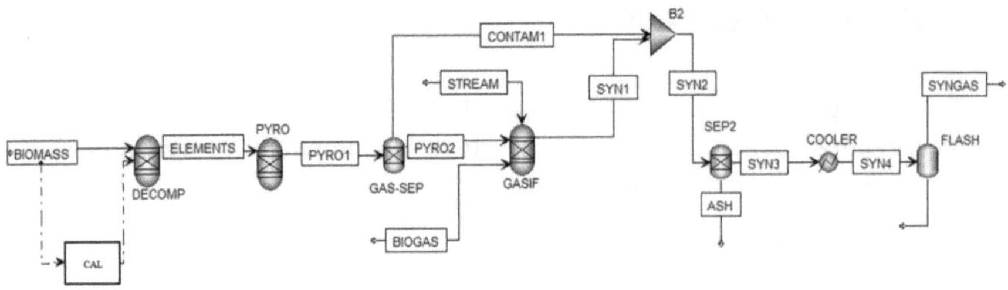

Fig. 6.1 Aspen Plus flowsheet for the process simulation of biomass gasification

SYNGAS2 by using the SEP block. The product gas from the biomass gasification process is cooled, and consequently, water is separated from the product gas using a Flash block.

In the case of biomass gasification with biogas co-feeding, the biogas is fed into the gasification step. Biogas from bagasse typically consists of 60–90% methane, 10–40% carbon dioxide, and trace amounts [9, 10]. This work used the treated biogas with a composition of 70% methane and 30% carbon dioxide for co-feeding in biomass gasification [11, 12]. The biomass-to-biogas ratio and steam-to-fuel ratio are described by Eqs. (6.1) and (6.2).

$$\text{biomass} - \text{to} - \text{biogas ratio} = \frac{\text{Inlet mass flow rate of biomass}}{\text{Inlet mass flow rate of biogas}} \tag{6.1}$$

$$\text{steam} - \text{to} - \text{fuel ratio} = \frac{\text{Inlet mass flow rate of steam}}{\text{Inlet mass flow rate of biomass and biogas}} \tag{6.2}$$

The system efficiency can be calculated from the heating value of the product gas to the total heating value of the input as shown in Eq. (6.3).

$$\text{system efficiency}\left(\%\right) = \frac{\dot{m}_{\text{syngas}}\text{LHV}_{\text{syngas}}}{\dot{m}_{\text{biomass}}\text{LHV}_{\text{biomass}} + \dot{m}_{\text{biogas}}\text{LHV}_{\text{biogas}} + Q} \tag{6.3}$$

where $\dot{m}_i$ is the inlet mass flow rate of i; LHV_i is the lower heating value of i; and Q is the heat consumption of the gasification process.

6.3 Model Validation

For the validation model, the simulation data were compared with the experimental data of Galindo et al. (2014). In the condition of the validation model, the biomass mass flow rate of 12 kg/h and the airflow rate of 18 Nm^3/h were introduced in the gasifier. The gasification and combustion temperatures were 577.16 and 743.99 °C, respectively. Table 6.2 compares syngas composition from gasification between the

Table 6.2 Validation model results [9]

Component	% Vol. (dry basis)		Error (%)
	Experimental data	Simulation data	
Hydrogen	14.88	14.72	1.08
Carbon monoxide	16.67	16.39	1.71
Carbon dioxide	15.39	16.18	5.15
Methane	2.05	5.18	60.41

simulation and the experimental results, presenting a good agreement between the simulation results and experimental data. The errors of the main products, i.e., hydrogen, carbon monoxide, and carbon dioxide, are less than 5.5%. The methane composition predicted by the model is slightly higher than the experimental values. However, the methane content remains relatively low compared to the concentrations of other components in the syngas. Therefore, the gasification model is accurate and can be reliably used for further analysis.

6.4 Results and Discussion

Adding Biogas in Biomass Gasification

The biomass gasification without and with biogas co-feeding is compared under the same conditions to investigate the effect of adding biogas to biomass gasification. In both cases, the fuel fed into the gasification process is specified at 20 kg/s. In the case of biomass gasification with biogas co-feeding, the biomass-to-biogas ratio is 2. The steam gasification was operated at 900 °C and 1 bar with a steam-to-fuel ratio of 1. Figure 6.2 shows the syngas products of biomass gasification with and without biogas co-feeding. It indicates that the feeding of biogas to biomass gasification leads to a 22.12% increase in hydrogen and 18.44% in carbon monoxide in the products, compared to biomass gasification without biogas, while the amount of carbon dioxide decreases. The addition of biogas in biomass gasification increases the methane and carbon dioxide in the gasification process, resulting in more reactions of methane with steam via steam reforming and methane with carbon dioxide via methane dry reforming reactions.

Effect of Steam to Fuel Ratio

To investigate the impact of the steam-to-fuel ratio, a biomass-to-fuel ratio of 2 and an operating temperature of 900 °C were specified. The influence of the steam-to-fuel ratio, ranging from 0.5 to 2.5, on the gas component and hydrogen-to-carbon ratio of the product gas is depicted in Fig. 6.3.

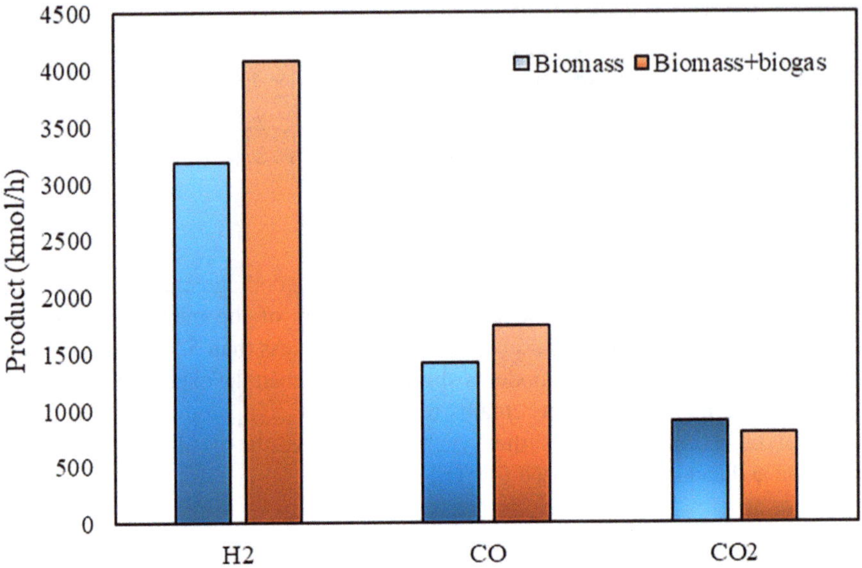

Fig. 6.2 Comparison of products between biomass gasification without and with biogas co-feeding

Figure 6.3a indicates that the hydrogen and carbon dioxide are enhanced by increasing the steam-to-fuel ratio, whereas the carbon monoxide decreases. The increase of steam in gasification drives more forward on the water-gas shift and methane reforming reactions, leading to an increase in hydrogen in the product gas. Moreover, more carbon monoxide is converted to carbon dioxide via the water-gas shift reaction with increasing steam in the feed. The increase of steam in gasification has more influence on the increment of hydrogen than the increase in carbon dioxide. Thus, the increase in steam in the gasification results in a rise in the hydrogen-to-carbon ratio in the product gas, as shown in Fig. 6.3b. However, an excessive increase in steam leads to higher energy consumption for steam gasification reactions due to its endothermic reaction, resulting in a decrease in system efficiency, as shown in Fig. 6.4.

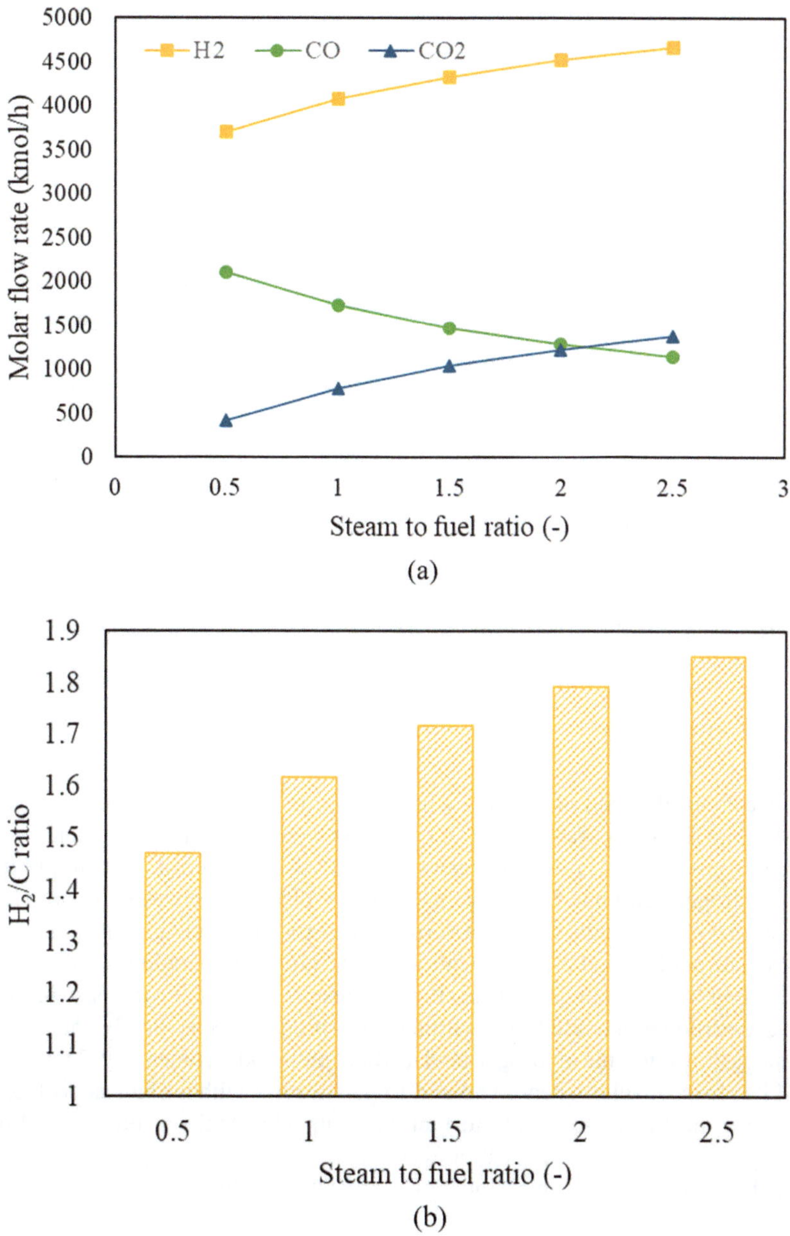

Fig. 6.3 Effect of steam-to-fuel ratio on (**a**) products and (**b**) H_2/C ratio in syngas

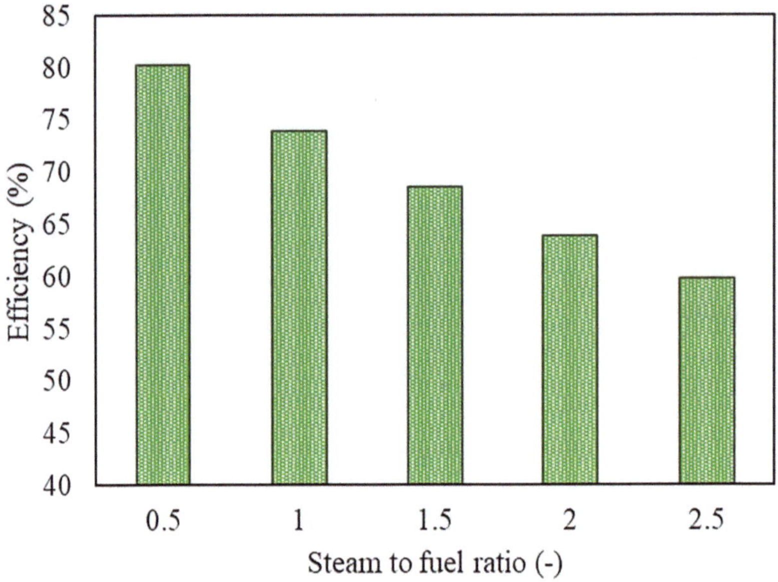

Fig. 6.4 Effect of steam-to-fuel ratio on the system efficiency in the gasification process

6.5 Conclusions

The steam gasification of biomass with biogas co-feeding was investigated in this work. The biomass gasification model was verified with experimental data. Biomass gasification is reliable because of a 1.08–5.15% error of main products, i.e., hydrogen, carbon monoxide, and carbon dioxide, between the simulation results and experimental data. The verified biomass gasification model was used to investigate adding biogas to gasification and the effect of operating conditions. The co-feeding of biogas in biomass gasification results in a notable increase in hydrogen and carbon monoxide levels within the gas products. Specifically, the presence of biogas elevates the hydrogen and carbon monoxide content by 22.12% and 18.44%, respectively, when compared to gasification without biogas co-feeding. Moreover, increasing the steam-to-fuel ratio can enhance the amount of hydrogen in syngas from biomass gasification with biogas co-feeding. However, supplying a suitable amount of steam in gasification is crucial, as adding too much can decrease the system efficiency.

Acknowledgments This work was supported by (i) Burapha University (BUU), (ii) Thailand Science Research and Innovation (TSRI), and (iii) National Science, Research and Innovation Fund (NSRF) (Fundamental Fund: Grant no.1.31/2568).

References

1. Nakyai T, Patcharavorachot Y, Arpornwichanop A, Saebea D (2020) Comparative exergoeconomic analysis of indirect and direct bio-dimethyl ether syntheses based on air-steam biomass gasification with CO_2 utilization. Energy 209:118332
2. Zhang Y, Zhou A, Li Z, Zhang H, Xiao R, Hu ZB, Wang X (2024) Numerical simulation analysis of biomass gasification and rich-H2 production process in a downdraft gasifier. J Energy Inst 114:101596
3. Shayan E, Zareb V, Mirzaee I (2018) Hydrogen production from biomass gasification; a theoretical comparison of using different gasification agents. Energy Convers. Manage. 159:30–41
4. Nakyai T, Aunthayanun S, Patcharavaracchot Y, Arpornwichanop A, Assabumrungrat S, Saebea D (2018) Exergoeconomics of hydrogen production from biomass air-steam gasification with methane co-feeding. Energy Convers Manag 140:228–239
5. Park S, Choe S, Lee H, Myung J (2024) Methylosinus trichosporium OB3b drives composition-independent application of biogas in poly(3-hydroxybutyrate) synthesis. Fuel 378:132730
6. Yang S, Liu Y, Wu N, Zhang Y, Svoronos S, Pellammanappallil P (2019) Low-cost, Arduino-based, portable device for measurement of methane composition in biogas. Renew Energy 138:224–229
7. Figueroa JEJ, Ardila YC, Lunelli BH, Filho RM, Maciel MRW (2013) Evaluation of pyrolysis and steam gasification processes of sugarcane bagasse in a fixed bed reactor. Chem Eng Trans 32:925–929
8. Moura CPC, Filho MAA, Vilardi HGD, Cavalcante RM, Yong AF (2023) Process simulation and economic evaluation of an integrated production plant for methanol, acetic acid and DME synthesis via sugarcane bagasse gasification. Energy Convers Manag 286:117051
9. Galindo AL, Lora ES, Andrade RV, Giraldo SY, Jaen RL, Cobas VM (2014) Biomass gasification in a downdraft gasifier with a two-stage air supply: effect of operating conditions on gas quality. Biomass Bioenergy 61:236–244
10. Prasad DY (2017) Biogas generation from bagasse-based paper mills by anaerobic digestion. J Chem Technol Biotechnol 51:515–524
11. Arelli V, Mamindlapelli NK, Begum S, Juntupally S, Shailaja S, Maddala RM, Anupoju GR (2022) Press mud and bagasse from sugar mill: cumulative effect of anaerobic co-digestion integrated with pre-treatment on overall performance. Biomass Bioenergy 166:106625–106625
12. Calbry-Muzyka A, Madi H, Rüsch-Pfund F, Gandiglio M, Biollaz S (2022) Biogas composition from agricultural sources and organic fraction of municipal solid waste. Renew Energy 181:1000–1007

Chapter 7
Sizing of Storage Components and Hydrogen Tank for Fuel Cell Hybrid Electric Vehicles

Bachir Bendjedia, Nassim Rizoug, and Moussa Boukhnifer

Abstract Fuel cell-hybrid electric vehicles have become one of the most promising alternatives for next-generation vehicles. The design of the hydrogen storage system is a very complex task that consists of defining the optimum of the weight, volume and cost. The influence of the hydrogen storage system size on the vehicle receives less attention in the literature. While the weight of the hydrogen storage tank increases the power and energy consumption, it is necessary to verify that the designed hybrid source can ensure the supply of the vehicle after the integration of the storage container and batteries. Also, these problems directly affect the vehicle's performance, safety, compactness and consumption. This paper focuses on a detailed study of the storage tank, the fuel cell system, and battery pack sizing when the influence of the battery element technologies is tested and validated. The proposed sizing algorithms show satisfactory results compared to actual vehicle data.

Keywords Batteries · Energy Storage System · Electric vehicle · Fuel cells · Hydrogen Tank · Sizing

7.1 Introduction

Air pollution is of serious concern in many countries, especially in populated cities. For this reason, hybrid electric vehicles have attracted many research institutions, industrial companies, and others in recent years. The future generation of vehicles will likely be hybrid fuel cells and/or battery-based vehicles. FCVs use a polymer

---

B. Bendjedia (✉)
LACoSERE Laboratory, University of Amar Telidji, Laghouat, Algeria
e-mail: b.bendjedia@lagh-univ.dz

N. Rizoug
ESTACA'LAB, ESTACA, Laval, France

M. Boukhnifer
Université de Lorraine, LCOMS, Metz, France

L. Chen (ed.), *Clean Energy Research*, Green Energy and Technology,
https://doi.org/10.1007/978-3-031-99312-1_7

electrolyte membrane fuel cell, which is powered by hydrogen gas. These vehicles are desirable because they produce no environmentally damaging emissions. In addition, once a sustainable method for H_2 gas production is developed, FCVs will have a constant fuel supply [1]. Unfortunately, at 0.0108 MJ/L, gaseous H_2 also has the lowest energy density by volume (over 3000 times smaller than gasoline), and it can explode violently when brought into contact with air. Consequently, it is needed to make FCVs commonly used, so that storing H_2 will be more compact, safe, and have high volumetric & weight density. In addition, there is limited space to store fuel in a vehicle tank that must hold enough hydrogen for the automobile to travel about 500 km between fills [2].

Regarding the vehicle's performance, it is necessary to minimize, as much as possible, the weight of all embedded sources and their auxiliaries to enhance performance, especially autonomy (fuel consumption). Major research works are not interested in this problem and it is assumed that the sizing concerns only the sources (e.g., for a fuel cell, only the cell number is considered) with an approximate percentage for their assistants. It is noted that some papers deal with the compressed hydrogen storage tank systems for automotive applications [2–4]. A few papers such as [5–7] are interested only in the influence of storage system sizing on battery lifetime. In [8], an energy source sizing approach based on the statistical description of driving cycles is used for hybrid fuel cell vehicles.

This paper focuses on the integrated sizing of the fuel cell system (stack and auxiliaries), hydrogen storage tank, and battery pack. A repeated Artemis driving cycle is used to simulate a vehicle range of 650 km. Recent data on hydrogen storage technologies and battery cell advancements are incorporated into the study. A simple energy management strategy, which splits power demand between the primary source (fuel cell) and the secondary source (battery), is proposed and applied. The hybrid source sizing is conducted and analysed using two developed algorithms. The results of these algorithms demonstrate good agreement with data from actual vehicles, confirming their reliability and practical relevance.

7.2 Technical Specifications

Regarding the Artemis cycle (Fig. 7.1), power loads involve important power exchanges in terms of storing or providing, which makes the batteries the most promising candidate for an energy storage unit. Then, the hybrid power sources, fuel cell/battery, combine the high energy density of fuel cells with the high-power density of batteries. The considered structure is shown in Fig. 7.2.

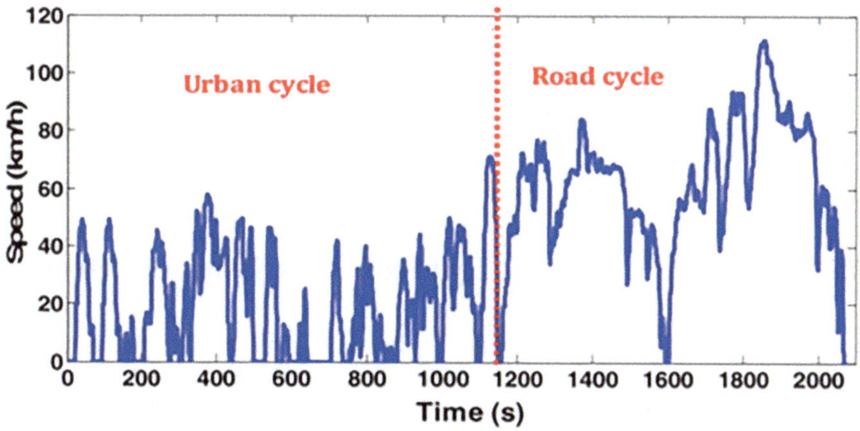

Fig. 7.1 Artemis cycle profile [11]

Fig. 7.2 Hybrid system topology

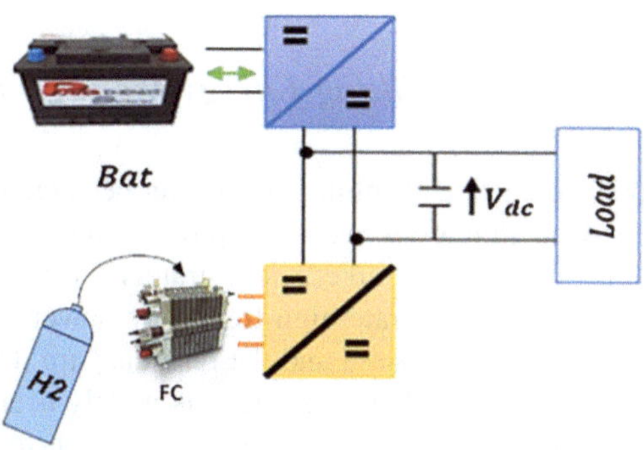

Driving Cycle (Mission)

In order to validate the proposed approaches, the speed profile must be defined. Currently, different missions exist (speed profile). These cycles were considered useful in comparing the effectiveness of HEVs in terms of cost, fuel consumption, and range in pure electric mode [9, 10]. However, the actual driving patterns are often much more complex and particular (speed, acceleration, etc.).

The ARTEMIS cycle has much closer driving dynamics to reality than other cycles. It is used in [11] to size a hybrid energy storage system. Then, in our case, the Artemis cycle with an average slope of 2.5% is used for sizing the fuel cell stack and its auxiliary storage tank. In the objective to achieve a distance of 650 km in this work, the ARTEMIS driving cycle is repeated 30 times.

Electric Vehicle Parameters

To estimate the power and energy needed for the proposed profile, a dynamic model of the vehicle is used under Matlab-Simulink software, where the vehicle's parameters are shown in Table 7.1 [12].

7.3 Fuel Cell, Battery and Hydrogen Tank Sizing Algorithm

In order to size the fuel cell system and the hydrogen tank, an algorithm is needed, which is based on a process of adding one cell if the calculated (needed) energy is not adequate compared to that found on the vehicle dynamic model. In this case a novel calculation of the hydrogen mass and the tank weight to be added to the novel vehicle mass, which is used again in the sizing process.

Frequency Splitting Energy Management Strategy

In the objective of splitting the power between two or more sources, an energy management strategy is needed to provide the proper power reference for each source. One of the simple and efficient strategies is the filtering management strategy based on a filter to separate high-frequency power components of ($\mathbf{P_v}$) to be sent to the batteries ($\mathbf{P_{bat}}$) system and the residuals are dedicated to the main source ($\mathbf{P_{FC}}$) as shown in Fig. 7.3. This strategy is widely used to manage the hybrid sources [13, 14], for sizing [12], and other strategies performances evaluation by adjusting the cut off frequency (F_C).

Battery Pack Sizing

The depth of discharge of batteries (DOD), which are used as a secondary power source, is 20%. This EMS defines the maximum power and energy for each source (battery and Fuel cell).

Table 7.1 Electric vehicle parameters [12]

Parameters	Values
Vehicle mass (kg)	860
Frontal area (m^2)	2.75
Air density (kg/m^3)	1.2
Penetration coefficient	0.3
Rolling resistance coefficient	$1.6 * 10^{-6}$

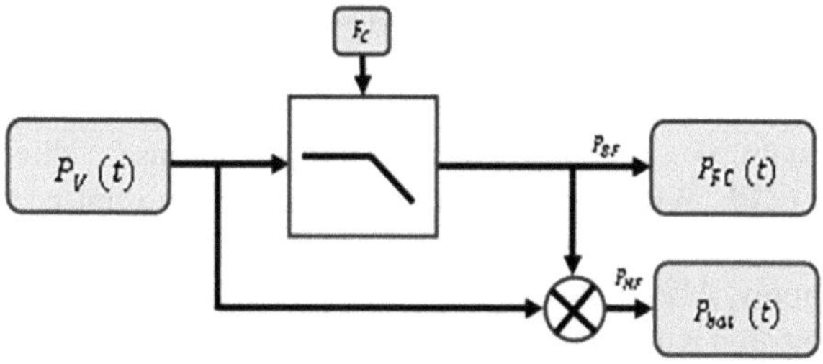

Fig. 7.3 Principle of separation frequency management strategy

The specification of these maximum powers and energy allows computing the number of cells in the battery pack. This last one must ensure all three conditions E_{bat_cons}, P_{bat_cons} and P_{bat_rec} mentioned in the sizing algorithm according to the chosen management strategy. Each condition imposes a specific number of cells. So, the number of battery cells can be expressed according to each condition in three ways:

1. The battery must generate a maximum of energy consumed E_{bat_cons} to ensure the necessary autonomy:

$$N^{E}_{Cel\_Bat} = \frac{E}{DOC * C_{Cel\_Bat} * 3600 * CU_{Cel\_Bat}} \quad (7.1)$$

2. The battery must generate the maximum consumed power P_{bat_cons}. In this case, the cell number will be expressed according to the discharge power as follows:

$$N^{Pcons}_{Cel\_Bat} = \frac{P_{Bat\_cons}}{P_{Cel\_Disc}} \quad (7.2)$$

3. The battery must generate the maximum recovered power P_{bat_rec}. In this case the cell number will be expressed according to the charge power as follows:

$$N^{Prec}_{Cel\_Bat} = \frac{P_{Bat\_rec}}{P_{Cel\_Char}} \quad (7.3)$$

In this work, for the battery sizing, the last three conditions must be ensured, so the number of cells must verify the formulas (7.1), (7.2) and (7.3):

$$N_{Cel\_Bat} = \max\left(N^{E}_{Cel\_Bat}, N^{Pcons}_{Cel\_Bat}, N^{Prec}_{Cel\_Bat} \right) \quad (7.4)$$

From the obtained results, it's observed that the frequency has a direct influence on the system sizing, weight, volume and other vehicle characteristics, and at the same time, the storage vessel and the fuel cell system are affected. Recent works have noted that the finest operating frequency of this type of fuel cell dedicated to automotive applications is around 50 mHz [14, 15]. For this reason, this frequency is considered to size the fuel cell system.

From Table 7.2, it is noted that sizing, taking into account the characteristics of the storage system, is a crucial constraint. Simply because it directly affects the vehicle weight (with about 60 kg), hydrogen consumption is raised by about 210 g, which makes the cost more expensive but enhances the vehicle's performance, especially its autonomy. Also, this quantity needed a special tank, which is heavier by about 15% compared with the simple case. In addition, the tank volume is higher by 7%, to ensuring the additional power needed only to transport the tank. Weight. As a result, the main source is too affected when its power rises by 1.7 kW and the weight increases by 10 kg.

7.4 Sizing Results

In the objective to size and optimize the hybrid source, four different technologies of battery cells are chosen to be tested, compared and lastly selected to be embedded in the hybrid source.

The different types of cells, as mentioned in Table 7.3, have diverse electrical specifications, such as capacities, charging/discharging rate and voltage. Also, they have dissimilar specifications, such as weight and volume, which are considered crucial parameters in this study to be minimized as much as possible to enhance the global system characteristics (drive range, cost, weight, etc.). The expected cost of an 80-kWnet automotive polymer electrolyte membrane (PEM) fuel cell system based on current technology, planned for commercialization in this year (2016), is approximately 280 $/kWnet when manufactured at a volume of 20,000 units/year [17, 18]. This cost ratio is used in this study to evaluate and compare the technical and economic efficiency of ESS.

The analysis of four hybrid source technologies reveals significant differences in weight, with KOKAM HP 7.5 being the heaviest (over 300 kg) due to its low

Table 7.2 Sizing results with $\mathbf{F_C = 50}$ **mHz**

Parameters	With taking into account tanks and fuel cell auxiliaries	Without taking into account tanks and fuel cell auxiliaries
Weight vehicle (kg)	1272.2	1213
Tank weight (kg)	57.9469	50.77
Fuel cell power (kW)	28.973	27.32
Fuel cell weight (kg)	141.1921	132.7
Tank volume (L)	75.9805	71.09
Cells number	1159	1093
H_2 masse: (kg)	3.19	2.98

Table 7.3 Battery cell's specifications [16]

KOK_7.2HP (7.1)	KOK_13HP (7.3)	MHI (7.2)	KOK_UHP (7.4)
$C_{\_elb}$ = 7.2 Ah	$C_{\_elb}$ = 13 Ah	$C_{\_elb}$ = 20 Ah	$C_{\_elb}$ = 0.45 Ah
$U_{\_elb}$ = 3.7 V	$U_{\_elb}$ = 3.7 V	$U_{\_elb}$ = 3.7 V	$U_{\_elb}$ = 3.7 V
$I_{\_elb\_Dis}$ = 14.4A	$I_{\_elb\_Dis}$ = 104 A	$I_{\_elb\_Dis}$ = 100 A	$I_{\_elb\_Dis}$ = 22.5 A
$I_{\_elb\_C}$ = 21.6A	$I_{\_elb\_C}$ = 39A	$I_{\_elb\_C}$ = 100A	$I_{\_elb\_C}$ = 22.5 A
$W_{\_elb}$ = 0.22 kg	$W_{\_elb}$ = 0.325 kg	$W_{\_elb}$ = 0.85 kg	$W_{\_elb}$ = 0.014 kg
V_elb = 0.12 L	V_elb = 0,174 L	V_elb = 0.42 L	V_elb = 0.0073 L
Cost = /	Cost = 30.36 euros	Cost = /	Cost = 4.6 Euros

power-to-weight ratio. The MHI technology achieves a lighter weight (290 kg) due to its high charging rate. KOKAM HP 13 offers the lightest source, reducing fuel cell and battery pack weight, resulting in a 5% reduction in total vehicle weight. KOKAM UHP performs best overall, minimizing the battery pack weight to 11.42 kg and reducing fuel cell weight by 10%, leading to a 62% improvement in total weight.

As shown in Fig. 7.4, in terms of volume, the fuel cell system is larger than the battery pack, despite the battery pack contributing more to the overall weight. Using the MHI and KOKAM HP 13 technologies, the obtained ESSs are nearly identical in terms of volume, while the KOKAM UHP technology achieves the smallest volume due to its compact size and high-power density, making the hybrid source more compact compared to other technologies.

Hydrogen consumption is a key parameter, with KOKAM UHP technology outperforming the others by reducing consumption by 11% (approximately 350 g), significantly enhancing vehicle drive range. In contrast, the MHI and KOKAM HP 13 technologies show similar hydrogen consumption levels, reflecting their comparable weights and system characteristics.

Cost analysis highlights the advantages of UHP technology, with the UHP battery pack being four times less expensive than the KOKAM HP 13 pack. This results in a 50% reduction in the total cost of the hybrid source. Coupled with its superior performance in weight, volume, and hydrogen consumption, UHP technology proves to be the optimal choice for a secondary battery.

7.5 Conclusion

This paper focuses on optimizing the design, cost, and performance of hybrid fuel cell electric vehicles (FC-HEVs) by improving the sizing of multiple energy sources. It proposes an energy management strategy (EMS) based on frequency splitting to minimize the size and weight of both the fuel cell system and its hydrogen storage tank. A dedicated sizing algorithm is developed to determine the optimal fuel cell configuration, ensuring a balance between efficiency and performance.

Fig. 7.4 Sizing results: 1—ESS weight; 2—ESS volume; 3—H2 consumption; 4—ESS cost

The study demonstrates that replacing KOKAM HP 13 batteries with Ultra High Power (UHP) batteries significantly reduces cost and weight while enhancing the overall performance of the hybrid energy source. The proposed solution integrates a fuel cell with UHP batteries, leading to improvements in mass, volume, cost efficiency, and vehicle autonomy. The effectiveness of this approach is validated through MATLAB/Simulink simulations.

References

1. Jorgensen SW (2011) Hydrogen storage tanks for vehicles: recent progress and current status. Curr Opin Solid State Mater Sci 15(2):39–43
2. Durbin DJ, Malardier-Jugroot C (2013) Review of hydrogen storage techniques for on board vehicle applications. Int J Hydrog Energy 38(34):14595–14617
3. Hua TQ, Ahluwalia RK, Peng J-K, Kromer M, Lasher S, McKenney K, Law K, Sinha J (2011) Technical assessment of compressed hydrogen storage tank systems for automotive applications. Int J Hydrog Energy 36(4):3037–3049

4. Ahluwalia RK, Hua TQ, Peng JK (2012) On-board and off-board performance of hydrogen storage options for light-duty vehicles. Int J Hydrog Energy 37(3):2891–2910
5. Schaltz E, Khaligh A, Rasmussen PO (2009) Influence of battery/ultracapacitor energy-storage sizing on battery lifetime in a fuel cell hybrid electric vehicle. IEEE Trans Veh Technol 58(8):3882–3891
6. Araujo RE, de Castro R, Pinto C, Melo P, Freitas D (2014) Combined sizing and energy management in EVs with batteries and supercapacitors. IEEE Trans Veh Technol 63(7):3062–3076
7. Pourabdollah M, Murgovski N, Grauers A, Egardt B (2013) Optimal sizing of a parallel PHEV powertrain. Veh Technol IEEE Trans 62(6):2469–2480
8. Ravey A, Watrin N, Blunier B, Bouquain D, Miraoui A (2011) Energy-source-sizing methodology for hybrid fuel cell vehicles based on statistical description of driving cycles. IEEE Trans Veh Technol 60(9):4164–4174
9. Silva C, Ross M, Farias T (2009) Evaluation of energy consumption, emissions and cost of plug-in hybrid vehicles. Energy Convers Manag 50(7):1635–1643
10. Graham L, Christenson M, Karman D (2007) Light duty hybrid vehicles—Influence of driving cycle and operating temperature on fuel economy and GHG emissions. 2006 IEEE EIC Clim. Chang. Technol. Conf. EICCCC 2006
11. Hammani A, Sadoun R, Rizoug N, Bartholomeüs P, Barbedette B, Le Moigne P (2012) Influence of the management strategies on the sizing of hybrid supply composed with battery and supercapacitor. in 2012 1st International Conference on Renewable Energies and Vehicular Technology, REVET 2012, pp. 1–7
12. Alloui H, Marouani K, Becherif M, Sid MN, Benbouzid MEH (2014) A control strategy scheme for fuel cell-vehicle based on frequency separation. 2014 First Int. Conf. Green Energy ICGE 2014, pp. 170–175
13. Azib T, Larouci C, Chaibet A, Boukhnifer M (2014) Online energy management strategy of a hybrid fuel cell/battery/ultracapacitor vehicular power system. IEEJ Trans Electr Electron Eng 9(5):548–554
14. Azib, T., Hemsas, K. E., & Larouci, C. (2014). Energy management and control strategy of hybrid energy storage system for fuel cell power sources. International Review on Modelling and Simulations (*IREMOS*), 7(6), 935.
15. Mesbahi T, Khenfri F, Rizoug N, Chaaban K, Bartholomes P, Le Moigne P (2016) Dynamical modeling of Li-ion batteries for electric vehicle applications based on hybrid particle swarm-Nelder-Mead (PSO-NM) optimization algorithm. Electr Power Syst Res 131:195–204
16. Burke A, Miller M (2009) Performance characteristics of Lithium-ion batteries of various chemistries for plug-in hybrid vehicles. EVS24 Int Batter Hybrid Fuel Cell Electr Veh Symp:1–13
17. Spendelow J, Marcinkoski J (2013) DOE Fuel Cell Technologies Office Record 14012: Fuel Cell System Cost—2013. vol. 14012, pp. 1–8
18. Li SG, Sharkh SM, Walsh FC, Zhang CN (2011) Energy and battery management of a plug-in series hybrid electric vehicle using fuzzy logic. IEEE Trans Veh Technol 60(8):3571–3585

Chapter 8
Mitigating Hydrogen Embrittlement Through Advanced Material Solutions

Haneen Barrak Aldossary and Gaydaa AlZohbi

Abstract Hydrogen embrittlement (HE) is a major challenge limiting the durability of high-strength materials exposed to hydrogen-rich environments. This paper presents a comprehensive analysis of advanced material strategies aimed at mitigating HE, including the development of multi-principal element alloys (MPEAs), nanostructured materials, and protective coatings. The study investigates the role of grain boundary engineering, carbide precipitation, and diffusion barriers in reducing hydrogen ingress and improving mechanical resilience. Experimental case studies on cadmium-plated fasteners and boiler tubes highlight the effectiveness of these approaches, with graphical analysis demonstrating significant reductions in ductility loss, crack propagation, and material degradation. The findings emphasize the necessity of integrating computational modeling, experimental validation, and industrial applications to enhance material performance in hydrogen-intensive sectors. This research provides a foundation for future advancements in hydrogen-resistant materials, contributing to the development of a safer and more reliable hydrogen infrastructure.

Keywords Hydrogen Embrittlement · High-Entropy Alloys · Nanostructured Materials

8.1 Introduction

Hydrogen embrittlement (HE) is a critical failure mechanism affecting high-strength materials exposed to hydrogen-rich environments. It causes a loss of ductility, strength, and toughness, leading to premature failure [1]. HE is a major concern in industries like aerospace, oil and gas, and hydrogen storage, where hydrogen can enter materials through corrosion, electrochemical reactions, or direct gas exposure

H. B. Aldossary · G. AlZohbi (✉)
Prince Mohammad bin Fahd University (PMU), Khobar, Saudi Arabia
e-mail: galzohbi@pmu.edu.sa

L. Chen (ed.), *Clean Energy Research*, Green Energy and Technology,
https://doi.org/10.1007/978-3-031-99312-1_8

[1]. To combat HE, researchers have explored advanced materials such as nanostructured metals, high-entropy alloys (HEAs), and protective coatings that limit hydrogen diffusion and enhance durability [2]. Early theories like the hydrogen pressure theory by Zapffe and Sims (1941) suggested that hydrogen accumulates at voids, generating internal pressure and cracking [3]. Later, the hydrogen-enhanced decohesion (HEDE) and hydrogen-enhanced localized plasticity (HELP) models explained embrittlement via weakened atomic bonds [4] and increased dislocation motion [5], respectively. Nanostructured materials reduce hydrogen diffusivity through engineered grain boundaries [6], while HEAs offer high mechanical strength and resistance to hydrogen-induced cracking [7]. Protective surface coatings, such as hydrogen-resistant ceramics, have also proven effective in minimizing hydrogen ingress and improving material longevity [8]. This review outlines key hydrogen embrittlement (HE) mechanisms and advanced mitigation strategies, combining theory and experimental findings. It highlights recent progress and existing gaps and supports future work in developing hydrogen-resistant materials.

8.2 Methodology

A literature review was done using major databases, focusing on studies from 2005 to 2025 with data on hydrogen embrittlement. Relevant works were grouped into HE mechanisms, techniques, material solutions, and case studies to highlight key advances and gaps.

8.3 Mechanisms of Hydrogen Embrittlement

Hydrogen embrittlement (HE) weakens high-strength materials as hydrogen enters, spreads, and damages the microstructure. A key explanation is the hydrogen pressure theory by Zapffe and Sims (1941), which suggests that hydrogen collects at defects like microvoids, forming molecular hydrogen and building pressure. When the pressure exceeds material strength, cracks form [9]. This explains "fish eye" fracture patterns seen in embrittled steels (Fig. 8.1) [9].

Hydrogen enters metals through gas exposure, corrosion, or electrochemical reactions. It sticks to the surface, breaks into atoms, and moves into the metal. This depends on factors like temperature and surface energy [9]. Sieverts' law says hydrogen solubility rises with the square root of its pressure. Temperature, stress, and defects like dislocations or voids affect how it spreads, weakening the metal and reducing strength [10].

The dissolved hydrogen concentration (C_{H}) in metals follows Sieverts' law, expressed as in Eq. (8.1):

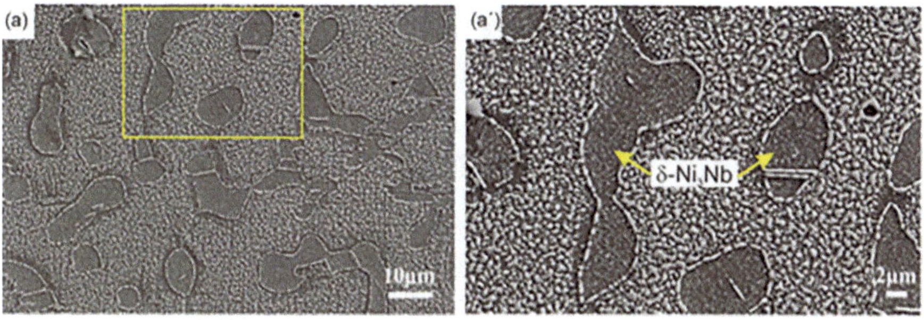

Fig. 8.1 Illustrates microstructural evolution during hydrogen charging contributing to defect formation and crack initiation [9]. Panel (**a**) shows a textured surface with irregular particles; panel (**a'**) is a magnified view of the boxed area in (**a**), showing d-Ni3Nb precipitates

Fig. 8.2 Hydrogen traps in steels—(**a**) interstitial sites, (**b**) surface traps, (**c**) subsurface traps, (**d**) grain boundary traps, (**e**) dislocation traps, (**f**) vacancy traps [12]

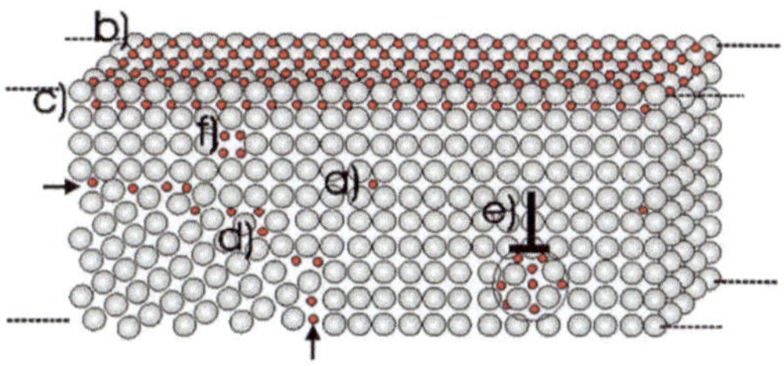

$$C_{\mathrm{H}} = S\sqrt{P_{\mathrm{H_2}}} \tag{8.1}$$

where (S) represents the solubility constant, which varies based on the alloy composition and temperature, while (P_{H_2}) denotes the hydrogen partial pressure.

8.4 Microstructural Changes

Once hydrogen has entered the material, it interacts with its microstructure, leading to the formation of hydrogen traps [11]. These traps influence how hydrogen accumulates and impacts material integrity. The atomic ratio of dissolved hydrogen in body-centered cubic (BCC) iron is theoretically low, approximately 2×10^{-8}, as predicted by Sieverts' law [12]. However, in practice, the actual hydrogen content in metals is significantly higher due to the presence of numerous hydrogen traps, such as vacancies, dislocations, grain boundaries, and surface traps, as illustrated in Fig. 8.2.

Hydrogen traps are grouped by activation energy (E_a) into reversible and irreversible types. Reversible traps ($E_a < 60$ kJ/mol), like dislocations, grain boundaries, and precipitates, allow hydrogen to move in and out [13, 14]. Irreversible traps ($E_a > 60$ kJ/mol), such as carbides and inclusions, hold hydrogen permanently [13, 15]. These traps affect hydrogen embrittlement (HE) by letting hydrogen build up at

Table 8.1 Activation energies for various hydrogen traps in steels [16]

Trapping sites	Activation energy (KJ **mol^{-1}**)
Reversible trapping sites	
Interstitial sites in iron	4–8
Dislocation	26.4–26.8
Lath boundary	17.8–18.6
Austenite/martensite	22
Grain boundary	17.8–18.6
NbC (coherent)	39–48
Irreversible trapping sites	
Ferrite/cementite interface	66.3–68.4
Fe_3C interface	84
Al_2O_3 interface	79
MnS interface	72
NbC (incoherent)	63–68

weak points, reducing strength and causing fractures. Reversible HE happens when hydrogen weakens bonds at stress spots, while irreversible HE involves gas buildup and cracking [16]. Alloys with more reversible traps tend to be more prone to HE [17]; please refare to Table 8.1 for activation energies of various hydrogen traps.

8.5 Theoretical Models

Hydrogen embrittlement (HE) is a well-known degradation mechanism in materials science, with multiple theoretical models explaining how hydrogen interacts with metallic structures under stress and environmental exposure. The hydrogen-enhanced decohesion (HEDE) model suggests that hydrogen accumulates at grain boundaries and defects, weakening atomic bonds and promoting crack initiation. In contrast, the hydrogen-enhanced localized plasticity (HELP) model proposes that hydrogen increases dislocation mobility, reducing resistance to dislocation motion and causing localized plastic deformation and crack growth. Experimental studies support HELP, showing reduced dislocation resistance in hydrogen-charged materials [17]. In some materials, hydrogen absorption triggers phase transformations that destabilize the microstructure. This is especially critical in high-strength steels and nickel-based alloys, where hydrogen stabilizes brittle phases, increasing the risk of cracking and internal stress buildup [18]. Additionally, in metals like titanium and zirconium, hydrogen forms brittle hydrides that act as crack starters and lower fracture toughness. Unlike HEDE and HELP, this involves a clear phase change that directly weakens the material [18]. Each HE model highlights different failure modes based on material, stress, and environment—key for effective prevention in high-risk industries.

8.6 Experimental Techniques and Testing Methods

Hydrogen Permeation Test (HPT)

Hydrogen permeation testing is a basic yet effective method for measuring diffusible hydrogen in steel, providing insight into its susceptibility to hydrogen embrittlement (HE) [19]. The test uses a double-cell setup: one side acts as the charging cell, and the other as the oxidation or exit cell, separated by a steel membrane [20, 21]. Hydrogen is introduced electrochemically into the charging cell and diffuses through the membrane into the oxidation cell. While simple, this method is often combined with other analytical techniques to improve accuracy and better evaluate hydrogen transport behavior in steels.

Thermal Desorption Spectroscopy (TDS)

Temperature desorption spectroscopy (TDS), also referred to as temperature-programmed desorption (TPD), is widely accepted for studying HE and hydrogen-induced failures [22–24]. Various methods exist to quantify diffusible hydrogen in steel and examine its role in material failure. Hydrogen mobility is another critical factor influencing HE [25]. TDS effectively qualifies and quantifies the presence of diffusible hydrogen. Figure 8.3 presents a schematic representation of the TDS setup.

Thermal desorption spectroscopy (TDS) measures hydrogen release from steel by heating it gradually. Hydrogen trapped in the metal is released once enough energy is supplied. The release temperature, called the desorption temperature, is measured using mass spectroscopy [26, 27]. TDS can detect very low hydrogen levels (as little as 0.1 mg/kg) [28, 29] and, when combined with other methods, gives useful insights into hydrogen trapping and embrittlement behavior.

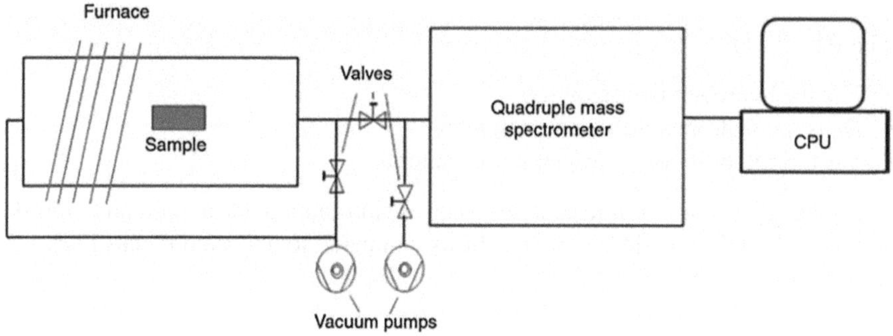

Fig. 8.3 Schematic of TDS testing apparatus [24]

8.7 Advanced Material Solutions

Novel Alloys

High-entropy alloys (HEAs) resist hydrogen embrittlement well because their complex atomic structures disrupt hydrogen movement and trapping [30]. Their high entropy interferes with hydrogen migration, as shown in Eq. (8.2). However, they often use expensive elements, making large-scale use difficult. Adjusting their composition and production methods can help reduce costs without losing performance [30]. HEAs usually need advanced techniques like vacuum arc melting or additive manufacturing, which offer precise control but aren't ideal for mass production [31]. Still, studies show they keep their resistance even after long exposure to hydrogen, making them great for aerospace and energy use [32].

The diffusion coefficient of hydrogen in metals follows Fick's second law:

$$\frac{\partial C}{\partial t} = D_H \frac{\partial^2 C}{\partial x^2} \tag{8.2}$$

where C is the hydrogen concentration, t is time, x is the spatial coordinate, and D_H is the diffusion coefficient of hydrogen.

Protective Barrier Coating

Protective coatings help block hydrogen from entering metals. Graphene and ceramic coatings are very effective [32]. Ceramics are strong but costly, while polymers are cheaper but weaker. Methods like PVD, CVD, and spraying apply these coatings based on needs [33]. They improve durability but can wear down over time [34].

The effectiveness of coatings can be evaluated using permeability Eq. (8.3):

$$P = D_H \times S \tag{8.3}$$

where

- P is the hydrogen permeability.
- D_H is the hydrogen diffusion coefficient
- S is the solubility of hydrogen in the material

Studies show that ceramic coatings, such as aluminum oxide and titanium nitride, significantly reduce hydrogen permeability compared to uncoated metals [34].

Nanostructured Materials

Nanostructured materials, like ultrafine-grained steels, resist hydrogen embrittlement well thanks to their many grain boundaries that trap hydrogen and stop cracks [35]. They can be made using methods like severe plastic deformation or powder metallurgy, but scaling up is still tough [36]. Thermomechanical processing helps, but controlling grain size is key [37]. These materials work well in hydrogen environments, though more research is needed to confirm their long-term stability under tough conditions [37].

The total number of hydrogen traps (N_t) in nanostructured materials can be approximated by Eq. (8.4):

$$N_t = \frac{V_{\text{GB}}}{V_{\text{total}}} \times N_{\text{sites}} \tag{8.4}$$

where

- V_{GB} is the grain boundary volume
- V_{total} is the hydrogen diffusion coefficient
- N_{Sites} is the number of trapping sites per unit volume.

Optimizing grain size and microstructure in nanomaterials boosts resistance to hydrogen embrittlement [38]. Ongoing advances in alloys and coatings support this, with future work aiming to enhance performance in hydrogen and aerospace applications.

8.8 Result and Discussion

Case Study: Cadmium-Plated Fasteners

Cadmium coatings are commonly used in aerospace and automotive sectors for their corrosion resistance and ability to limit hydrogen ingress in high-strength steel fasteners. However, Jha et al. (2010) found that these fasteners can still suffer from hydrogen embrittlement (HE) under stress in hydrogen-rich conditions [39]. Tests showed delayed fractures, and SEM analysis revealed intergranular cracking along with ductile zones, indicating HE followed by tearing (Fig. 8.4). Hydrogen sources included heat treatment at high temperatures, cadmium plating, and poor post-baking. To improve resistance, alternatives like zinc-nickel and ceramic coatings have shown better performance and durability in harsh environments.

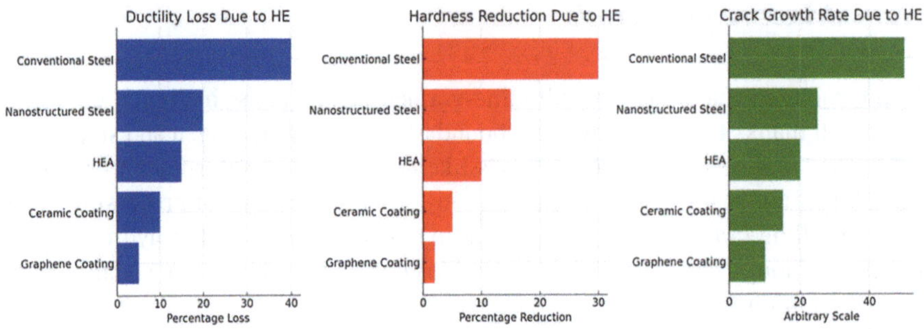

Fig. 8.4 Comparative analysis of hydrogen embrittlement effects on different materials

Case Study: Boiler Tubes in Power Plants

Boiler tubes in power plants face hydrogen embrittlement due to high-pressure hydrogen exposure. Djukic et al. (2015) found that microstructural changes in failed tubes allowed hydrogen to spread and cause cracking [40]. Analysis showed hydrogen buildup at grain boundaries and quasi-cleavage fractures, confirming HE (Fig. 8.4). The main issue was high-temperature hydrogen attack (HTHA), worsened by HEDE and HELP, leading to lower ductility and strength. A "window" fracture pattern pointed to hydrogen damage. Resistance depended on hydrogen levels and operating conditions. To reduce failures, nanostructured steels and oxide coatings were used—fine grains trap hydrogen, and coatings block it. Figure 8.4 shows that these materials perform better than standard steels.

Practical Implications and Future Work

These case studies show that advanced materials like nanostructured alloys and high-performance coatings offer better protection against hydrogen embrittlement than traditional options. Future work should focus on cost-effective coatings, microstructure tuning, and long-term testing to boost durability in hydrogen environments.

8.9 Conclusion

Hydrogen embrittlement (HE) remains a major challenge for structural materials in hydrogen-rich environments like aerospace, energy, and transportation. Mechanisms such as hydrogen-enhanced decohesion (HEDE) and localized plasticity (HELP) highlight the need for advanced materials. Solutions like high-entropy alloys (HEAs), nanostructured metals, and protective coatings have shown strong potential by reducing hydrogen diffusion and improving mechanical strength. Microstructural strategies, including grain boundary control and carbide formation, further enhance resistance. Coatings like graphene and ceramics also help limit hydrogen ingress. Graphical data supports these improvements, showing less ductility loss and slower crack growth. As hydrogen infrastructure expands, addressing HE will require integrated efforts in material development, process optimization, and predictive modeling to ensure safer, longer-lasting components.

References

1. Dwivedi SK, Vishwakarma M (2018) Hydrogen embrittlement in different materials: a review. Int J Hydrog Energy 43(46):21603–21616
2. Badenes B et al. (2022) Development of advanced materials for borehole heat exchangers guided by numerical simulations. Energy and Buildings
3. Zapffe CA, Sims CE (1941) Hydrogen embrittlement in steel. Journal of the Iron and Steel Institute
4. Lynch SP (2012) Hydrogen embrittlement mechanisms and models. Corros Sci 53(1):310–330
5. Robertson IM et al (2015) Hydrogen embrittlement understood via multiscale modeling. Nat Mater 14(2):129–135
6. Gong P et al (2020) Hydrogen embrittlement in nanoprecipitation-strengthened steels. Sci Adv
7. Li X et al (2022) Hydrogen embrittlement in multi-principal element alloys: a review. J Mater Sci Technol 122:20–32
8. Wetegrove M et al (2023) Barrier coatings for the hydrogen economy. Hydrogen 4(2):307–322
9. Sezgin J-G et al (2017) Hydrogen interactions in structural materials. Int J Hydrog Energy 42:15403
10. Liu Q, Atrens A (2013) Hydrogen embrittlement and corrosion review. Corros Rev 31:85
11. Toribio J, Kharin V (2018) Hydrogen and phase transformations in high-strength steels. Metals 8(8):602
12. Kiuchi K, McLellan RB (1983) The solubility and diffusivity of hydrogen in well-annealed iron. Acta Metall 31(7):961–984
13. Pressouyre GM (1979) A classification of hydrogen traps in steel. Metall Trans A 10(11):1571–1573
14. Nagumo M et al (2001) Hydrogen absorption, diffusion and trapping in high-strength steels. Metall Mater Trans A 32(2):339–347
15. Oriani RA (1987) Hydrogen—the versatile embrittler. Corrosion 43(7):390–397
16. Lynch SP (2012) HE and liquid-metal embrittlement in alloys. Metall Mater Trans A 43(7):2385–2397
17. Gangloff RP (2003) Hydrogen-assisted cracking of high-strength alloys. In: Comprehensive structural integrity, vol 6, pp 31–101
18. Robertson IM et al (2015) Hydrogen embrittlement understood. J Mater Res 30:5
19. Devanathan MAV, Stachurski Z (1962) The adsorption and diffusion of electrolytic hydrogen in palladium. Trans Faraday Soc 58:246
20. Danford MD (1987) Hydrogen trapping and the interaction of hydrogen with metals. NASA Technical Paper 2744
21. Mertens G, et al. (2007) Hydrogen absorption and desorption in steel by electrolytic charging. Advanced
22. Saini N et al (2017) Effect of diffusible hydrogen content on embrittlement of P92 steel. Int J Hydrog Energy 42(27):17328–17338

23. Thomas S et al (2016) The effect of absorbed hydrogen on the dissolution of steel. Heliyon 2(12):e00209
24. Pundt A, Kirchheim R (2006) Microstructural aspects. Annu Rev Mater Res 36(1):555–608
25. Song J, Curtin WA (2013) Atomic mechanism and prediction of hydrogen embrittlement in iron. Nat Mater 12(2):145–151
26. Kishi A, Takano N (2010) Effect of hydrogen cathodic charging on fatigue fracture of type 310S stainless steel. J Phys Conf Ser 240(1):012050
27. Mine Y (2007) Fatigue crack growth behavior and hydrogen penetration in stainless steels. Tetsu-to-Hagane 93(3):47–56
28. Vergani L, et al. (2014) Hydrogen effect on fatigue behavior of quenched and tempered steel. Procedia
29. Kim YS, Kim JG (2017). Electroplating of reduced-graphene oxide on austenitic stainless steel to prevent hydrogen embrittlement. International Journal of Hydrogen Energy, 42(44), 27428–27437.
30. Venezuela J et al. (2016) Hydrogen embrittlement of martensitic advanced high-strength steels. Corrosion
31. Koyama M et al (2017) Hydrogen embrittlement in high-Mn steels. Int J Hydrog Energy 42(17):12706–12723
32. Zamanzade M, Barnoush A (2014) Hydrogen embrittlement of iron aluminides. Procedia Mater Sci 3:2016–2023
33. Abhay K et al (2008) Failure analysis of cadmium-plated 35NCD16 fasteners due to hydrogen embrittlement. Eng Fail Anal 16(1):125–133
34. Djukic MB et al (2019) Hydrogen embrittlement. Eng Fail Anal 100:111–121
35. Lynch SP (2012) Techniques for characterizing hydrogen-assisted cracking of metals. In Gaseous
36. Gangloff RP, Somerday BP (2012) Hydrogen embrittlement: mechanisms and modeling. Gaseous
37. Djukic MB et al (2019) Synergistic action of hydrogen embrittlement mechanisms in steels. Engineering
38. Troiano AR (1960) The role of hydrogen in the embrittlement. Metall Trans 52:54–80
39. Beachem CD (1972) Model for hydrogen-assisted cracking in steels. Metall Trans 3(2):441–455
40. Jack TA et al (2017) Hydrogen-induced cracking in steel. Int J Hydrog Energy 42(22):15411–15415

Chapter 9
Hydrogen Energy Prospectus and Challenges in Kingdom of Saudi Arabia

Feroz Shaik, Fouad Mohammad Jastaniah, and Ahmed Gazi Alharbi

Abstract This paper provides a comprehensive analysis of the potential for hydrogen energy to become the leading energy source in the Kingdom of Saudi Arabia (KSA). As hydrogen emerges as a viable energy source, demand and investment are increasing significantly, particularly in nations with vast deposits of renewable energy resources. Moreover, green energy is manufactured using renewable sources like wind and solar. During electrolysis, water molecules are split into hydrogen and oxygen without carbon dioxide emission. The prospect of green hydrogen generation in Saudi Arabia is necessary because of its abundant energy resources. Moreover, the Saudi government has already made substantial investments in the renewable energy sector, and its determined renewable energy goals show a solid dedication to sustainable development. Saudi Arabia benefits from having extensive desert lands that can be utilized for building massive solar farms. The demand for hydrogen energy is at its initial stages and is expected to increase due to the demand for green energy. There is a need to assess economies of scale and environmental sustainability while considering the storage of green hydrogen. Nonetheless, the production of green hydrogen still faces obstacles, such as the current high production costs and political and regulatory challenges. Therefore, this paper provides a historical overview of hydrogen energy in Saudi Arabia, its demand, potential resources for its production, and current challenges.

Keywords Hydrogen · Renewable energy · Green energy · Energy resources · Energy challenges

F. Shaik (✉) · F. M. Jastaniah · A. G. Alharbi
Department of Mechanical Engineering, Prince Mohammad Bin Fahd University, Al Khobar, Kingdom of Saudi Arabia
e-mail: fshaik@pmu.edu.sa

L. Chen (ed.), *Clean Energy Research*, Green Energy and Technology,
https://doi.org/10.1007/978-3-031-99312-1_9

9.1 Introduction

Advancing cleaner energy sources is a worldwide imperative, as it supports climate change initiatives. Presently, hydrogen is a potential source of clean energy that can reduce carbonic acid gas and address climate change. It is imperative to assert that fossil fuels such as coal, oil, and electricity are still widely in use in the Kingdom of Saudi Arabia [1]. Further, Saudi Arabia has the second-largest oil reserves worldwide and is the second-largest producer of oil, producing about 12 million barrels a day [2]. Despite having extensive oil and gas deposits, Saudi Arabia has yet to make efforts to support the generation and utilization of hydrogen energy. Nonetheless, there is still doubt concerning the prospects of hydrogen energy and its contribution to determining the future. It is necessary to explore the possibility of hydrogen energy becoming a substantial aspect of the nation's energy mix. Apart from examining the historical pattern and the potential for hydrogen to contribute to a development role in the local and international setting, this paper explores Saudi Arabia's present energy demand, preparedness, and the possible challenges to its adoption.

Historical Overview of Hydrogen Energy in Saudi Arabia

Saudi Arabia has supported low-carbon development for over 10 years by implementing local and international initiatives. The objectives include reducing overreliance on oil and gas and incorporating energy-saving initiatives [1]. Though the government launched its all-inclusive plan in 2020, its participation in hydrogen energy initially surfaced in 2010 when national hydrogen research was initiated. The research and development was focused on carbon emissions reduction with an alternate approach towards hydrogen energy utilization [3]. Since the government has been under intense political pressure from local and external forces to mitigate climate change, it has started investing in clean hydrogen energy. Many nations, including Saudi Arabia, have substantially invested in research and development to support green hydrogen technology. Apart from the mega development projects in the eastern parts of the country, the government, companies, and stakeholders had invested approximately 900 million dollars in hydrogen initiatives by 2020 [4]. The magnitude of the funded programs shows that Saudi Arabia is shifting from investigation and protests toward substantial development. Moreover, its Vision 2030 strives to make the country's energy sector more effective. As seen in Table 9.1, the country has clear plans and milestones towards appreciating green energy.

Financial achievement and creating more job opportunities are essential assurances made in communications supporting the hydrogen economy. Even though countries like Australia have committed huge investments and demonstrated political will, there are many challenges to developing hydrogen as an alternative energy source [5]. The hydrogen economy is not thriving since its steady supply is only guaranteed with an increased market demand. Potential manufacturers are only

Table 9.1 Clean energy ambitions in the Saudi Arabia Countrywide Vision 2023 [2]

Aspiration	Formulations in the vision document (Page Number)	Implementation (Stand 2021)
Localization of renewables	"Localize a significant portion of the renewable energy value chain in the Saudi economy" (P.49)	Evidence of localization progress so far provided in Alotaibi et al.
Renewable capacity	"Initial target of generating 9.5 gigawatts of renewable energy (P.49)	Target raised in 2021 to 50% of capacity by 2030
Renewable investments	"Launch the King Salman Renewable Energy Initiative" (P. 49)	Initiative used since 2016 as a banner for renewable projects
Renewable energy sector regulation and participation	"Review the legal and regulatory framework that allows the private sector to buy and invest in the renewable energy sector" and "encourage public-private partnerships" (P.49)	Renewable projects being realized by the domestic company ACWA power (single buyer) through purchase agreements (PPAs) with consortia involving private sector
Energy liberalization	"Guarantee the competitiveness of renewable energy through the gradual liberalization of the fuels market" (P.49). "Free market prices shall, in the long term, stimulate productivity and competitiveness among utility companies and open the door to investment and diversification of the energy mix in the kingdom. We will also seek to set clear subsidy criteria" (P.51)	Energy subsidy reforms introduced in 2016 and a citizen account program established to provide more targeted support to eligible citizens

willing to invest in such initiatives with a promising market trend. Identifying environmentally friendly and cheapest methods to produce large quantities of sustainable hydrogen is necessary to establish a booming hydrogen economy [6]. Many companies still need to identify the most appropriate way of producing affordable and ecologically friendly hydrogen.

9.2 Hydrogen Demand in Saudi Arabia

Even though the demand for green hydrogen worldwide is currently at its initial stages, it is advancing rapidly. Moreover, the cost of manufacturing green hydrogen is decreasing, and it is projected to continue going down in the coming years due to increased demand from different sectors [2]. Many countries are currently supporting the development of green hydrogen to create a suitable environment for its market growth. Since the cost of renewable energy is increasingly becoming low and the demand for clean sources of energy increases, green hydrogen is likely to become the key contributor to the global energy market [7]. Many institutions provide different hydrogen demand projections because of the relative age of the hydrogen industry. For instance, the International Energy Agency (IEA) predicts

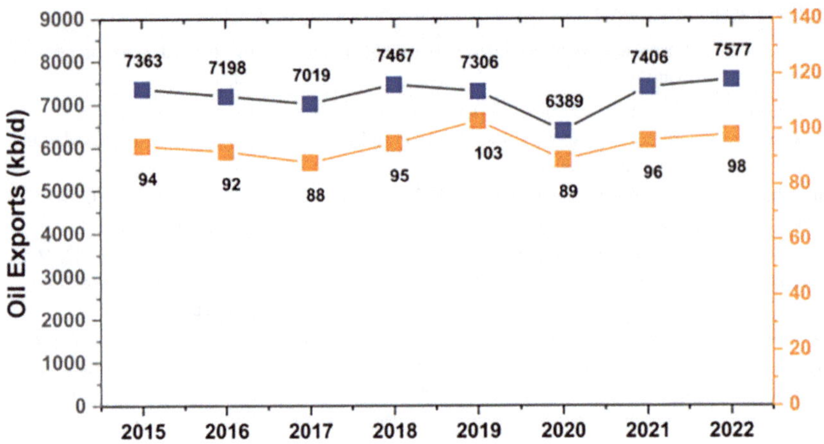

Fig. 9.1 Saudi Arabia's oil and gas exports between 2015 and 2022 [4]

that global hydrogen usage will be more than 35 exajoules (EJ) by 2050. The demand is projected to increase to 66 EJ, particularly with net-zero situations. Further, the demand is projected to triple by 2050 to 78 EJ. The increase in hydrogen supply is projected to meet more than 22% of the total market for energy sectors globally, such as transport, electricity generation, and commercial energy. Figure 9.1 shows that the production and consumption of natural gas is steadily rising. Specifically, the energy-rich nation exported 0.3 trillion cubic feet of natural gas in 2029 [4].

9.3 Possible Resources to Generate Green Energy in Saudi Arabia

Some potential resources for the country to generate green hydrogen include solar and wind energy. The country has one of the leading solar energy abilities globally. The government is dedicated to improving its renewable energy industry, especially solar power, to reduce its overreliance on oil and gas [8]. Moreover, green energy can be generated from renewable energy sources like solar energy. The geographic information system (GIS) map in Fig. 9.2 demonstrates the possibility of solar radiation in Saudi Arabia, an essential aspect in establishing the perspective for solar energy production.

As Fig. 9.2 shows, many parts of the nation exhibit high solar radiation, especially along the central and northern regions. In some cities like Riyadh, 2200 (kWh/m^2) and Hail (2500 kWh/m^2), which record high solar radiation, can generate green hydrogen [4]. For instance, they can create huge volumes of solar energy for the generation of green hydrogen, which can assist Saudi Arabia in reducing its

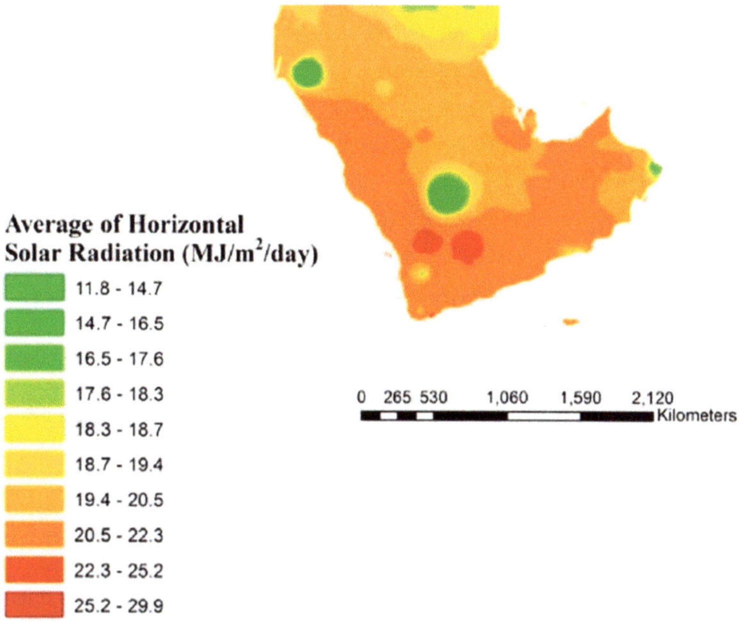

Fig. 9.2 GIS map showing Saudi Arabia's possible solar radiation [4]

greenhouse gas emissions and overreliance on oil and gas. Saudi Arabia is also recognized for its ability to generate wind energy, especially along the western and southwestern regions. Figure 9.3 shows the projected wind speed in Saudi Arabia, essential in establishing wind energy production.

As demonstrated in Fig. 9.3, the western and southwestern parts of Saudi Arabia have the highest concentration of wind energy, with some having a density of more than 500 W/m^2. Moreover, some of these cities, like Jeddah, Mecca, and Abha, can be utilized to generate green hydrogen since they have the potential to create vast volumes of wind power [4].

9.4 Possible Technologies to Generate Green Hydrogen in Saudi Arabia

Modern technology is a great catalyst for the efficient production of Hydrogen energy. Currently, the volume of Green Hydrogen production globally only accounts for 0.1% of overall hydrogen amounts [8]. Businesses and individual investors have visualized the possibility of plugging the intermittency of wind and solar energy to power operations in the form of natural gas. It is possible to improve the output quantity in KSA using various technologies. They include digital twins and artificial intelligence. Increased levels of knowledge concerning optimum designs are

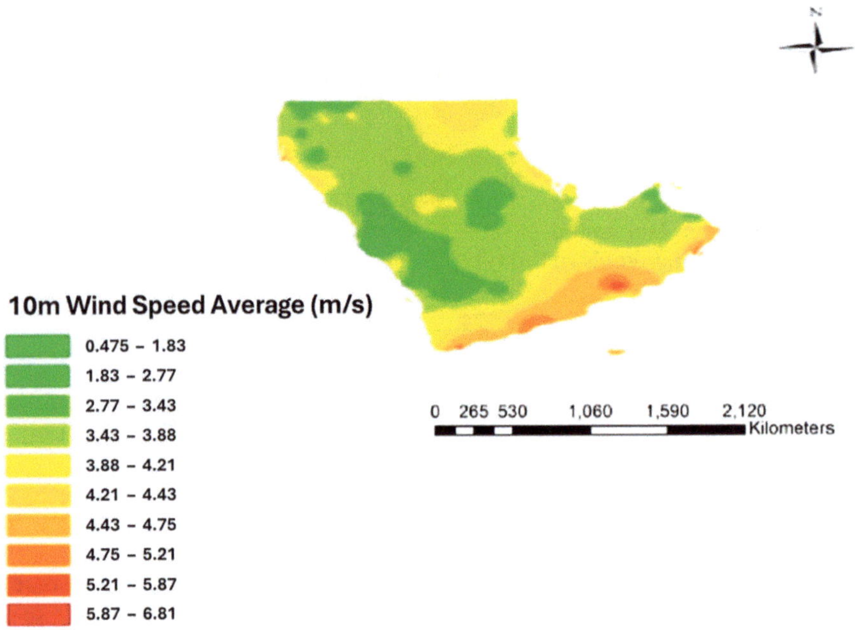

Fig. 9.3 GIS map showing Saudi Arabia's possible wind speed [4]

necessary to ensure the innovations facilitate the reduction of carbon footprints and maximize returns on investments.

Digital twins allow for system configuration before the commitment of capital to Green Hydrogen production projects. The digital twins technology allows developers to model different scenarios in advance, as illustrated in Fig. 9.4.

The key considerations include the weather, demand volatility, off-takers and the local infrastructure. As a result, it is possible to minimize risks while promoting high returns on investment. Research indicates that the digital twin technology optimizes capital by up to 15%. At the same time, it reduces potential risk by 30–50%. KSA can use the technology to ensure that current and future projects for Green Hydrogen production will produce the anticipated quality and quantity of output effectively. Artificial intelligence can be used to facilitate advanced monitoring and control activities. The developers need to accurately track energy consumption, production rates, Hydrogen purity, storage, and overall plant performance. The Internet of Things technology will provide reliable data through the installation of sensors on assets for monitoring key performance indicators, cloud-based remote assessment away from control units, as well as alarms. The technology provides real-time feedback, hence the improvement of efficiency and cost reductions [9].

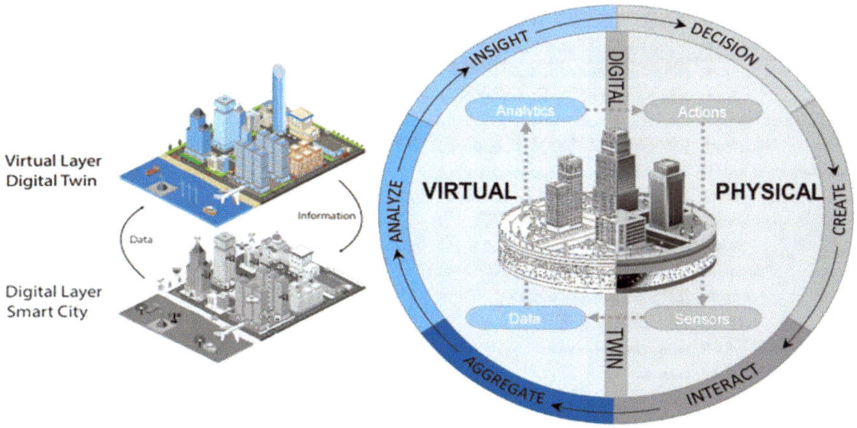

Fig. 9.4 Scenarios for hydrogen production [9]

9.5 Green Hydrogen Storage and Transition

Notably, the only two significant scales of pure hydrogen storage are in the gas and liquid phases. Large volumes of hydrogen gas should be stored at pressures of more than 100 bars in above-ground storage containers but 200 bars in underground vessels. This approach is to take care of the material and operating cost elements. The three primary metallic containers for safe gaseous hydrogen storage include gas holders, spherical pressure vessels, and pipe storage. Regarding liquid storage, the hydrogen gas must first be liquefied to minimize evaporation when stored. The vessels for liquefied hydrogen are normally double-layered, with an existing vacuum between the materials. Despite the complexity of the liquid gas storage containers, there is an indication that this approach is relatively cheaper than storing hydrogen in gas form [8]. Moreover, storing hydrogen in the gas state needs high-pressure tanks, while keeping it in a liquid form needs cryogenic temperatures since its boiling point at room temperature is −252.8 °C [10]. Energy generated from wind farms can be stored as cold/cryo-compressed gas. In this context, the power storage system has an electrolyte that converts the point to green hydrogen.

The storage density of hydrogen gas commonly affects its investment costs because volumetric storage density typically regulates the size of the vessels. In contrast, gravimetric hydrogen accommodation dictates the quantity of the materials required per unit mass of the gas stockpiled. As seen in Fig. 9.5, storing hydrogen gas has the lowest volumetric storage density across all vessel technologies, including high pressures of 700 bar. The trend depicts that greater volumetric storage mass correlates with extreme volumetric storage density [8].

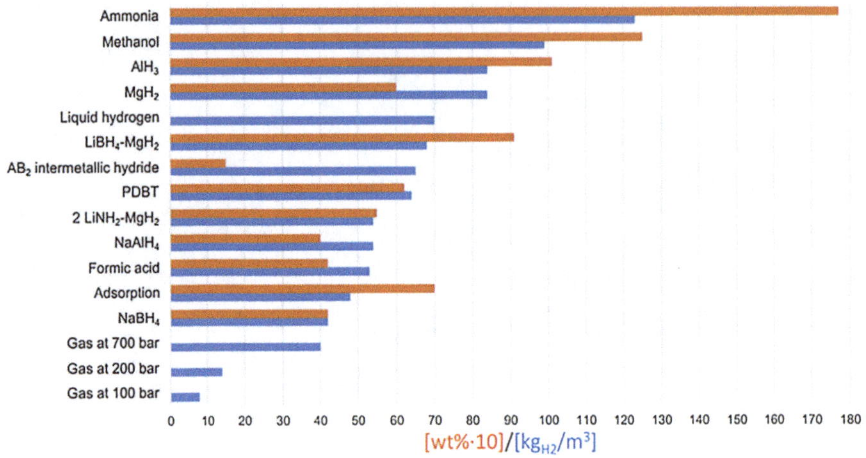

Fig. 9.5 Volumetric and gravimetric hydrogen storage densities of preferred technologies [8]

Green hydrogen storage and transportation are substantial challenges for creating a sustainable energy economy. Nonetheless, it faces challenges like technological advancements, efficiency, and profitability. Presently, most countries generate and use hydrogen locally. However, there will be a financial appraisal of large-scale hydrogen storage because its use as an alternative clean energy source is increasingly becoming common [11]. Assessing the economies of storing green hydrogen in massive amounts will ensure that it comes with environmental advantages and supports the sustainability of its entire supply chain. Many governments, including Saudi Arabia, and research organizations have acknowledged and researched to evaluate the financial aspects of green hydrogen storage systems.

9.6 Present Barriers to the Generation of Green Hydrogen in Saudi Arabia

High Cost of Production

The generation of green hydrogen needs substantial capital investment in the development of its infrastructure and technology. Unlike natural gas, the cost of constructing renewable energy plants like wind and solar systems is still high [12]. Since Saudi Arabia is still trying to improve its renewable energy sector, the government might be forced to spend vast amounts of money to create the needed infrastructure to generate green hydrogen. Green hydrogen is generated through the electrolysis process, which is comparatively expensive and inefficient. For instance, the process requires more energy to convert water into hydrogen and oxygen, which is costly, particularly with high electricity generation. The country's electricity generation cost is high because of its overdependence on oil and natural gas for energy production [13]. Therefore, the increased cost of generating green hydrogen in Saudi Arabia is a substantial challenge that requires more infrastructure and technology investment to improve its demand. Since the country strives to improve its renewable energy industry and support the production of green hydrogen, the associated costs are projected to reduce, making it a suitable alternative for different sectors.

Political and Regulatory Challenges

Some of the political and regulatory challenges include limited local and international markets, dependency on oil and natural gas, and the need for proper regulations. The manufacture of green hydrogen is likely to come with challenges because of the limited local market for hydrogen. Energy usage in most parts of Saudi Arabia is concentrated in the oil and gas industry, leaving other sectors with few calls and access to green hydrogen. The international market for green hydrogen is still developing, and its demand is still low globally. Therefore, this can affect Saudi Arabia's goal of exporting its green hydrogen [14]. Moreover, the country depends on its oil and natural gas, and the sectors are recognized as substantial contributors to its economy. Therefore, the government might be reluctant to invest in green energy, which could threaten the country's existing oil and gas industry.

In Saudi Arabia, the government still needs to create an all-inclusive regulatory framework for generating and supplying green hydrogen, creating doubt for investors and derailing the development of the industry [15–18]. The political and regulatory challenges of the nation's green hydrogen generation are substantial problems that must be solved for the industry to grow. Therefore, the government must create an all-inclusive regulatory framework for the generation of green hydrogen and improve the needed infrastructure to move the industry forward. To address these potential problems, the government should support the growth of the green hydrogen industry. However, the probable advantages of reducing Saudi Arabia's overdependence on oil and gas make it valuable.

9.7 Conclusions

Apart from having the ability to become one of the leading contributors to the international green hydrogen market, Saudi Arabia is presently ranked among the countries producing vast amounts of conventional hydrogen. Since the world is increasingly shifting towards reducing greenhouse gas emissions, green hydrogen is becoming essential since it is a sustainable and environmentally friendly option to the old methods of generating hydrogen. Nonetheless, the country also faces various challenges in its change towards the production of green hydrogen, such as poor infrastructure and increased cost of production. To address these problems, the government should support research and development and create partnerships with other international organizations with enough experience in producing green hydrogen. Suppose the country successfully creates a sustainable green hydrogen sector; in that case, the government will control its reliance on oil and gas and become a significant contributor to the international green hydrogen market. Though Saudi Arabia faces substantial problems in its advancement towards the generation of green hydrogen, it has a viable opportunity to become one of the top producers, thus effectively contributing to the global fight to control climate change.

References

1. Kotowicz J, Uchman W, Jurczyk M, Sekret R (2023) Evaluation of the potential for distributed generation of green hydrogen using metal-hydride storage methods. Appl Energy 344:1–32
2. Al-Saidi M (2022) Energy transition in Saudi Arabia: Giant leap or necessary adjustment for a large carbon economy? Energy Rep 8:312–318
3. Khan MI, Sami GA (2023) Hydrogen economy for sustainable development in GCC countries: a SWOT analysis considering current situation, challenges, and prospects. Int J Hydrog Energy 48(28):10315–10344
4. Hassan Q, Aws ZS, Hayder MS, Marek J, Mohammed A, Mohammad A (2023) Energy futures and green hydrogen production: is Saudi Arabia trend? Results Eng 18:1–10
5. Kar SK, Akhoury SKS, Rohit B, Bahman S, Sidhartha H (2023) Overview of hydrogen economy in Australia. Wiley Interdiscip Rev Energy Environ 12(1):457
6. Li Y, Beni S, Jianjun Y, Junyi F, Adhityo G, Baskoro S (2023) A strategic roadmap for ASEAN to develop hydrogen energy: economic prospects and carbon emission reduction. Int J Hydrog Energy 48(30):11113–11130
7. Rangel CM, Fernandes VR, António JG (2022) Metal hydride-based hydrogen production and storage system for stationary applications powered by renewable sources. Renew Energy 197:398–405
8. Andersson J, Stefan G (2019) Large-scale storage of hydrogen. Int J Hydrog Energy 44(23):11901–11919
9. Caprari G, Castelli G, Montuori M, Camardelli M, Malvezzi R (2022) Digital twin for urban planning in the green deal era: a state of the art and future perspectives. Sustainability 14(10):6263
10. Hazrat MA, Rasul M, Jairul MI, Chowdhury A (2022) Techno-economic analysis of recently improved hydrogen production pathway and infrastructure. Energy Rep 8:836–844
11. Ma N, Weihua Z, Wenzhong W, Haiqin Z (2023) Large scale of green hydrogen storage: opportunities and challenges. Int J Hydrog Energy 1-37
12. Khan MA, Al-Shankiti I, Zianj A, Idriss H (2021) Demonstration of green hydrogen production using solar energy at 28% efficiency and evaluation of its economic viability. Sustain Energy Fuels 5(4):1085–1094
13. Crotogino F (2022) Large-scale hydrogen storage. Storing Energy 613-632
14. Camargo L, Daniel C, Yulineth CE, Anibal AM, Gaylord CC, Ivan P (2022) Bibliometric analysis of global trends around hydrogen production based on the scopus database in the period 2011–2021. Energies 16(1):87
15. Balabel A, Munner SA, Marwan SA, Abdullah OA, Mohammed MA, Hisham A (2023) Potential of solar hydrogen production by water electrolysis in the NEOM green city of Saudi Arabia. World J Adv Eng Technol Sci 8(1):29–52

16. Al-Sulaiman FA, Al-Shehri DA, Al-Mutairi AM, Al-Mutairi AM (2023) The clean hydrogen economy and Saudi Arabia: domestic developments and international perspectives. Stanford Hydrogen Initiative
17. Yamani Z, Al-Sulaiman FA, Al-Shehri DA, Al-Mutairi AM, Al-Mutairi AM (2023) Advancing green hydrogen production in Saudi Arabia. SCIEPublish J
18. Hasan, S., Al-Fattah, S.M., Al-Mutairi, S.H., Al-Shehri, D.A., Al-Mutairi, A.M., Al-Mutairi, A.M.: The Economics and Resource Potential of Hydrogen Production in Saudi Arabia. KAPSARC Discussion Paper (2022)

Chapter 10
Drivetrain Motion Analyses of Eight Floating Wind Turbines

Mohamed Maktabi and Eugen Rusu

Abstract This work considers drivetrain motion analyses of eight floating wind turbines. From this perspective, his work will provide the six-degree-of-freedom drivetrain motions. This work is a continuation of previous work concerning the floater responses of the same floating wind turbines. The drivetrain analyses carried out in this paper are five semi-submersible floating wind concepts (OO-Star, CSC, WindFloat, Volturn-US, and INO-WINDMOOR), two spar-buoy floating wind concepts (DTU Spar 1 and DTU Spar 2) and one TLP floating wind concept.

Keywords Offshore renewable energy · Floating wind · Drivetrain motion

10.1 Introduction

This work could be considered as a continuation of the work carried out in [1–3]. From this perspective, previous work considered support structure motion analyses of eight floating wind turbines. In this paper, we will carry on the same analysis but in terms of the drivetrain motion instead of the support structure motion. The drivetrain analysis is important to see what effects the floater motion has on drivetrain motion, which will affect the blades and the nacelle function, to mention just a few

M. Maktabi (✉)
Department of Mechanical Engineering, 'Dunarea de Jos' University of Galati, Galati, Romania

Department of Marine Technology, Norwegian University of Science and Technology (NTNU), Trondheim, Norway

Department of Law and Economics, Islamic University of Madinah, Madinah, Saudi Arabia
e-mail: mm307@student.ugal.ro

E. Rusu
Department of Mechanical Engineering, 'Dunarea de Jos' University of Galati, Galati, Romania
e-mail: eugen.rusu@ugal.ro

L. Chen (ed.), *Clean Energy Research*, Green Energy and Technology,
https://doi.org/10.1007/978-3-031-99312-1_10

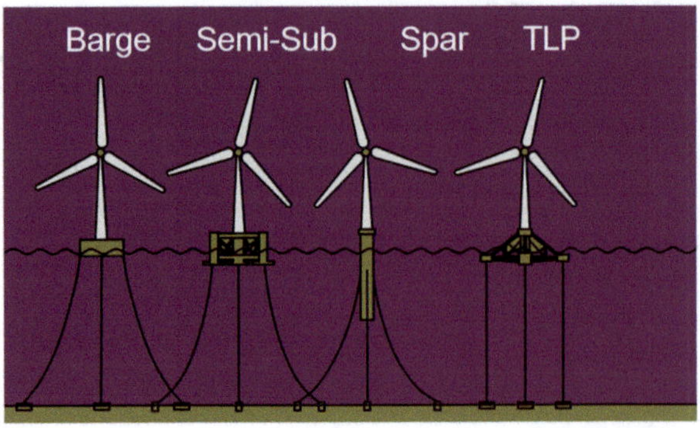

Fig. 10.1 Floating wind turbine types considered in the drivetrain response analyses in this paper. The authors designed the figure in accordance with the information presented in [3]

aspects. However, the most important output data will be given in Sect. 10.3. Section 10.2 will only cover input data concerning only one of the floating wind turbines (OO-Star) due to the length limit of this paper. However, input data concerning all the floating wind turbines analyzed in this paper are mentioned and can be referred to in [3]. Figure 10.1 shows the four generic floating wind turbines, three of which are used in the analysis of this paper, namely, semi-submersible, spar and TLP. More information on each concept and illustrative figures can be found in [3] and is not provided in this paper due to the same reason as the paper length limit. The drive-train response offset data presented in Sect. 10.3 are average values corresponding to 10,000 s time series, excluding the transients. The analysis timestep was set as 0.005 s. The time increment was set as 0.1 s to ensure correct analysis solutions. Papers [4–7] present innovative aspects of relevance to floating wind turbine drive-trains. Figure 10.1 presents the four most popular floating wind turbine types. Three of them (spar, semi-submersible, and TLP) are covered under the eight floating wind turbine concepts studied in this paper. The analysis conducted in this paper is a dynamic analysis with stationary uniform wind and JONSWAP wave.

10.2 Materials and Methods

Sima software implements the blade element momentum (BEM) theory for calculating wind loads. In terms of the wave loads, potential flow theory is implemented in the Sima software. It has assumptions of irrotational flow and incompressible and inviscid fluid, as expressed in the following equation [8].

$$\nabla^2 \phi = 0 \tag{10.1}$$

According to [3, 9], the time-series equation of motion implemented in Sima software has the following expression.

$$\left(m+A_{\infty}\right)\ddot{x}+D_1\dot{x}+Kx+\int_0^t h\left(t-\tau\right)\dot{x}\left(\tau\right)\mathrm{d}\tau=q\left(t,x,\dot{x}\right) \tag{10.2}$$

where m is the mass, A_{∞} is the infinite-frequency added mass, D_1 is the external linear damping matrix, K is the hydrostatic restoring matrix, h is the retardation matrix, and q is the external loads vector including first and second-order loads, Morison viscous loads, tower loads, and mooring forces.

Furthermore, the Sima software uses control theory to control the dynamic loads and elastic theory to calculate the mooring system dynamic loads. Further information about the detailed background behind Sima software can be found in [3, 9].

Table 10.1 presents the nine sea state conditions that are analyzed separately for each of the eight floating wind turbines considered in this paper. The implemented wind speed in the analyses is the rated wind speed. It is 11.4 m/s for 10 MW floating wind turbines and 10.6 m/s for 12 and 15 MW floating wind turbines. The MW corresponding values of each of the floating wind turbine concepts are mentioned in the Tables in Sect. 10.3.

Table 10.2 illustrates the input data of the OO-Star floating wind turbine analyzed in this paper. The input data of the other seven floating wind turbines are not presented here and can be found in [3].

10.3 Results and Discussions

This section will present the drivetrain response offset results of the eight floating wind turbines considered in this study (See Tables 10.1, 10.2, 10.3, 10.4, 10.5, 10.6, 10.7 and 10.8). Some of the corresponding results will be compared with the results mentioned in external references.

Table 10.3 presents OO-Star drivetrain's six degrees of freedom. The drivetrain surge response offset is 0.3–65.8 m. The drivetrain sway response offset is 0.003–0.602 m. The drivetrain heave response offset is 0.43–1.13 m. The drivetrain roll response offset is 0.0015–0.434°. The drivetrain pitch response offset is 0.27–5.9°. The drivetrain yaw response offset is 0.0002–0.13°.

Table 10.4 presents CSC drivetrain's six degrees of freedom. The CSC drivetrain surge response offset is 2.143–52.17 m. The drivetrain sway response offset is 0.0014–0.766 m. The drivetrain heave response offset is 0.03–1.57 m. The drivetrain roll response offset is 0.0004–0.473°. The drivetrain pitch response offset is 0.315–7.79°. The drivetrain yaw response offset is 0.0002–0.112°.

Paper [10] has analyzed surge, sway and heave motions on a semi-submersible floating wind turbine. This paper reports nacelle surge motion as around 8 m. It reports nacelle sway motion as around 0.2 m. It reports nacelle heave motion as around 86 m. Note that the wind turbines and their wind and wave conditions are different from the wind turbine analyzed in Table 10.3.

Table 10.5 presents DTU Spar 1 drivetrain's six degrees of freedom. The DTU Spar 1 drivetrain surge response offset is 2.16–82.7 m. The drivetrain sway response

Table 10.1 Environmental wind and wave conditions of the nine sea state analyses carried out for the eight floating wind turbines studied in this paper

Condtions								
1	2	3	4	5	6	7	8	9
Only wind	Only wave	Only wave	Only wave and current	Only wave and current	Only wind and wave	Only wind and wave	Combined wind, wave and current	Combined wind, wave and current
Hs = 0.01 m	Hs = 4 m	Hs = 2.5 m	Hs = 4 m	Hs = 2.5 m	Hs = 4 m	Hs = 2.5 m	Hs = 4 m	Hs = 2.5 m
Tp = 10 s	Tp = 12 s	Tp = 10 s	Tp = 12 s	Tp = 10 s	Tp = 12 s	Tp = 10 s	Tp = 12 s	Tp = 10 s

Table 10.2 OO-Star floating wind turbine input characteristics [3]

OO-Star	
Support structure's height [m]	**Total height until the hub [m]**
27	131.63
Accumulated mass of support structure [kg]	
2.1709×10^7	
Support structure's accumulated volume [m³]	
8199.1	
Length of each blade [m]	
86.365	
Additional masses at hub [kg]	
Nacelle: 1, hub: 1.0502×10^5	
Stress-free length of each mooring line [m]	
703	
Water depth [m]	
130	

Table 10.3 OO-Star drivetrain response offset for different conditions and degrees of freedom

OO-Star [m] [°]—10 MW									
Conditions	1	2	3	4	5	6	7	8	9
Surge	40.8	0.6	0.3	39.6	38.8	42.3	41.8	65.8	65.4
Sway	−0.5	−0.005	−0.003	0.006	0.004	−0.55	−0.54	−0.601	0.602
Heave	1.13	0.43	0.43	0.53	0.53	0.53	1.13	0.83	0.83
Roll	0.4	0.0017	0.0015	0.003	0.002	0.434	0.433	0.417	0.419
Pitch	5.9	−0.28	−0.27	−1.99	−1.96	5.8	5.9	3.8	3.9
Yaw	0.1	0.0005	0.0002	0.003	0.004	0.127	0.13	0.123	0.124

Table 10.4 CSC drivetrain response offset for different conditions and degrees of freedom

CSC [m] [°]—10 MW									
Conditions	1	2	3	4	5	6	7	8	9
Surge	29.5	2.143	2.148	25.2	25.03	29.5	29.3	52.12	52.17
Sway	0.7	0.0017	0.0014	0.014	−0.012	−0.74	−0.74	−0.766	−0.764
Heave	1.57	0.03	0.07	0.27	0.27	1.47	1.47	1.37	1.37
Roll	0.4	0.0005	0.0004	0.00304	0.0034	0.473	0.472	0.427	0.426
Pitch	7.7	−0.316	−0.315	−3.059	−3.054	7.79	7.74	5.43	5.49
Yaw	−0.08	0.0003	0.0002	0.044	−0.043	0.08	−0.09	−0.112	0.1101

offset is 0.0037–0.97 m. The drivetrain heave response offset is 0.02–3.88 m. The drivetrain roll response offset is 0.00097–0.328°. The drivetrain pitch response offset is 0.2205–9.15°. The drivetrain yaw response offset is 0.000083–0.37°.

Paper [11] shows that the maximum spar nacelle surge motion is 40 m. The maximum nacelle pitch rotation is 22°.

Table 10.5 DTU Spar 1 drivetrain response offset for different conditions and degrees of freedom

DTU Spar 1 [m] [°]—10 MW									
Conditions	1	2	3	4	5	6	7	8	9
Surge	58.1	2.16	2.17	43.7	43.6	57.09	57.6	82.5	82.7
Sway	−0.92	−0.015	0.0037	−0.011	−0.0045	−0.97	−0.95	0.967	0.962
Heave	2.83	0.03	0.03	0.08	0.02	2.73	2.83	3.88	3.88
Roll	0.32	0.0038	0.00097	0.0035	0.0013	0.328	0.326	0.325	0.324
Pitch	8.774	−0.221	−0.2205	0.64	0.62	8.58	8.67	9.109	9.15
Yaw	−0.32	0.00034	−0.0032	0.000083	0.00039	−0.37	−0.35	−0.33	−0.32

Table 10.6 DTU Spar 2 drivetrain response offset for different conditions and degrees of freedom

DTU Spar 2 [m] [°]—10 MW									
Conditions	1	2	3	4	5	6	7	8	9
Surge	56.5	1.964	1.968	32.2	32.1	55.7	56.01	76.1	76.5
Sway	−1.09	−0.02	0.0048	−0.01	−0.005	−1.13	−1.12	1.124	−1.11
Heave	4.12	0.02	0.02	0.01	0.01	4.02	4.02	5.21	5.21
Roll	0.38	0.0047	0.00101	0.0044	0.0015	0.393	0.391	0.387	0.386
Pitch	11.5	−0.29	−0.28	0.37	0.34	11.3	11.4	11.5	11.6
Yaw	−0.27	0.00021	0.0027	0.000042	0.00027	−0.32	0.305	−0.28	−0.27

Paper [12] analyses surge, roll and pitch response offsets on a spar floating wind turbine. The results show that the surge motion is around 30 m. The roll rotation is around 0.2°. The pitch rotation is around 4.5°. Note that this is a support structure response analysis and not a drivetrain response analysis, as it is the topic of this paper. Also note that the floating wind turbine input data and wind and wave conditions differ from what this paper considers.

Table 10.6 presents DTU Spar 2 drivetrain's six degrees of freedom. The DTU Spar 2 drivetrain surge response offset is 1.964–76.5 m. The drivetrain sway response offset is 0.0048–1.13 m. The drivetrain heave response offset is 0.01–5.21 m. The drivetrain roll response offset is 0.00101–0.393°. The drivetrain pitch response offset is 0.28–11.6°. The drivetrain yaw response offset is 0.000042–0.305°.

Further external references are given as follows. Paper [13] carries out a dynamic analysis of relevance to a spar floating wind turbine. Paper [14] presents a spar multibody system (MBS) dynamic analysis of relevance to floating wind turbine drivetrain. The paper presents [15] drivetrain dynamic analysis of a multibody system (MBS) spar floating wind turbine. Paper [16] presents a spar floating wind turbine dynamic analysis.

Table 10.7 presents Volturn-US drivetrain's six degrees of freedom. The Volturn-US drivetrain surge response offset is 3.42–34.08 m. The drivetrain sway response offset is 0.0028–1.24 m. The drivetrain heave response offset is 0.03–0.53 m. The drivetrain roll response offset is 0.00027–0.507°. The drivetrain pitch response offset is 1.93–4.37°. The drivetrain yaw response offset is 0.0017–0.32°.

Table 10.7 Volturn-US drivetrain response offset for different conditions and degrees of freedom

Volturn-US [m] [°]—15 MW									
Conditions	1	2	3	4	5	6	7	8	9
Surge	30.49	−3.42	−3.47	3.66	3.59	30.1	30.3	33.8	34.08
Sway	−1.24	−0.0056	0.0028	0.00704	−0.0039	−1.216	−1.22	−1.218	−1.23
Heave	0.53	0.03	0.03	0.03	0.03	0.43	0.43	0.43	0.53
Roll	0.506	0.00096	0.00027	0.001	0.00062	0.503	0.504	0.506	0.507
Pitch	4.37	−1.95	−1.93	−2.31	−2.29	4.24	4.307	3.91	3.98
Yaw	−0.32	0.017	−0.0017	0.016	0.0097	−0.31	−0.32	−0.25	−0.25

Table 10.8 INO-WINDMOOR drivetrain response offset for different conditions and degrees of freedom

INO-WINDMOOR [m] [°]—12 MW									
Conditions	1	2	3	4	5	6	7	8	9
Surge	42.09	39.7	39.6	74.9	74.7	42.4	42.3	77.03	76.8
Sway	0.0075	0.0058	0.0037	0.0031	−0.0023	0.0075	0.0074	0.011	0.0057
Heave	0.39	0.01	0.01	1.11	1.11	0.39	0.39	0.81	0.81
Roll	0.0016	0.00104	0.00059	0.00086	0.00063	0.00165	0.00163	0.0027	0.0024
Pitch	3.017	2.38	2.39	−0.13	−0.106	2.99	3	0.49	0.52
Yaw	0.0068	0.0047	0.0031	0.0015	−0.0032	0.00667	0.00668	0.013	0.008

Table 10.9 WindFloat drivetrain response offset for different conditions and degrees of freedom

WindFloat [m] [°]—10 MW									
Conditions	1	2	3	4	5	6	7	8	9
Surge	80.4	53	52.9	70.8	70.5	79.8	80.1	100.1	100.2
Sway	−0.39	0.00086	0.00055	0.00064	0.00043	−0.5006	−0.45	−0.64	−0.61
Heave	3.38	5.48	5.48	5.68	4.68	3.08	3.18	2.08	2.28
Roll	0.35	0.00046	0.00026	0.00047	0.00027	0.352	0.3508	0.307	0.305
Pitch	12.9	5.22	5.21	4.16	4.17	12.6	12.81	11.84	12
Yaw	0.087	0.00082	0.00039	0.00077	0.00038	0.022	0.0522	−0.05208	−0.035

Table 10.8 presents INO-WINDMOOR drivetrain's six degrees of freedom. The INO-WINDMOOR drivetrain surge response offset is 39.6–77.03 m. The drivetrain sway response offset is 0.0023–0.0075 m. The drivetrain heave response offset is 0.01–1.11 m. The drivetrain roll response offset is 0.00059–0.00165°. The drivetrain pitch response offset is 0.106–3.017°. The drivetrain yaw response offset is 0.0015–0.0068°.

Table 10.9 presents WindFloat drivetrain's six degrees of freedom. The WindFloat drivetrain surge response offset is 52.9–100.2 m. The drivetrain sway response offset is 0.00043–0.64 m. The drivetrain heave response offset is 2.08–5.68 m. The drivetrain roll response offset is 0.00026–0.352°. The drivetrain pitch response offset is 4.16–12.9°. The drivetrain yaw response offset is 0.00038–0.087°.

Table 10.10 presents TLP drivetrain's six degrees of freedom. The TLP drivetrain surge response offset is 2.65–8.05 m. The drivetrain sway response offset is

Table 10.10 TLP drivetrain response offset for different conditions and degrees of freedom

TLP [m] [°]—10 MW									
Conditions	1	2	3	4	5	6	7	8	9
Surge	5.8	2.66	2.65	4.67	4.64	5.82	5.83	8.03	8.05
Sway	−0.079	−0.00013	−0.000064	0.00009705	0.00005009	−0.077	−0.078	−0.0622	−0.0628
Heave	0.18	0.02	0.02	0.02	0.02	0.18	0.18	0.08	0.08
Roll	0.098	0.000021	0.0000105	−0.0000207	−0.0000106	0.0982	0.0984	0.09853	0.09856
Pitch	0.46	−0.0563	−0.0566	−0.055	−0.056	0.4603	0.462	0.48	0.49
Yaw	−0.11	0.000077	0.0000071	0.00033	0.00015	−0.11	−0.11	−0.101	−0.1008

0.000064–0.079 m. The drivetrain heave response offset is 0.02–0.18 m. The drivetrain roll response offset is 0.0000105–0.09856°. The drivetrain pitch response offset is 0.055–0.49°. The drivetrain yaw response offset is 0.000071–0.11°.

Further external references are given as follows. Paper [17] presents dynamic drivetrain analysis on a barge floating wind turbine. Paper [18] presents a further drivetrain dynamic analysis of a floating wind turbine. Paper [19] analyses the forces acting on a wind turbine drivetrain. Paper [20] presents dynamic load analysis on spar, semi-submersible and TLP floating wind turbine drivetrains. Paper [21] presents dynamic load analysis on a barge floating wind turbine and drivetrain aspects of relevance.

10.4 Conclusions

This paper has analyzed the drivetrain response offsets of eight floating wind turbines.

The results presented in Sect. 10.3 show the drivetrain response offset in descending order as follows. For the surge drivetrain motion, WindFloat has maximum amplitude (100.2 m), then comes DTU Spar 1 (82.7 m), then INO-WINDMOOR (77.03 m), then DTU Spar 2 (76.5 m), then OO-Star (65.8 m), then CSC (52.17 m), then Volturn-US (34.08 m), then TLP (8.05 m).

For the sway drivetrain motion, Volturn-US has maximum amplitude (1.24 m), then comes DTU Spar 2 (1.13 m), then DTU Spar 1 (0.97 m), then CSC (0.766 m), then WindFloat (0.64 m), then OO-Star (0.602 m), then TLP (0.079 m), then INO-WINDMOOR (0.0075 m).

For the heave drivetrain motion, WindFloat has maximum amplitude (5.68 m), then comes DTU Spar 2 (5.21 m), then DTU Spar 1 (3.88 m), then CSC (1.57 m), then OO-Star (1.13 m), then INO-WINDMOOR (1.11 m), then Volturn-US (0.53 m), then TLP (0.18 m).

For the roll drivetrain motion, Volturn-US has maximum amplitude (0.507°), then comes CSC (0.473°), then OO-Star (0.434°), then DTU Spar 2 (0.393°), then WindFloat (0.352°), then DTU Spar 1 (0.328°), then TLP (0.09856°), then INO-WINDMOOR (0.00165°).

For the pitch drivetrain motion, WindFloat has maximum amplitude (12.9°), then comes DTU Spar 2 (11.6°), then DTU Spar 1 (9.15°), then CSC (7.79°), then OO-Star (5.9°), then Volturn-US (4.37°), then INO-WINDMOOR (3.017°), then TLP (0.49°).

For the yaw drivetrain motion, DTU Spar 1 has maximum amplitude (0.37°), then comes Volturn-US (0.32°), then DTU Spar 2 (0.305°), then OO-Star (0.13°), then CSC (0.112°), then TLP (0.11°), then WindFloat (0.087°), then INO-WINDMOOR (0.0068°).

From the above analysis results, the surge drivetrain motion (maximum of 100.2 m) has significantly higher amplitudes compared to the other degrees of freedom. This will affect the turbine's functionality, and if it is subjected to the same conditions for an extended period, it might result in collapse and accidents with other neighboring structures.

References

1. Maktabi M, Rusu E (2024) A review of perspectives on developing floating wind farms. Inventions 9(2):24. https://doi.org/10.3390/inventions9020024
2. Maktabi M, Rusu E (2024) Constant wind analyses on eight floating wind turbines. Preprint. https://doi.org/10.20944/preprints202411.0362.v1
3. Maktabi M, Rusu E (2025) Constant wind analyses on eight floating wind turbines. MDPI [wind]. Paper under peer review
4. Nejad AR, Torsvik J (2021) Drivetrains on floating offshore wind turbines: lessons learned over the last 10 years. Forsch Ingenieurwes 85:335–343. https://doi.org/10.1007/s10010-021-00469-8
5. Moan T, Gao Z, Bachynski EE, Nejad AR (2020) Recent advances in integrated response analysis of floating wind turbines in a reliability perspective. ASME J Offshore Mech Arct Eng 142(5):052002. https://doi.org/10.1115/1.4046196
6. Nejad AR, Keller J, Guo Y, Sheng S, Polinder H, Watson S, Dong J, Qin Z, Ebrahimi A et al (2022) Wind turbine drivetrains: state-of-the-art technologies and future development trends. Wind Energy Sci 7(1):387–411. https://doi.org/10.5194/wes-7-387-2022
7. Torsvik J (2020) Dynamic analysis in design and operation of large floating offshore wind turbine drivetrains. Doctoral theses, NTNU
8. Tian Z, Shi W, Li X, Park Y, Jiang Z, Wu J (2025) Numerical simulations of floating offshore wind turbines with shared mooring under current-only conditions. Renew Energy 238:121918. https://doi.org/10.1016/j.renene.2024.121918. ISSN 0960-1481
9. Souza CE, Engebretsen E, Bachynski-Polić E, Lene E, Berthelsen PA, Haslum H (2021) Definition of the INO WIND-MOOR 12 MW base case floating wind turbine
10. Yu Z, Zhu C, Tan J, Song C, Wang Y (2022) Fully-coupled and decoupled analysis comparisons of dynamic characteristics of floating offshore wind turbine drivetrain. Ocean Eng 247:110639. https://doi.org/10.1016/j.oceaneng.2022.110639. ISSN 0029-8018
11. Xing Y, Karimirad M, Moan T (2014) Modelling and analysis of floating spar-type wind turbine drivetrain. Wind Energy 17(4):565–587
12. Jiang Z, Yang L, Gao Z, Moan T (2022) Integrated dynamic analysis of a spar floating wind turbine with a hydraulic drivetrain. Renew Energy 201(Part 1):608–623. https://doi.org/10.1016/j.renene.2022.10.104. ISSN 0960-1481
13. Skaare B, Hanson T, Nielsen F, Yttervik R, Hansen A, Thomsen K, Larsen T (2007) Integrated dynamic analysis of floating offshore wind turbines. In: European wind energy conference and exhibition
14. Viadero F, del Rincon AF, Liaño E, Serna MA, Diaz MA (2012) Dynamic analysis of an offshore wind turbine drivetrain on a floating support. Springer, Berlin
15. Viadero F (2014) Non-stationary dynamic analysis of a wind turbine power drivetrain: offshore considerations. Appl Acoust 77:204–211
16. Wang S, Nejad A, Moan T (2024) Dynamic behaviour comparison of a 10-MW drivetrain in a bottom-fixed monopile and a spar floating wind turbines
17. Xie S, Jin X, He J (2019) Structural vibration control for the offshore floating wind turbine including drivetrain dynamics analysis. J Renew Sustain Energy 11(2):023304. https://doi.org/10.1063/1.5079427
18. Li Z, Wen B, Wei K, Yang W, Peng Z, Zhang W (2020) Flexible dynamic modeling and analysis of drive train for offshore floating wind turbine. Renew Energy 145(C):1292–1305

19. Li Z, Wen B, Peng Z, Dong X, Qu Y (2020) Dynamic modeling and analysis of wind turbine drivetrain considering the effects of non-torque loads. Appl Math Model 83:146–168. ISSN 0307-904X
20. Nejad A, Bachynski E, Kvittem MI, Luan C, Gao Z, Moan T (2015) Stochastic dynamic load effect and fatigue damage analysis of drivetrains in land-based and TLP, spar and semi-submersible floating wind turbines. Mar Struct 42:137–153. https://doi.org/10.1016/j.marstruc.2015.03.006. ISSN 0951-8339
21. Jonkman JM (2007) Dynamics modeling and loads analysis of an offshore floating wind turbine. Technical report NREL/TP-500-41958

Chapter 11
Inference of the Levelized Cost of Energy for a Future Offshore Wind Farm Located in the Western Area of the Black Sea

Marius Manolache, Alexandra-Ionelia Manolache, and Gabriel Andrei

Abstract The levelized cost of energy (LCOE) has been widely used as a criterion for the comparison of electrical energy supply options from different sources of energy, generally in the comparison of a renewable energy source against the conventional way of obtaining the same electrical energy from a fossil source. Also, LCOE is used as a criterion for making investment decisions in the expansion of coverage of the electricity service. This research estimates LCOE using historical prices and costs. For the research, a Black Sea farm with 72 wind turbines was used. The Siemens Gamesa SWT-8.4-167 wind turbine power curve data model and observed wind data were used to predict energy production at this location. The analysis also considers water depth, grid connection expenses, equipment costs, operating and maintenance costs, and other wind farm lifecycle costs. This study excludes some financial parameters, incentives, taxes, fiscal policies, and even some basic accounting concepts, which affect their accuracy and uncertainty.

Keywords Black Sea · Wind turbine · Wind speed · LCOE · CAPEX · OPEX

11.1 Introduction

Massive fossil fuel consumption has exacerbated environmental issues and hastened global warming, which has sparked the growth of alternative energy sources. The best renewable energy option for meeting emission reduction goals is thought to be wind power, a cost-effective and sustainable energy source. By the end of 2023, 73 GW of offshore wind capacity had been installed globally [1]. Forty-two new offshore wind farms were put in operation globally in 2022, with 29 of those being installed in China, 5 in Vietnam, 2 in Japan, 1 in each of the UK, South Korea, Italy, France, Spain, and Germany, and 5 each in South Korea and Vietnam. Offshore

M. Manolache (✉) · A.-I. Manolache · G. Andrei
"Dunărea de Jos" University of Galați, Galați, Romania
e-mail: marius.manolache@ugal.ro

L. Chen (ed.), *Clean Energy Research*, Green Energy and Technology,
https://doi.org/10.1007/978-3-031-99312-1_11

wind farms are better able to fully maximize power generation efficiency because they lack many of the limiting variables of onshore wind farms (such as noise, visual impacts, and land use). Almost 45% of the gas that the European Union buys for use in industry, residences, and energy generation comes from Russia. We calculate that between 100 and 200 TWh of the European Union's natural gas-based power is supplied by Russia, considering country-level supply dependencies. On the other hand, the incremental growth of renewable electricity output up to 180 TWh from 2021 to 2023, almost equivalent to the peak value of gas-fired generation, depends on Russia. Considering current deployment patterns, the rise of wind and solar PV throughout the European Union has the potential to dramatically lessen the need for Russian gas for energy use.

The levelized cost of energy (LCOE) is the theoretical price at which electricity should be sold in order to break even. As a result, it is a critical parameter in determining the economic viability of an energy project and serves as a standardized method for comparing the costs of various energy sources. The LCOE of a project is made up of two major components: the costs and the energy produced. Expressions based on site-specific variables, such as distance to shore and water depth, may be used to estimate project costs [2]. A cost breakdown like this is useful for identifying cost-cutting opportunities. In terms of energy production, it is typically computed using average wind climate conditions at specific locations [3]. However, the wind climate has been shown to significantly influence a number of economic indices [4], and this is only emphasized when considering the impact of climate change on wind energy [5]. To summarize, an accurate, site-specific estimation of energy production is required to identify potential areas for the deployment of renewable energy. The global offshore wind benchmark is now $3/MWh lower than coal and $18/MWh lower than gas. This is the first time that the benchmark undercuts fossil fuels. High commodity prices continue to raise the cost of coal and gas projects. Our benchmark gas LCOE is now more than double that of solar and onshore wind, at $93/MWh globally. It anticipates that fossil fuel prices will remain high in the short to medium term.

The LCOE for a location in the Black Sea will be calculated in this paper. A case study that simulates the placement of a wind farm with 72 Siemens Gamesa SWT-8.4-167 turbines and a total capacity of 608 MW using a floating structure. The simulated project will have a 20-year lifespan and will analyze wind speed data to better characterize the location environment. Furthermore, site-specific estimation of electricity production provides a more accurate assessment of energy costs.

11.2 Methods and Materials

Target Area

The Black Sea is a semi-enclosed sea in the Atlantic basin that is a hydrological remnant of the Sarmatian Sea. It is located in the extreme south-east of Europe.

A lot of significant factors should be considered when selecting a location. The depth to which the structure is functional; in this case, a spar-type structure operates to depths of more than 100 m. Another issue is the distance from the coast, which must be far enough not to be visibly disturbed but also to avoid noise [6]. There are several restrictions related to these issues that vary by country. For countries with operable wind farms in the North Sea, for example, there are restrictions of 12 nautical miles (NM) legally set or not (as in the case of Belgium, Holland, and Scotland) or 2 km from the shore (as in Norway) [7]. They must be located roughly 11 km from the shore in Greece and 12 km [8] from the shore in Spain. A final factor considered in this analysis is maritime transport. As illustrated in Fig. 11.1 the largest number of transports for the east of the Black Sea are seen around ports such as Romania's Constanta port. After analyzing this statistic and considering all the previously stated factors, we will select the place with the lowest traffic density (Fig. 11.2).

The levelized cost of electricity is the ratio of all costs incurred during the life of an energy project to the amount of electricity produced. It can be thought of as the price at which electricity would have to be sold to break even on costs, and it is thus an important criterion to consider when assessing the feasibility of energy projects. The LCOE can be written as [10, 11]:

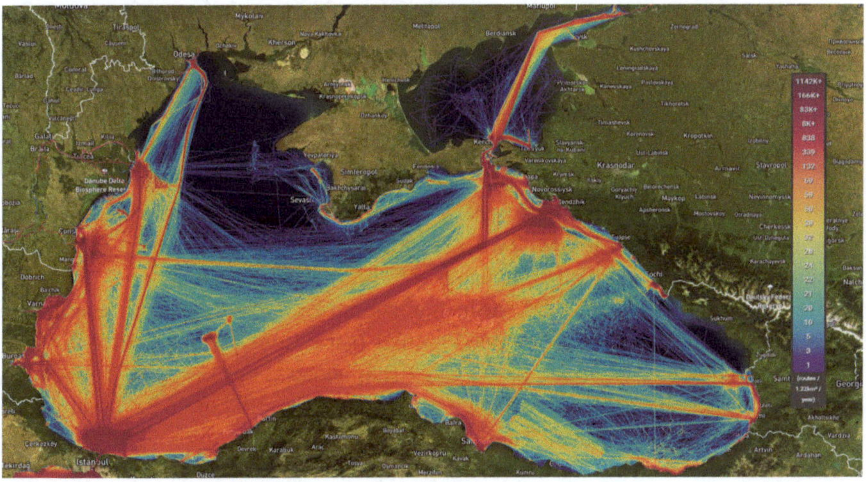

Fig. 11.1 Map with maritime traffic in the Black Sea [9]

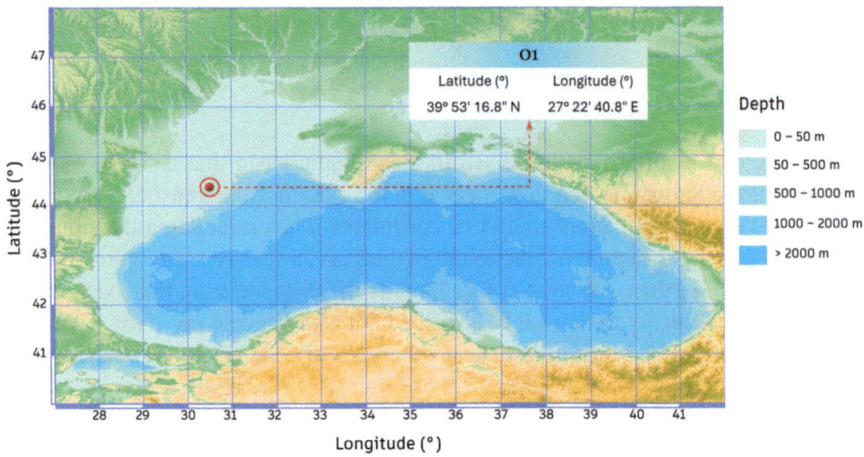

Fig. 11.2 The chosen location in the Black Sea for the study

$$\mathrm{LCOE} = \frac{\sum_{i=1}^{T}\left(\mathrm{CAPEX}_i + \mathrm{OPEX}_i\right)\left(1+r\right)^{-i}}{\sum_{i=1}^{T}\mathrm{AEP}_i\left(1+r\right)^{-i}} \tag{11.1}$$

where CAPEX represents the capital expenditure [12], OPEX is the operational expenditure [12], AEP is the annual energy production, T is the life of the project, and r is the discount rate.

- The capital expenditure of an offshore wind power project includes engineering expenses for equipment procurement and installation, construction engineering fees, sea area use fees, project management fees, and interest throughout construction. The literature states that wind power project equipment procurement, installation engineering, and construction engineering expenditures account for 75–80% of overall project investment. Location, water depth, coast distance, sea conditions, and others affect offshore wind energy project costs.
- Offshore wind farms' LCOE includes 23% of total investment cost for O&M charges, compared to 5% for onshore turbines [13]. Thus, minimizing O&M costs helps reduce LCOE. Maintenance expenses are more important than operating costs for LCOE management. Equipment costs dominate O&M spending, followed by revenue losses. Two expenditures are directly related to maintenance: the direct cost of maintenance and the cost of not maintaining. Over the lifespan of an offshore wind farm, maintenance affects downtime length and LCOE. Engineering structures need regular inspections and repairs to fix problems or replace problematic parts. OWT maintenance costs vary by foundation type and location. High maintenance expenses hinder offshore wind plant development. Ageing equipment causes wind farm downtime, although reasonable and effective maintenance may decrease it [14].

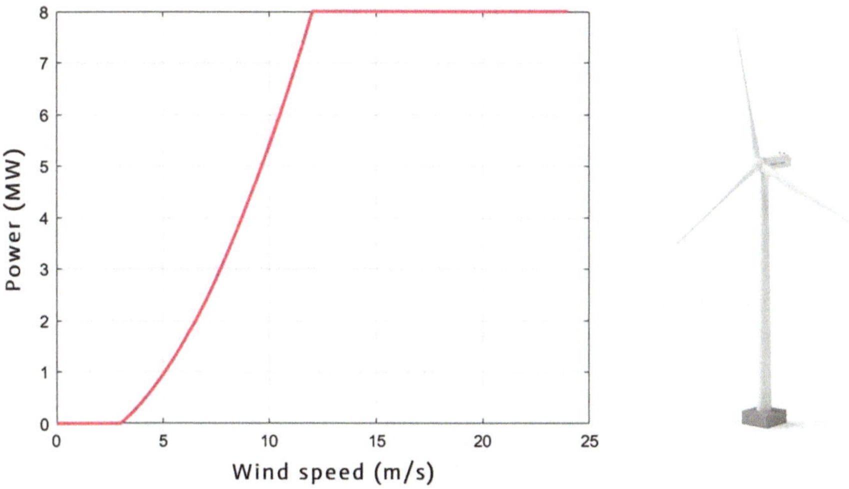

Fig. 11.3 Siemens Gamesa 8 MW turbine power curve

- The LCOE calculation requires the AEP, which describes the total energy generated annually for a certain turbine type and can be determined using the formula [15]:

$$\text{AEP} = T \times \int_{\text{cut-in}}^{\text{cut-out}} f(u) P(u) du \tag{11.2}$$

- where $T = 8760$ h/year is the turbine's annual operating time. The wind value at which the wind turbine begins to operate is referred to as the cut-in wind speed; the turbine continues to rotate until it achieves maximum efficiency at the cut-out wind speed (in m/s for both parameters) [16]. $P(u)$ denotes the power curve of a particular wind turbine (Fig. 11.3) and $f(u)$ is the Weibull distribution function, which can be determined using the formulas from [17, 18]. The wind speed used was at a height of 90 m and was determined using the wind profile's logarithmic expression [19].

11.3 Results and Discussion

For this work, a location in the Black Sea was chosen to simulate a wind turbine farm, considering all the aspects presented in Sect. 11.2. To obtain the CAPEX, the following costs were considered:

- For development and consent, the costs are approximately €210,000/MW, of which 70% development services, 15% seabed survey, 7.5% met mast and 7.5% environmental survey [20].

- The average cost of a generic wind turbine for the year 2022 is set at €1,200,000/MW.
- The structure has a total cost of €3,947,800 per turbine, which is estimated based on material consumption and manufacturing process estimation.
- The transmission can be HVAC (high voltage alternating current) or HVDC (high voltage direct current). For wind farms that are located at distances greater than 80 km, the latter is recommended. These costs are made up of the costs for the export cables (for HVDC this price is estimated at €1,168,000/km), the costs for the offshore substation (€142,750,000/substation) and the onshore substation (€84,350,000/substation) [20].
- The anchoring system for a spar-type structure is made of a combination of steel wires and chains. The anchoring system depends a lot on both the number of lines needed and the depth of the water. Thus, for waters with a depth of 100 m, the length of the anchor line must be 500 m, to which 50 m of chains are added. The cost for 1 m of steel wire is estimated at 45 euros, and for a chain, 250. In addition to this, there is also the baseline for the anchor, which costs 342,000 euros per turbine [21].
- The installation of an offshore wind farm is a challenging one and a multitude of aspects and problems that may arise must be considered. To simplify the algorithm, the price for installing a wind farm is approximated at €100,000,000.
- Decommissioning consists of a reverse process to the installation and for this work it will be estimated at approximately €250,000/MW [20].

Table 11.1 summarizes the total CAPEX results for the reference scenario. The distance to shore has a significant impact on capital expenditure, which influences transmission and installation prices. Interestingly, the expenses directly related to water depth are seen to be less substantial when compared to those influenced by distance to the coast. Just the length of the mooring lines is impacted by water depth in the technique provided in this paper.

The costs associated with operation and maintenance are provided by daily maintenance, the replacement of some small parts and, in some cases, large components, the repair of cables, the operation on and to the site, the provision of staff housing and port facilities, and other costs. These costs were approximately €1475 million in the case of this study.

Table 11.1 The results of the costs considered in CAPEX

Parameters	Value	Unit of measurement
Development and consenting	120.96	(M€)
Turbine	699.29	(M€)
Structure	284.24	(M€)
Mooring	32.184	(M€)
Transmission	744.68	(M€)
Installation	100	(M€)
Decommissioning	144	(M€)
Total CAPEX	2125.35	(M€)

By computing the wind data, we obtained a maximum wind speed of 27.31 m/s and an average speed of approximately 8.05 m/s. Comparing the average speed with the wind classes from [17], we can conclude that the chosen location falls into wind class 4, which is good for exploiting wind resources but cannot produce extraordinary amounts of energy. When considering the operating period of the turbine, it is limited to 65% and will have a capacity factor of 44.7 due to the wind speed not exceeding 12 m/s. The annual energy production of a turbine was computed using the Siemens turbine's power curve and wind speed; the result was an AEP of 35249.2 MWh.

In this study, the LCOE has a value of €74/MWh, which is close to the real value of this parameter. For example, in the case of the United States, the LCOE has a value in the range of €61–116/MWh. Comparing the result obtained with the one from [22] where the LCOE is €75/MWh for the most favorable case, we can say that the results obtained are close, the areas studied in both works being close in terms of wind speeds. Another study that analyzed the LCOE is the one in [20] for which values between €95/MWh were obtained for areas with the best wind resources such as the UK, Ireland and NW Spain and €160/MWh for areas with lower resources, the Gulfs of Biscay and Cadiz. Similarly, the study in [23] obtained LCOE values between €82/MWh and €236.7/MWh, the difference in LCOE in this case being given by the type of foundation used. We note that compared to both studies, the results obtained in this case are below those in the evaluated studies but are comparable to those obtained in this study.

11.4 Conclusion

The LCOE for a floating offshore wind farm based on a floating spar foundation located in the Black Sea was evaluated in this study. A breakdown of all costs incurred during the wind farm's lifetime was presented using expressions based on site-specific variables such as distance to shore and water depth. Furthermore, based on wind speed data and the power curve of an exemplary wind turbine, the annual electricity production was estimated. This allowed for a more accurate site-specific estimation of theoretical electricity generation.

For the floating wind farm, several key cost drivers have been identified. It is possible to distinguish between site-dependent and thus predictable aspects and uncertain aspects. The discount rate, distance from shore, farm size, and depth are the most sensitive to the LCOE of the predictable aspects. The accuracy of the load factor and the variation in steel price are two of the most influential factors in distinguishing the foundation concepts from the more uncertain aspects. Given that the wind turbine and substructure account for most of the costs, the computed LCOE values have a high potential for cost reduction. Furthermore, given the high reliance of the LCOE on the turbine on the wind resource, future optimization and standardization of different turbines for specific wind climates may significantly reduce energy costs. In other words, as turbine technology develops and adapts to specific wind climates, areas with high energy costs in this work may have greater potential in the future. More research is needed to implement cost-saving potential and scaling effects, particularly for mass-produced components such as turbines, mooring lines, and anchors. Other studies can be conducted to investigate the effects of extending the turbine's lifetime. Future studies should evaluate the impact of uncertainty on the LCOE estimations. The data utilized in this analysis are limited by the relatively new nature of floating offshore wind technology. The more accurately these uncertainties are measured, the more pertinent and beneficial LCOE estimations are for the sector in formulating strategic decisions.

References

1. Manolache AI, Andrei G (2024) A comprehensive review of multi-use platforms for renewable energy and aquaculture integration. Energies 17:4816. https://doi.org/10.3390/en17194816
2. Benabadji AS, Rahmoun K, Bahar FA, Dahani A, Martinez A, Iglesias G (2025) Geospatial LCOE analysis for floating offshore wind energy in SW Mediterranean Sea. Renew Energy 245:122797. https://doi.org/10.1016/j.renene.2025.122797
3. Martinez A, Iglesias G (2024) Levelized cost of energy to evaluate the economic viability of floating offshore wind in the European Atlantic and Mediterranean. e-Prime Adv Electr Eng Electron Energy 8:100562. https://doi.org/10.1016/j.prime.2024.100562
4. Castro-Santos L, Diaz-Casas V (2015) Sensitivity analysis of floating offshore wind farms. Energy Convers Manag 101:271–277. https://doi.org/10.1016/j.enconman.2015.05.032
5. Serri L, Airoldi D, Lanni F, Naldi R, Castorrini A, Rispoli F, Soukissian T, Castro Santos L, Le Boulluec M, Maisondieu C (2024) Technical and economic challenges for floating offshore wind deployment in Italy and in the Mediterranean Sea. WIREs Energy Environ 13:e533. https://doi.org/10.1002/wene.533
6. Díaz H, Guedes Soares C (2020) Review of the current status, technology and future trends of offshore wind farms. Ocean Eng 209:107381. https://doi.org/10.1016/j.oceaneng.2020.107381
7. SEANSE (2019) Planning criteria for offshore wind energy—North Sea region overview
8. Melissas D, Vasilakos N (2022) An optimal strategy for initiating and efficiently developing the floating offshore wind energy sector in Greece
9. Marine Traffic. https://www.marinetraffic.com
10. Johnston B, Foley A, Doran J, Littler T (2020) Levelised cost of energy, a challenge for offshore wind. Renew Energy 160:876–885. https://doi.org/10.1016/j.renene.2020.06.030
11. Xu Y, Yang K, Yuan J (2021) Levelized cost of offshore wind power in China. Environ Sci Pollut Res 28:25614–25627. https://doi.org/10.1007/s11356-021-12382-2
12. Kang T, Gbadago DQ, Hwang S (2024) Techno-economic analysis and comparison of hydrogen supply using offshore wind power sources on the coast of the Korean peninsula. Int J Hydrog Energy 71:1201–1213. https://doi.org/10.1016/j.ijhydene.2024.05.188
13. Dinwoodie I, Mcmillan D (2014) Operation and maintenance of offshore wind farms. Eng Technol Ref 1
14. Staffell I, Green R (2014) How does wind farm performance decline with age? Renew Energy 66:775–786. https://doi.org/10.1016/j.renene.2013.10.041
15. Tumse S, Bilgili M, Yildirim A, Sahin B (2024) Comparative analysis of global onshore and offshore wind energy characteristics and potentials. Sustainability 16:6614. https://doi.org/10.3390/su16156614
16. Nizamani Z, Muhammad AK, Ali MOA, Wahab MA, Nakayama A, Ahmed MM (2024) Renewable wind energy resources in offshore low wind speeds regions near the equator: a review. Ocean Eng 311:118834. https://doi.org/10.1016/j.oceaneng.2024.118834

17. Diaconita AI, Rusu L, Andrei G (2021) A local perspective on wind energy potential in six reference sites on the western coast of the Black Sea considering five different types of wind turbines. Inventions 6:44. https://doi.org/10.3390/inventions6030044
18. Diaconita A, Andrei G, Rusu L (2021) New insights into the wind energy potential of the west Black Sea area based on the North Sea wind farms model. Energy Rep 7:112–118. https://doi.org/10.1016/j.egyr.2021.06.018
19. Manolache AI, Chirosca A-M, Rusu L (2024) Assessment of wind and wave climate dynamics in the Mediterranean and Black Seas for renewable energy potential analysis. In: Chen L (ed) Trends in clean energy research: selected papers from the 9th international conference on advances on clean energy research (ICACER 2024). Springer Nature Switzerland, Cham, pp 83–90. https://doi.org/10.1007/978-3-031-67987-2_8
20. Martinez A, Iglesias G (2022) Mapping of the levelised cost of energy for floating offshore wind in the European Atlantic. Renew Sust Energ Rev 154:111889. https://doi.org/10.1016/j.rser.2021.111889
21. Bjerkseter C, Ågotnes A (2013) Levelised costs of energy for offshore floating wind turbine concepts
22. Martinez A, Iglesias G (2021) Multi-parameter analysis and mapping of the levelised cost of energy from floating offshore wind in the Mediterranean Sea. Energy Convers Manag 243:114416. https://doi.org/10.1016/j.enconman.2021.114416
23. Myhr A, Bjerkseter C, Ågotnes A, Nygaard TA (2014) Levelised cost of energy for offshore floating wind turbines in a lifecycle perspective. Renew Energy 66:714–728. https://doi.org/10.1016/j.renene.2014.01.017

Chapter 12
Assessment of Offshore Wind Potential in the Six Countries with Access to the Black Sea

Adriana Silion, Liliana Rusu, and Alexandra-Ionelia Manolache

Abstract Renewable energy sources are vital for reducing pollution and combating global warming as fossil fuel supplies diminish and energy consumption rises. Due to its vast, untapped wind resources, the Black Sea region presents a significant opportunity for the development of offshore wind energy exploitation. This study investigates the offshore wind potential at six locations within the bordering nations: Romania, Bulgaria, Turkey, Ukraine, Russia, and Georgia. The aim is to identify the best sites for wind farm installation. Geographic distribution, wind exposure, and economic relevance were taken into consideration when selecting the sites, which were close to major ports and strategically important in marine trade. Key parameters analyzed include mean wind speed at 100 m altitude, wind variability, power density, and predominant wind direction, utilizing a decade of ERA5 reanalysis data from 2015 to 2024. Results highlight that locations like A1, A2, and A6 exhibit superior wind speeds and power density, making them the most suitable for development. A seasonal analysis reveals that wind energy potential is highest in winter and lowest in summer, influencing energy output predictability. This research underscores the critical role of offshore wind in bolstering energy security, reducing dependence on fossil fuels, and increasing renewable energy integration within the national grids of Black Sea coastal states.

Keywords Offshore wind energy · Wind potential · Renewable energy · Black Sea

12.1 Introduction

In recent decades, the adverse consequences of global warming have not been slow to appear and are becoming increasingly visible [1]. The 2011–2020 decade was the warmest on record, and 2024 set a new benchmark by exceeding pre-industrial

A. Silion (✉) · L. Rusu · A.-I. Manolache
"Dunărea de Jos" University of Galați, Galați, Romania
e-mail: adriana.patilea@ugal.ro

L. Chen (ed.), *Clean Energy Research*, Green Energy and Technology,
https://doi.org/10.1007/978-3-031-99312-1_12

levels by 1.60 °C, as noted by the Copernicus Climate Change Service (C3S) [2], considering data going back to 1850. These trends highlight the accelerating pace of climate change and its measurable impacts. In the context of the energy transition and the need to reduce greenhouse gas emissions, offshore wind energy continues to be the fastest-growing source of renewable energy [3]. Maritime regions offer significant advantages over terrestrial ones due to higher and more consistent wind speeds, which lead to increased wind turbine efficiency [4]. Offshore wind energy is one of the most promising renewable energy sources, with Europe holding a leading role in this sector [5]. In this regard, the Black Sea represents a major area of interest for future offshore wind project developments, having considerable untapped potential [6].

The main objective of the present study is to assess the offshore wind potential in six locations within the Black Sea, belonging to the riparian countries. The first step of the analysis is to identify the most favorable locations for offshore wind farm implementations across the region. By evaluating the average wind speed (at 100 m altitude), variability, density, and predominant wind direction, the study seeks to provide an overall picture based on fundamental data that can support investment decisions in offshore wind energy in the Black Sea.

All six Black Sea coastal countries, namely Romania, Bulgaria, Turkey, Ukraine, Russia, and Georgia, are interested in the development of the utilization of renewable energy sources in the region [7]. The construction of offshore wind farms can become a significant factor in the energy security of the countries, in the reduction of the consumption of fossil fuels, and in the increase of the share of green energy in the national energy sector [8].

In this context, the present study aims to provide a clear overview of the Black Sea offshore wind potential, along with the challenges and opportunities of this emerging industry.

12.2 Methods and Materials

Given the interest in renewable energy production in the Black Sea region, six locations have been selected to analyze the potential of offshore wind resources. The locations proposed for the analysis are rather evenly distributed along the Black Sea, corresponding to each country of the region. The uniform distribution of these sites confirms the intention to perform a general study of the region and not to give priority to a specific zone. The locations were selected near the most important ports in each country, as indicated in Table 12.1. All chosen sites correspond to strategic port cities that play a crucial role in maritime trade and the economic infrastructure of their respective nations. The selection aims to balance geographic distribution, exposure to strong winds, and the economic significance of the ports.

For this research, wind speed analysis data was sourced from the European Centre for Medium-Range Weather Forecasts (ECMWF) ERA5 reanalysis database [9, 10]. The most recent atmospheric reanalysis, ERA5, integrates meteorological

Table 12.1 The characteristics of the chosen sites

Site	Country	Port	Longitude	Latitude
A1	Romania	Constanta	30°15′ E	44°05′ N
A2	Bulgaria	Burgas	28°25′ E	42°45′ N
A3	Turkey	Zonguldak	33°18′ E	42°04′ N
A4	Georgia	Poti	41°36′ E	42°08′ N
A5	Russia	Novorossiisk	37°42′ E	44°36′ N
A6	Ukraine	Sevastopol	33°00′ E	44°45′ N

observations into an advanced computational model to provide geographical and temporal data with accuracy [11, 12]. This resource is essential for climatological, meteorological, and energy research since extensive data on atmospheric and surface factors is provided [13].

The u (west-east) and v (south-north) components of wind speed at an elevation of 10 m above sea level were acquired for this investigation. The data, characterized by a spatial resolution of approximately 31 km [14], encompasses a decade-long timeframe from 2015 to 2024, with measurements taken at 2-h intervals.

In this study, to assess the wind speed and power production at the specific height of a wind turbine, wind data must be recalibrated utilizing a logarithmic law [15–17].

$$U_z = U_{zref} \frac{\ln(z / z_0)}{\ln(z_{ref} / z_0)} \tag{12.1}$$

where U_z denotes the wind speed at height z (specifically, z is 100 m), U_{zref} signifies the known wind speed at height z_{ref} (specifically, 10 m), and z_0 indicates the roughness length, which is set at 0.0002 m.

The equation for wind power density is expressed as [18–20]:

$$P_{wind} = \frac{1}{2} \rho_{air} U_z^3 \tag{12.2}$$

where ρ_{air} represents the air density, which is quantified at 1.225 kg/m^3.

The fluctuations in renewable energy resources, particularly wind power, over various timescales significantly influence their incorporation into the power grid. Locations exhibiting moderate but stable energy outputs tend to be more attractive for wind farm development compared to sites with greater resources that experience variability. The findings indicate that enclosed basins exhibit a higher degree of variability in comparison to open seas, as highlighted in various studies [21, 22].

Seasonal variability (SV) and monthly variability (MV) indexes have been established to more accurately define these swings. These indexes offer a quantitative assessment of the disparities in energy availability between the most and least energetic intervals. They are defined as follows [18, 23]:

$$Monthly = \frac{P_{Mwind_{\max}} - P_{Mwind_{\min}}}{P_{wind_{average}}}$$
$$Seasonal = \frac{P_{Swind_{\max}} - P_{Swind_{\min}}}{P_{wind_{average}}} \quad (12.3)$$

The analysis presents the maximum and minimum wind power values observed across all months, denoted as $P_{Mwind_{\max}}$ and $P_{Mwind_{\min}}$. The values $P_{Swind_{\max}}$ and $P_{Swind_{\min}}$ denote the highest and lowest wind power outputs observed across all seasons. The average wind power value of the entire dataset is represented by $P_{wind_{average}}$.

12.3 Results and Discussion

Figure 12.1 shows that all locations exhibit a similar seasonal variation: wind speed is higher in the autumn-winter months (November–December and January–February) and lower in summer (June–July). The highest wind speed values are recorded in the first month of the year, reaching approximately 9 m/s, but only in three locations: A1, A2, and A6. From February onward, wind speed decreases more sharply in most locations, except for A4, followed by A3, where the decline is smaller or more gradual. The lowest wind speed values are recorded in June and July, dropping to 4–5 m/s in locations such as A3 and A4. A1, A2, and A6 show a rapid increase in wind speed at the beginning of the cold season, while A5 experiences a moderate recovery but does not reach the winter peaks.

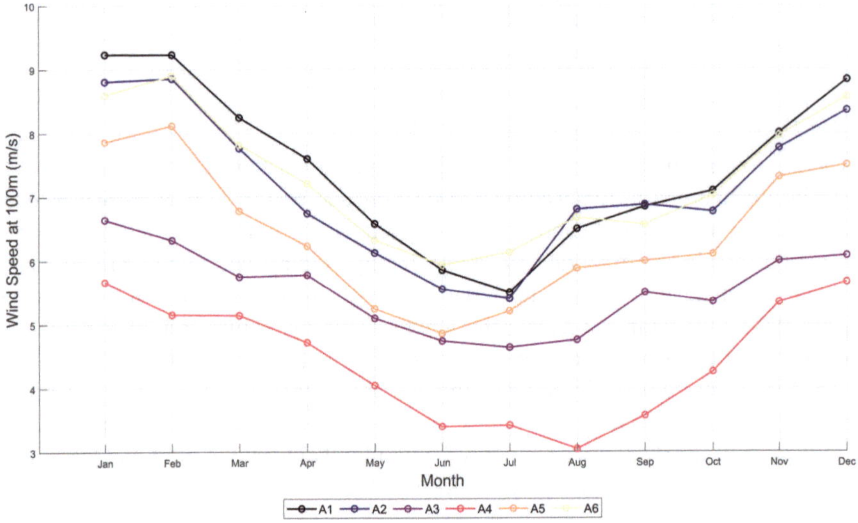

Fig. 12.1 Wind speed at 100 m by month for locations A1–A6

Figure 12.2 presents the average wind speed at 100 m altitude for each location (A1–A6) depending on the season (winter, spring, summer, autumn). It is observed that winter records the highest wind speed values for all analyzed locations, especially in areas A1, A2, and A6, where wind speed reaches approximately 9 m/s. In contrast, summer shows the lowest values, particularly at location A4, where wind speed is just over 3 m/s. In spring, there is a slight decrease in wind speed, but locations A1, A2, and A6 continue to record significant values of around 7 m/s, while A5 has lower values, around 6 m/s. In autumn, the recorded values are similar to those in spring.

The following figure (Fig. 12.3) is regarding wind speed distribution. A6 exhibits the strongest and most consistent winds, along with A1, which also has high wind speeds with a high frequency of moderate and strong winds. On the opposite end, A4 records the weakest winds compared to the other locations. A5 also experiences episodes of strong winds (>15 m/s), while A2 and A3 tend to have moderate to strong winds. Figure 12.3 is about the dominant wind direction. Five out of six locations primarily experience winds from the North, with additional dominant influences from the North-East in A1, A2, and A5, and from the North-West in A3 and A6. The only location without a clear dominant orientation, influenced by winds from the West and South-West (W-SW), is A4.

To strengthen our understanding of the wind energy potential in each area, the wind power density analysis was made, and the results are presented in Fig. 12.4. It was observed that the highest wind power densities, exceeding 400 W/m^2, are recorded in A1, A6, and A2. These are followed by A5 and A3, which show more moderate values in the range of 200–350 W/m^2, while A4 ranks last, with values below 150 W/m^2.

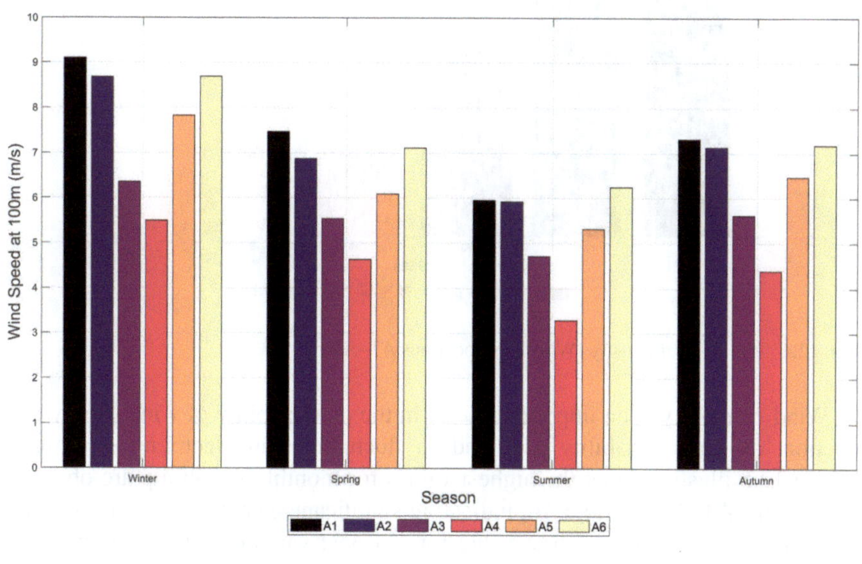

Fig. 12.2 Average wind speed at 100 m by season for locations A1–A6

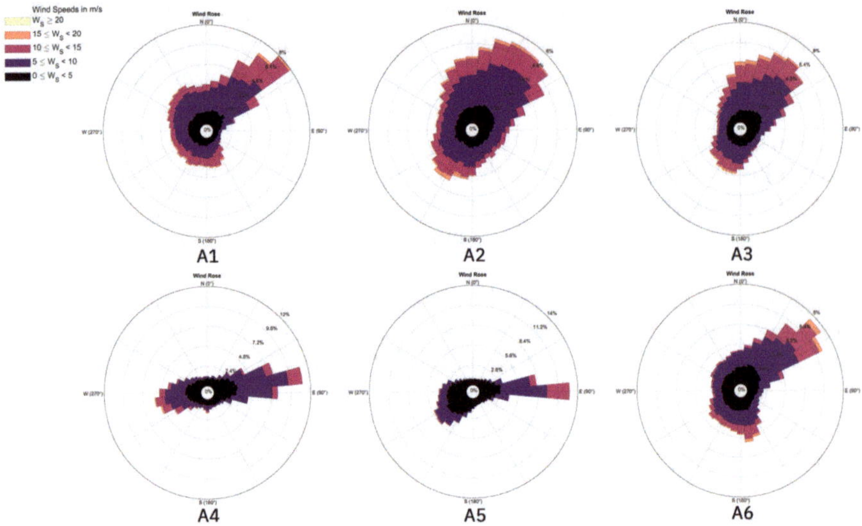

Fig. 12.3 Wind rose diagram for locations A1–A6

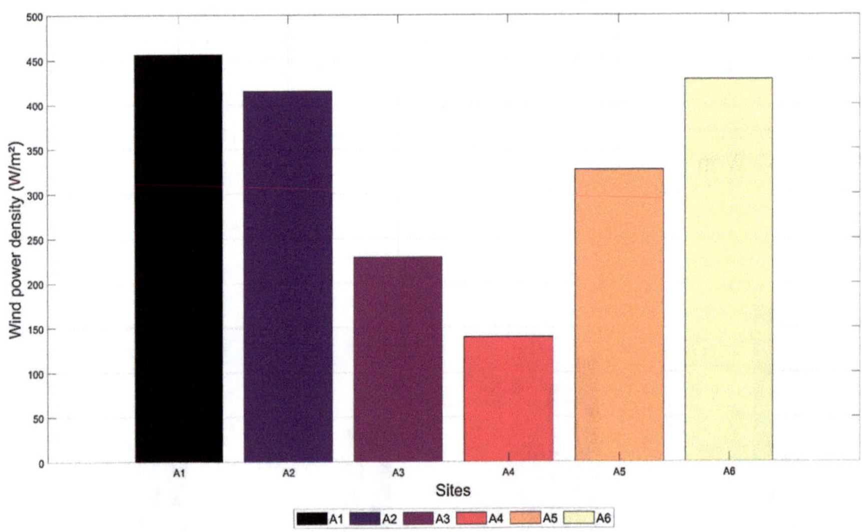

Fig. 12.4 Wind power density (W/m^2) for locations A1–A6

Wind variability is an important factor in the predictability of wind energy production, as high variability can lead to fluctuations in electricity generation. Figure 12.5 illustrates that the highest values for monthly variability are observed in locations A1, A2, A5, and A6, indicating significant short-term instability, while location A4 records the highest value for seasonal variability, highlighting long-term instability. A3 also shows higher seasonal variability compared to monthly

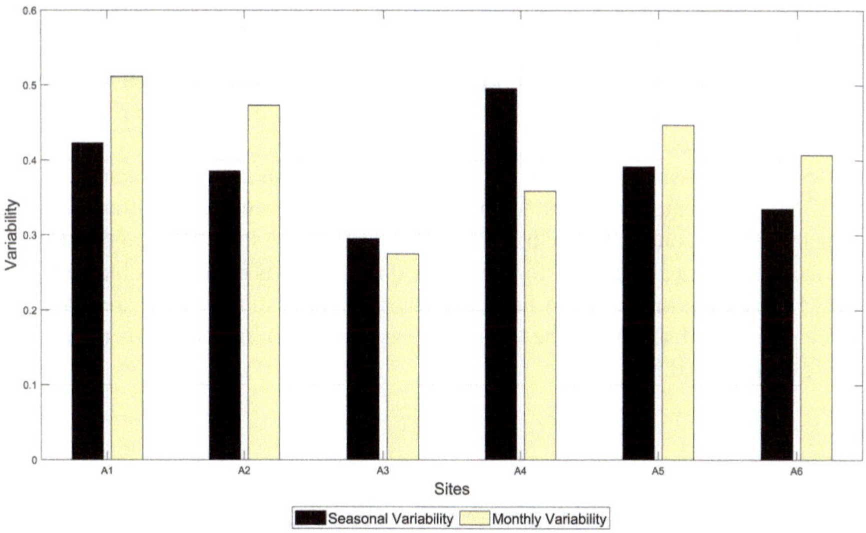

Fig. 12.5 Wind variability (seasonal and monthly) for locations A1–A6

variability, but it still stands out as the most stable location, with values remaining below 0.3.

12.4 Conclusion

The selected study sites are directly exposed to the Black Sea, making them ideal for wind analysis. To gain a comprehensive understanding of wind conditions and energy potential at each location, analyses have been conducted to provide complementary information on wind speed, direction, and offshore wind potential.

The monthly and seasonal wind speed charts demonstrate that wind speed is highest in winter and lowest in summer across all locations. Locations with high wind speeds are more suitable for wind energy, while wind variability must be considered, as large fluctuations can impact energy production.

Locations with consistent predominant wind directions offer better predictability for wind energy, such as A1, A2, and A6, compared to A4, which has variable directions, making it less suitable for such investments. For a wind farm to be efficient, not only does wind speed matter but also its power density.

A high seasonal variability (A4 and A5) means the wind speed fluctuates significantly between winter and summer, making energy production more unpredictable. Energy generation will be more stable if variability is moderate, as observed in locations A1, A2, and A6. However, the high monthly variability in A1 and A2 may lead to short periods of fluctuating production.

It should be noted that sites A1, A2, and A6 are at the top of the list of locations with the greatest offshore wind potential due to their high wind speeds, high power density, and moderate variability Meanwhile, A5 has potential but struggles with variability issues. A4 ranks last, having weak winds, high seasonal variability, and low power density.

The principal strength lies in the analysis of the offshore wind potential in key locations represented by the main ports of the basin that are vital for strategic maritime trade and economic growth. Nevertheless, one of the limitations of the study is the lack of real measurements for the selected locations, as the values obtained from ERA5 are approximate. In the future, it would be interesting to compare real measurements for at least one of the locations to assess the difference between them and the results derived from ERA5.

References

1. Farooq S, Ozturk I, Majeed MT, Akram R (2022) Globalization and CO2 emissions in the presence of EKC: a global panel data analysis. Gondwana Res 106:367–378. https://doi.org/10.1016/j.gr.2022.02.002
2. Goessling HF, Rackow T, Jung T (2025) Recent global temperature surge intensified by record-low planetary albedo. Science 387:68–73. https://doi.org/10.1126/science.adq7280
3. Taylor JW, Jeon J (2018) Probabilistic forecasting of wave height for offshore wind turbine maintenance. Eur J Oper Res 267:877–890. https://doi.org/10.1016/j.ejor.2017.12.021
4. Majidi Nezhad M, Neshat M, Azaza M, Avelin A, Piras G, Astiaso Garcia D (2023) Offshore wind farm layouts designer software's. e-Prime Adv Electr Eng Electron Energy 4:100169. https://doi.org/10.1016/j.prime.2023.100169
5. Soukissian TH, Denaxa D, Karathanasi F, Prospathopoulos A, Sarantakos K, Iona A, Georgantas K, Mavrakos S (2017) Marine renewable energy in the Mediterranean Sea: status and perspectives. Energies 10:1512. https://doi.org/10.3390/en10101512
6. Rusu L, Stratigaki V (2023) Offshore renewables for a transition to a low carbon society. J Mar Sci Eng 11:1185
7. Rusu L (2024) An overview of the renewable energy potential in the coastal environment of the Black Sea. J Eng Sci Innov 9(2):169–182
8. Koçak D, Çalışkan A (2023) The renewable energy performances of Black Sea and Balkan countries: an application of grey principal component analysis. Ege Acad Rev 23:617–626
9. Bell B, Hersbach H, Simmons A, Berrisford P, Dahlgren P et al (2021) The ERA5 global reanalysis: preliminary extension to 1950. Q J R Meteorol Soc 147:4186–4227. https://doi.org/10.1002/qj.4174
10. Hersbach H, Bell B, Berrisford P, Hirahara S, Horányi A, Muñoz Sabater J, Nicolas J, Peubey C, Radu R, Schepers D, Simmons A, Soci C, Abdalla S, Abellan X, Balsamo G, Bechtold P, Biavati G, Bidlot J, Bonavita M, Thépaut J-N (2020) The ERA5 global reanalysis. Q J R Meteorol Soc 146:1999. https://doi.org/10.1002/qj.3803
11. Gleixner S, Demissie T, Diro GT (2020) Did ERA5 improve temperature and precipitation reanalysis over East Africa? Atmosphere 11:996
12. Lavers DA, Simmons A, Vamborg F, Rodwell MJ (2022) An evaluation of ERA5 precipitation for climate monitoring. Q J R Meteorol Soc 148:3152–3165. https://doi.org/10.1002/qj.4351
13. Olauson J (2018) ERA5: the new champion of wind power modelling? Renew Energy 126:322–331

14. Molina MO, Gutiérrez C, Sánchez E (2021) Comparison of ERA5 surface wind speed climatologies over Europe with observations from the HadISD dataset. Int J Climatol 41:4864–4878. https://doi.org/10.1002/joc.7103
15. Onea F, Diaconita A, Ganea D (2022) Assessment of the Black Sea high-altitude wind energy. J Mar Sci Eng 10:1463. https://doi.org/10.3390/jmse10101463
16. Diaconita A, Andrei G, Rusu L (2021) New insights into the wind energy potential of the west Black Sea area based on the North Sea wind farms model. Energy Rep 7:112–118
17. Stull RB (2015) Practical meteorology: an algebra-based survey of atmospheric science. University of British Columbia, Vancouver
18. Diaconita AI, Rusu L, Andrei G (2021) A local perspective on wind energy potential in six reference sites on the western coast of the Black Sea considering five different types of wind turbines. Inventions 6:44. https://doi.org/10.3390/inventions6030044
19. Rusu L (2019) The wave and wind power potential in the western Black Sea. Renew Energy 139:1146–1158. https://doi.org/10.1016/j.renene.2019.03.017
20. Rusu L, Ganea D, Mereuta E (2018) A joint evaluation of wave and wind energy resources in the Black Sea based on 20-year hindcast information. Energy Explor Exploit 36:335–351. https://doi.org/10.1177/0144598717736389
21. Kalogeri C, Galanis G, Spyrou C, Diamantis D, Baladima F, Koukoula M, Kallos G (2017) Assessing the European offshore wind and wave energy resource for combined exploitation. Renew Energy 101:244–264. https://doi.org/10.1016/j.renene.2016.08.010
22. Bernardino M, Rusu L, Guedes Soares C (2021) Evaluation of extreme storm waves in the Black Sea. J Oper Oceanogr 14(2):114–128. https://doi.org/10.1080/1755876X.2020.1736748
23. Rusu E, Onea F (2019) An assessment of the wind and wave power potential in the island environment. Energy 175:830–846. https://doi.org/10.1016/j.energy.2019.03.130

Chapter 13
Mooring Line Analyses of Eight Floating Wind Turbines

Mohamed Maktabi and Eugen Rusu

Abstract This work aims to study the mooring line axial forces of eight floating wind turbines. From this perspective, this paper will compare the mooring line axial forces of the different floating wind turbines under nine different wind and wave environmental conditions. The floating wind turbines studied in this paper are two spar buoys, five semi-submersibles and one TLP. This paper also provides a study of the displacements of the different mooring lines under different wind and wave environmental conditions. An additional analysis covering torsional moment will only be provided for the TLP floating wind turbine. This is due to its negligible values for the other floating wind concepts.

Keywords Offshore renewable energy · Floating wind · Mooring force

13.1 Introduction

This paper is a continuation of the work conducted in [1–3]. The same eight floating wind turbines are analyzed under similar wind and wave conditions but considering different aspects of the floaters' response. In this paper, we are focusing on the mooring line axial force and mooring line displacements. The data presented in this

M. Maktabi (✉)
Department of Mechanical Engineering, 'Dunarea de Jos' University of Galati, Galati, Romania

Department of Marine Technology, Norwegian University of Science and Technology (NTNU), Trondheim, Norway

Department of Law and Economics, Islamic University of Madinah, Madinah, Saudi Arabia
e-mail: mm307@student.ugal.ro

E. Rusu
Department of Mechanical Engineering, 'Dunarea de Jos' University of Galati, Galati, Romania
e-mail: eugen.rusu@ugal.ro

L. Chen (ed.), *Clean Energy Research*, Green Energy and Technology,
https://doi.org/10.1007/978-3-031-99312-1_13

paper are dynamic analysis results. The analysis type is constant wind analysis. Sima 4.6.4 software was used to conduct the analyses. Due to the length limit of this paper, the figures from Sima software will not be presented here. However, they were previously presented in [2, 3]. Also, the characteristics of floating wind turbines are mentioned in the same reference. In this paper, only the characteristics of one of the floating wind turbines (CSC) will be presented due to the page limit of this paper (See Table 13.1).

The analyses conducted in this paper are of aero-hydro-servo-elastic type. The analysis type is dynamic. The implemented wind type is stationary and regular (constant). The wave loads are based on a three-parameter JONSWAP wave spectrum.

In terms of the reliability of the Sima software, paper [4] concludes that the Sima software is efficient for carrying out coupling analysis for floating wind turbines, including giving reliable results of mooring line tension and axial force.

The significance of this work is that it is directly related to floating wind turbine mooring line axial forces, which will determine the appropriate mooring lines to use. Also, the studied floating wind turbine mooring line motions will contribute to future efforts to reduce and optimize them due to their direct connections to unwanted floating wind support structure motions, which leads to problems during their lifetime.

Paper [5] presents innovative aspects relevant to floating wind turbine response analysis. Figure 13.1 illustrates the most popular floating wind turbine types. Three of which are covered in the eight floating wind concepts analyzed in this paper.

Table 13.1 CSC floating wind turbine input characteristics [2]

CSC	
Support structure's height [m]	**Total height [m] (without hub and blades)**
30	135.63
Support structure's accumulated mass [kg]	
1.2642×10^7	
Support structure's accumulated volume [m^3]	
2355.9	
Length of each blade [m]	
86.365	
Additional masses at hub [kg]	
Nacelle: 4.4604×10^5, hub: 1.0552×10^5	
Stress-free length of each mooring line [m]	
880	
Water depth [m]	
200	

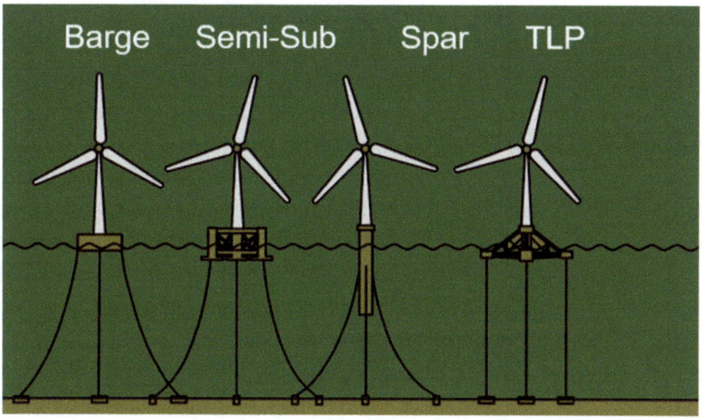

Fig. 13.1 Most popular floating wind turbine types (used in the analyses of this paper except Barge). The authors designed the figure in accordance with the information presented in [1]

13.2 Materials and Methods

The Sima software used to obtain the results presented in this study uses potential flow theory, which is based on velocity potential assumptions of inviscid and incompressible fluid and irrotational flow and has the following equation [6].

$$\nabla^2 \phi = 0 \tag{13.1}$$

The time domain equation implemented in Sima software is as follows [7].

$$(m + A_\infty)\ddot{x} + D_1\dot{x} + Kx + \int_0^t h(t-\tau)\dot{x}(\tau)\mathrm{d}\tau = q(t,x,\dot{x}) \tag{13.2}$$

where m is the mass, A_∞ is the added mass, D_1 is external linear damping matrix, K is restoring stiffness, h is retardation matrix, and q is external load vector including first- and second-order loads, mooring forces, Morison viscous loads, and tower loads.

Furthermore, the aerodynamic theory implemented in Sima software is based on Blade Element Momentum (BEM) theory [3, 7].

Furthermore, [8, 9] present the methodology behind floating wind turbine mooring line frequency and time domain analysis implemented in commercial software.

Reference [10] presents aspects related to aero-hydro mooring line system optimization. As well as commercial software related aspects to mooring lines.

The analyses conducted in this paper cover nine different environmental conditions. Condition 1 contains only wind (without wave and current). Condition 2 contains only a wave with Hs = 4 m and Tp = 12 s. Condition 3 contains only a wave with Hs = 2.5 m and Tp = 10 s. Condition 4 contains only wave and current (without wind) with Hs = 4 m and Tp = 12 s. Condition 5 contains only wave and current (without wind) with Hs = 2.5 m and Tp = 10 s. Condition 6 contains only wind and wave with Hs = 4 m and Tp = 12 s. Condition 7 contains only wind and wave with Hs = 2.5 m and Tp = 10 s. Condition 8 contains combined wind, wave, and current with Hs = 4 m and Tp = 12 s. Condition 9 contains combined wind, wave and current with Hs = 2.5 m and Tp = 10 s. The corresponding results of each of these conditions are presented in Sect. 13.3. The wind speed used in the analyses is rated speed (11.4 m/s for 10 MW floating wind turbines and 10.6 for 12 and 15 MW floating wind turbines, respectively). The MW corresponding values for each of the floating wind turbines are mentioned in the tables in Sect. 13.3.

Table 13.1 presents the characteristics of CSC floating wind turbines. As was mentioned above, the characteristics of the other floating wind turbines studied in this paper can be found in [3] and could not be presented here due to paper's length limit.

13.3 Results and Discussions

The results presented in this section correspond to only the maximum values of the three mooring lines corresponding to each of the eight floating wind turbines studied in this paper. This is to reduce the complexity and the amount of information and make it presentable. The displacements of the mooring lines in x-, y-, and z-directions refer to the motions of the nodes connecting the mooring lines to the floating wind support structures.

Table 13.2 presents the OO-Star mooring line axial force and responses. Table 13.2 shows that the OO-star mooring line axial force is 2.822–4.738 MN. The mooring line x-direction displacement is 0.79–58.8 m. The mooring line y-direction displacement is 0.005–0.115 m. The mooring line z-direction displacement is 0.1978–1.8902 m.

Reference [11] reports that the OO-Star axial force is 3.75–4 MN and corresponds to 43–55 m displacement. Note that the environmental conditions and the input values of both analyses are different.

Paper [12] reports the OO-Star mooring line tension of about 3.2 MN. The paper also presents the floater's motion and other aspects of relevance. Note that the environmental conditions and the input values of both analyses are different.

Table 13.2 OO-Star mooring line axial force and displacements

OO-Star (10 MW) mooring lines									
Conditions	1	2	3	4	5	6	7	8	9
Axial force [MN]	3.418	2.827	2.822	3.759	3.724	3.432	3.429	4.738	4.692
Response in x [m]	29.9	1.17	0.79	43.27	42.43	31.52	30.97	58.8	58.3
Response in y [m]	0.115	0.005	0.005	0.005	0.005	0.105	0.115	0.045	0.045
Response in z [m]	1.8702	0.1978	0.2218	1.0502	1.0102	1.8902	1.8802	1.0502	1.0502

Table 13.3 presents CSC mooring line axial force and responses. Table 13.3 shows that the CSC mooring line axial force is 2.878–6.39 MN. The mooring line x-direction displacement is 0.94–37.739 m. The mooring line y-direction displacement is 0.001–0.229 m. The mooring line z-direction displacement is 0.13–5.97 m.

Paper [13] reports CSC axial effective tension of about 1.14–1.3 MN. The same reference also reports CSC shared mooring tension and other aspects of relevance. Note that the environmental conditions and the input values of both analyses are different.

Table 13.4 presents DTU Spar 1 mooring line axial force and responses. Table 13.4 shows that the DTU Spar 1 mooring line axial force is 1.927–4.351 MN. The mooring line x-direction displacement is 0.0551–27.669 m. The mooring line y-direction displacement is 0.0008–0.0728 m. The mooring line z-direction displacement is 0.004–2.201 m.

Paper [6] reports the Spar mooring line force as around 1.8–2.5 MN. The paper also reports shared mooring line force and tension and other respects of relevance. As well as the platform's response under different mooring line lengths. Note that the environmental conditions and the input values of both analyses are different.

Table 13.5 presents the DTU Spar 2 mooring line axial force and responses. Table 13.5 shows that the DTU Spar 2 mooring line axial force is 2.161–4.405 MN. The mooring line x-direction displacement is 0.0422–20.3173 m. The mooring line y-direction displacement is 0.0008–0.0928 m. The mooring line z-direction displacement is 0.003–2.561 m.

Paper [14] reports Spar mooring line tension as about 0.8–1 MN. The paper also reports other aspects of relevance. Note that the environmental conditions and the input values of both analyses are different.

Table 13.3 CSC mooring line axial force and displacements

CSC (10 MW) mooring lines									
Conditions	1	2	3	4	5	6	7	8	9
Axial force [MN]	3.549	2.879	2.878	5.023	4.996	3.547	3.542	6.39	6.369
Response in x [m]	9.66	2.01	0.94	29.259	29.079	9.679	9.6	37.739	37.649
Response in y [m]	0.229	0.001	0.001	0.009	0.009	0.229	0.229	0.109	0.109
Response in z [m]	5.97	0.13	0.15	1.51	1.51	5.93	5.91	4.5	4.55

Table 13.4 DTU Spar 1 mooring line axial force and displacements

DTU Spar 1 (10 MW) mooring lines									
Conditions	1	2	3	4	5	6	7	8	9
Axial force [MN]	2.706	1.927	1.927	3.24	3.236	2.684	2.696	4.347	4.351
Response in x [m]	27.6693	0.0551	0.0562	0.321	0.272	27.2491	27.4692	12.501	12.542
Response in y [m]	0.0728	0.0038	0.0008	0.0008	0.0008	0.0648	0.0678	0.0498	0.0508
Response in z [m]	1.296	0.004	0.004	0.031	0.031	1.266	1.286	2.191	2.201

Table 13.5 DTU Spar 2 mooring line axial force and displacements

DTU Spar 2 (10 MW) 0020mooring lines									
Conditions	1	2	3	4	5	6	7	8	9
Axial force [MN]	2.932	2.161	2.161	3.328	3.324	2.918	2.924	4.397	4.405
Response in *x* [m]	20.3173	0.0431	0.0422	0.253	0.203	20.0671	20.1372	11.753	11.833
Response in *y* [m]	0.0928	0.0098	0.0018	0.0038	0.0008	0.0918	0.0908	0.0718	0.0718
Response in *z* [m]	1.483	0.003	0.003	0.031	0.021	1.463	1.473	2.541	2.561

Table 13.6 Volturn-US mooring line axial force and displacements

Volturn-US (15 MW) mooring lines									
Conditions	1	2	3	4	5	6	7	8	9
Axial force [MN]	4.004	2.404	2.399	2.874	2.862	3.998	4.003	4.61	4.607
Response in *x* [m]	18.448	0.608	0.498	8.668	8.538	18.428	18.448	22.998	23.038
Response in *y* [m]	0.152	0.01	0.002	0.002	0.002	0.132	0.142	0.11	0.122
Response in *z* [m]	3.283	0.013	0.003	0.403	0.383	3.203	3.243	3.123	3.163

Paper [15] reports that the Spar mooring line fairlead tension is about 0.775–1.309 MN. Paper [16] reports that the Spar mooring line tension is about 0.69 MN. The paper also reports other related aspects of relevance. Note that the environmental conditions and the input values of these analyses are different from those of our analysis.

Table 13.6 presents Volturn-US mooring line axial force and responses. Table 13.6 shows that the Volturn-US mooring line axial force is 2.399–4.61 MN. The mooring line x-direction displacement is 0.498–23.038 m. The mooring line y-direction displacement is 0.002–0.152 m. The mooring line z-direction displacement is 0.003–3.283 m.

Paper [17] reports the Volturn-US mooring line tension to be 0.75–3 MN. The paper also reports the floater's six degrees of freedom and other aspects of relevance. Note that the environmental conditions and the input values of both analyses are different.

Paper [18] reports that the VolturnUS-S mooring line dynamic tension is 2.647–3.485 MN. The paper also reports mooring line static tension, platform response offset and other aspects of relevance. Note that the environmental conditions and the input values of both analyses are different.

Paper [19] reports that the VolturnUS-S mooring line axial force is 2.175–2.9 MN. The paper also reports the floater's degrees of freedom and other aspects of relevance. Note that the environmental conditions and the input values of both analyses are different.

Table 13.7 presents the INO-WINDMOOR mooring line axial force and responses. Table 13.7 shows that the INO-WINDMOOR mooring line axial force is

Table 13.7 INO-WINDMOOR mooring line axial force and displacements

INO-WINDMOOR (12 MW) mooring lines									
Conditions	1	2	3	4	5	6	7	8	9
Axial force [MN]	1.108	1.064	1.061	4.749	4.704	1.12	1.117	4.884	4.838
Response in x [m]	0.811	0.27	0.33	40.41	40.18	1.11	1.03	41.15	40.94
Response in y [m]	0.01	0.01	0.01	0.01	0.01	0.01	0.01	0.01	0.01
Response in z [m]	0.4528	0.0091	0.0152	0.21979	0.23383	0.4381	0.4422	0.9289	0.9143

Table 13.8 WindFloat mooring line axial force and displacements

WindFloat (10 MW) mooring lines									
Conditions	1	2	3	4	5	6	7	8	9
Axial force [MN]	3.472	2.757	2.756	4.054	4.023	3.46	3.467	5.544	5.514
Response in x [m]	7.99	1.67	1.69	18.649	18.319	7.95	7.99	30.179	29.939
Response in y [m]	0.225	0.005	0.005	0.005	0.005	0.225	0.225	0.085	0.085
Response in z [m]	11.009	5.349	5.349	4.629	4.659	10.779	10.889	10.209	10.349

1.061–4.884 MN. The mooring line x-direction displacement is 0.27–41.15 m. The mooring line y-direction displacement is 0.01 m. The mooring line z-direction displacement is 0.0091–0.9289 m.

Reference [20] reports WINDMOOR mooring line tension as about 0.7–4 MN. The reference also represents other information of relevance to Sima and OrcaFlex software used to conduct numerical simulations related to floating wind turbines.

Table 13.8 presents WindFloat mooring line axial force and responses. Table 13.8 shows that the WindFloat mooring line axial force is 2.756–5.544 MN. The mooring line x-direction displacement is 1.67–30.179 m. The mooring line y-direction displacement is 0.005–0.225 m. The mooring line z-direction displacement is 4.629–11.009 m.

Furthermore, Paper [16] reports DeepCWind Semi-submersible mooring line fairlead tension as about 1.030–1.232 MN. The paper also reports other aspects of relevance. Note that the floating wind turbines and their environmental conditions and the input values of both analyses are different.

Paper [21] studies DeepCWind Semi-submersible mooring line tension and floater's degrees of freedom and other related aspects.

Table 13.9 presents the TLP mooring line axial force and responses. Table 13.9 shows that the TLP mooring line axial force is 28–29.87 MN. The mooring line x-direction displacement is 0.001–4.819 m. The mooring line y-direction displacement is 0.01–0.04 m. The mooring line z-direction displacement is 0.001–0.181 m.

Furthermore, the TLP mooring line torsional moment is 0.003–0.043 Nm.

Table 13.9 TLP mooring line axial force and displacements

TLP (10 MW) mooring lines									
Conditions	1	2	3	4	5	6	7	8	9
Axial force [MN]	29.87	28	28	28	28.01	29.86	29.87	29.83	29.84
Response in x [m]	2.631	0.006	0.001	2.006	1.989	2.626	2.639	4.806	4.819
Response in y [m]	0.03	0.01	0.01	0.01	0.01	0.03	0.03	0.04	0.03
Response in z [m]	0.181	0.001	0.001	0.011	0.011	0.171	0.181	0.071	0.071
Torsional moment [Nm]	0.014	0.003	0.003	0.003	0.001	0.043	0.027	0.01	0.004

13.4 Conclusions

The mooring line analysis results of the eight floating wind turbines studied in this paper are as follows. Regarding the mooring line axial force, the TLP floating wind turbine has 29.87 MN. Then comes CSC with 6.39 MN. Then comes WindFloat with 5.544 MN. Then comes INO-WINDMOOR with 4.884 MN. Then comes OO-star with 4.738 MN. Then comes Volturn-US with 4.61 MN. Then comes DTU Spar 2 with 4.405 MN. Then comes DTU Spar 1 with 4.351 MN.

Regarding the mooring line x-direction displacement, the OO-star floating wind turbine has a maximum response offset of 58.8 m. Then comes INO-WINDMOOR with 41.15 m. Then comes CSC with 37.739 m. Then comes WindFloat with 30.179 m. Then comes DTU Spar 1 with 27.669 m. Then comes Volturn-US with 23.038 m. Then comes DTU Spar 2 with 20.3173. Then comes TLP with 4.819 m.

Regarding the mooring line y-direction displacement, the CSC floating wind turbine has a maximum response offset of 0.229 m. Then comes WindFloat with 0.225 m. Then comes Volturn-US with 0.152 m. Then comes OO-Star with 0.115. Then comes DTU Spar 2 with 0.0928 m. Then comes DTU Spar 1 with 0.0728 m. Then comes TLP with 0.04. Then comes INO-WINDMOOR with 0.01 m.

Regarding the mooring line z-direction displacement, the WindFloat floating wind turbine has a maximum response offset of 11.009 m. Then comes CSC with 5.97 m. Then comes Volturn-US with 3.283 m. Then comes DTU Spar 2 with 2.561 m. Then comes DTU Spar 1 with 2.201 m. Then comes OO-Star with 1.890 m. Then comes INO-WINDMOOR with 0.9289 m. Then comes TLP with 0.181 m.

Moreover, when carrying out the analyses, it was observed that only the TLP mooring lines seem to have a torsional moment. However, it is anyway low and has a maximum value of 0.05 Nm.

The significance of this work is due to its relevance to the floating wind turbine mooring line motions and axial forces. This makes it important because the mooring line axial forces will determine which mooring lines to choose. Furthermore, the mooring line motions will directly influence the floating wind turbine motions. This work will contribute to optimizing both aspects.

References

1. Maktabi M, Rusu E (2024) A review of perspectives on developing floating wind farms. Inventions 9(2):24. https://doi.org/10.3390/inventions9020024
2. Maktabi M, Rusu E (2024) Constant wind analyses on eight floating wind turbines. Preprints. https://doi.org/10.20944/preprints202411.0362.v1
3. Maktabi M, Rusu E (2025) Constant wind analyses on eight floating wind turbines. Preprints. Wind. Under peer review
4. DNV GL (2018) Integrated analysis of floating wind turbines. White paper
5. Moan T, Gao Z, Bachynski EE, Nejad AR (2020) Recent advances in integrated response analysis of floating wind turbines in a reliability perspective. ASME J Offshore Mech Arct Eng 142(5):052002
6. Tian Z, Shi W, Li X, Park Y, Jiang Z, Wu J (2025) Numerical simulations of floating offshore wind turbines with shared mooring under current-only conditions. Renew Energy 238:121918. https://doi.org/10.1016/j.renene.2024.121918
7. Souza CE, Engebretsen E, Bachynski-Polić E, Lene E, Berthelsen PA, Haslum H (2021) Definition of the INO WIND-MOOR 12 MW base case floating wind turbine
8. Connolly P, Hall M (2019) Simulation methods for floating offshore wind turbine farms with shared moorings. In: EERA deepwind'19. Poster
9. Pegalajar-Jurado A, Borg M, Bredmose H (2018) An efficient frequency-domain model for quick load analysis of floating offshore wind turbines. Wind Energy Sci 3:693–712. https://doi.org/10.5194/wes-3-693-2018
10. Langeard O, Le Floc'h C (2019) Floating offshore wind: a simplified approach of aero-hydrodynamic coupling to optimize the mooring system design. MCEDD. Powerpoint presentation
11. Jan Ove K (2020) Mooring systems for floating wind turbine farms in deep water. Master thesis, NTNU
12. Payenda MA, Wang S, Jiang Z, Prinz A (2024) Prediction of mooring dynamics for a semi-submersible floating wind turbine with recurrent neural network models. Ocean Eng 313:119490. https://doi.org/10.1016/j.oceaneng.2024.119490
13. Munir H, Lee C, Ong M (2021) Global analysis of floating offshore wind turbines with shared mooring system. Mater Sci Eng 1201:012024
14. Bjørni F, Lien S, Midtgarden T, Kulia G, Verma A, Jiang Z (2021) Prediction of dynamic mooring responses of a floating wind turbine using an artificial neural network. IOP Conf Ser Mater Sci Eng 1201:012023
15. Trubat P, Molins C, Gironella X (2020) Wave hydrodynamic forces over mooring lines on floating offshore wind turbines. Ocean Eng 195:106730. https://doi.org/10.1016/j.oceaneng.2019.106730. ISSN 0029-8018
16. Li W, Du G, Lin S, Chuang Z, Wang F (2024) Dynamic response assessment of floating offshore wind turbine mooring systems with different in-line tensioner configurations based on fully coupled load calculations. J Mar Sci Eng 12:2138. https://doi.org/10.3390/jmse12122138
17. Lozon E, Hall M, McEvoy P, Kim S, Ling B (2022) Design and analysis of a floating-wind shallow-water mooring system featuring polymer springs. NREL. Preprint. NREL/CP-5000-83342
18. Hall M, Brian D, Ericka L (2024) Streamlined loads analysis of floating wind turbines with fiber rope mooring lines. NREL. NREL/CP-5000-87481. Preprint
19. Liu S, Chuang Z, Qu Y, Li X, Li C, He Z (2022) Dynamic performance evaluation of an integrated 15 MW floating offshore wind turbine under typhoon and ECD conditions. Front Energy Res 10:874438. https://doi.org/10.3389/fenrg.2022.874438

20. Rogne ØY (2024) Faster simulations of floating wind turbines with shared mooring. SINTEF blog Link. https://blog.sintef.com/energy/faster-simulations-of-floating-wind-turbines-with-shared-mooring. Accessed 2 Mar 2025
21. Bertozzi A, Niosi F, Jiang X, Jiang Z (2024) Numerical calibration of the mooring system for a semi-submersible floating wind turbine model. J Offshore Mech Arct Eng 146:1–16. https://doi.org/10.1115/1.4065551

Chapter 14
Integrated Desalination and CO_2 Capture System: A Pilot Study on Brine Treatment and Gas Utilization Efficiency

Maisa El Gamal, Ameera F. Mohammad, Ali H. Al-Marzouqi, Ayisha Siddiqua, Sawsan Dagher, and Noran Mousa

Abstract This study presents an innovative integrated system designed for the simultaneous desalination of industrial brine and the capture of carbon dioxide (CO_2), offering a sustainable solution to two critical environmental challenges. The system consists of three key components: a feed preparation unit, a multistage reactor system for chemical and electrocoagulation treatment, and a product collection section. This process treats concentrated brine to reduce salinity while CO_2 is efficiently sequestered in solid carbonates. Optimization experiments were carried out to identify the ideal operating conditions, including brine concentration, gas and liquid flow rates, and pH levels. Results demonstrated that the system achieved a 39% reduction in brine salinity over 24 h while capturing 0.5 g of CO_2 per gram of solid product. The precipitated solid by-products, primarily composed of carbonates of calcium and magnesium, were found to have high purity and potential for reuse in industrial applications. In addition to improving CO_2 capture efficiency, the system's integration with brine treatment presents an economically viable approach by transforming waste into valuable resources, minimizing environmental impact. These findings highlight the system's potential as a scalable solution for industries dealing with high volumes of brine waste and CO_2 emissions, contributing to circular economy efforts and sustainable industrial practices.

Keywords Desalination · CO_2 capture · Brine treatment · The multi-stage reactor · Sustainable industrial practices

---

M. El Gamal (✉)
College of Natural and Health Sciences, Zayed University, Abu Dhabi, UAE
e-mail: Maisa.elgamal@zu.ac.ae

A. F. Mohammad · A. H. Al-Marzouqi · N. Mousa
College of Engineering, UAE University, Al Ain, UAE

A. Siddiqua
Shiv Nadar Institution of Eminence Deemed to be University, Greater Noida, India

S. Dagher
Abu Dhabi Polytechnic, Abu Dhabi, UAE

L. Chen (ed.), *Clean Energy Research*, Green Energy and Technology,
https://doi.org/10.1007/978-3-031-99312-1_14

14.1 Introduction

The global rise in desalination processes and industrial CO_2 emissions underscores the need for sustainable solutions addressing resource efficiency and environmental protection.

The brine, a hypersaline concentrate, a waste byproduct of the desalination plant, has significant problems with its disposal or use in the circular economy. The brine from the desalination plants is hazardous as it contains many dangerous pretreatment chemicals, organics, and heavy metals, which lead to pH fluctuations, eutrophication, and an increase in heavy metals in the marine environment, e.g., $FeCl_3$, $AlCl_3$, H_2SO_4 [1–3].

The global brine production is 142 million m^3/day from 15,906 operational desalination plants, which produce around 95 million m^3/day of potable water. Middle Eastern countries like Saudi Arabia, the United Arab Emirates (UAE), Kuwait, and Qatar account for 55% of the global production [4]. The disposal of brine is expensive and has negative environmental impacts. Disposal happens via sea, surface water discharge land, well injection, deep electrodialysis, and sewage disposal. Because of the presence of hazardous chemicals, strategies of minimal liquid discharge (MLD)and zero liquid discharge (ZLD) are being applied [5, 6]. These strategies maximize the liquid brine treatment, which seems promising to eliminate the wastewater discharge while also recovering extra freshwater and valuable materials such as salts. The brine also has essential minerals like Lithium, Rubidium, and cesium, which can be extracted [7].

As the brine has high salt concentrations and foulants, it needs pretreatment before it can go through the final steps of ZLD. Generally, for ZLD, the pretreatment processes include softening, filtration, and pH adjustments. Then, the liquid is evaporated in a brine concentrator and crystallized or released in an evaporation pond. This process requires a considerable input of energy. There are membrane-based solutions, but they are also limited by highly saline brine and fouling. Another method is pumping hazardous brine into the deep sea, which could lead to long-term removal of CO_2. This desalination-assisted carbon concentration and carbon dioxide removal (CDR) could remove 3.8 Mton CO_2/year globally [8]. The sustainability of desalination plants is very important [9]. Brine management is also done by capturing CO_2 in ammoniated brine [10]. The process showed a more than 90% reduction in CO_2 concentration [11].

The desalination plants' CO_2 emission in reverse osmosis (RO), multi-effect distillation (MED), and multistage flash distillation (MSF) systems are between 1.75–2.79, 7.01–17.6, and 9.41–25 kg-CO_2/m^3, respectively [12]. The Middle East is highly dependent on thermal desalination (MED, MSF), which is 75% fossil fuel-based, and the remaining 25% is RO-based. Thermal desalination methods like MED and MSF have ten times higher GHG emissions than RO-based ones. The amount of renewable energy utilized in these desalination methods is only 1% of the total energy usage [13].

The carbon footprint of UAE desalination plants from producing one cubic meter of freshwater was assessed to be 2.7 kg CO_2 for the MSF method, 1.2 CO_2 for the MED method, and 2.3 kg for the RO method [14].

For desalination plants to become sustainable and reduce their carbon footprint, recent studies have explored the optimization of CO_2 capture and brine desalination using alkaline materials in modified Solvay processes. Researchers have investigated various parameters, including temperature, pressure, gas flow rate, and alkaline concentrations, to maximize CO_2 uptake and ion removal [15, 16]. Optimized conditions have demonstrated significant CO_2 capture efficiencies, ranging from 86% to 0.58 g/g KOH, alongside substantial removal of ions such as Na^+, Cl^-, Ca^{2+}, and Mg^{2+} [15, 17]. Integrating amine carriers with modified Solvay processes has shown promise in simultaneously addressing CO_2 emissions and brine treatment [18]. While these processes have achieved notable salinity reductions and valuable product recovery, challenges remain in achieving brine quality suitable for agricultural and industrial applications. Ongoing research aims to enhance process efficiency and overcome limitations through innovative technologies and multistage treatment strategies.

Sustainable brine management should shift from removing brine to reusing and recovering essential minerals. This research focuses on developing a prototype to effectively remove the salts and hazardous materials before the brine is let out in the ocean or for zero liquid discharge treatment. This study introduces an integrated system that combines desalination with CO_2 capture, thereby transforming brine and gaseous effluents into manageable and potentially valuable products. Using a mixture of brine and alkaline industrial solid waste, this system aims to optimize CO_2 capture while reducing the salinity of treated brine, aligning with resource recovery and pollution reduction goals.

14.2 System Design and Components

Figure 14.1 shows a schematic diagram for the desalination and CO_2 capture prototype, and Fig. 14.2 shows detailed schematic diagrams of each component, including the front, plan, and back views.

Feed Preparation Part

The feed preparation unit consists of a 100-L feed tank that stores fresh brine and alkaline industrial solid waste, which is used as an alkaline agent to enhance CO_2 capture efficiency. Mixer 01 provides controlled agitation (2–10 RPS) to achieve homogeneity in the brine-alkaline solid waste mixture. A liquid pump, connected via Tube 01 to the feed tank, maintains a specific volumetric flow rate to control the brine volume for each batch. Tube 02 transfers the prepared brine-alkaline solid

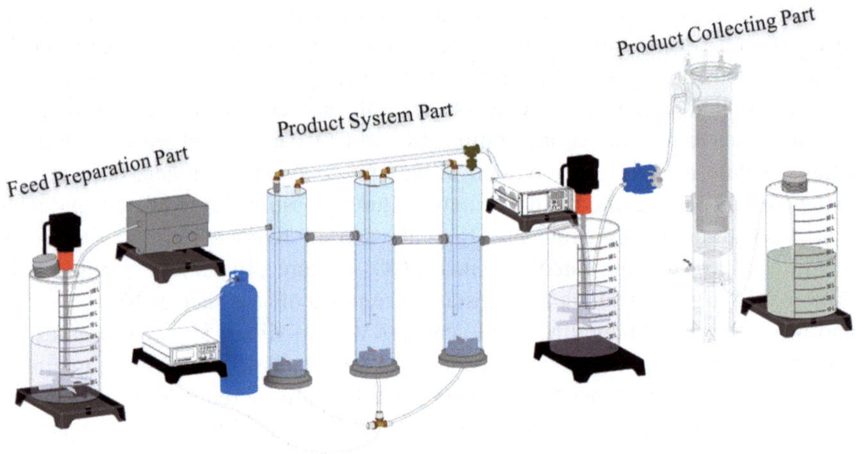

Fig. 14.1 Schematic diagram for the desalination and CO_2 capture prototype

waste mixture to the reactor system, ensuring a steady material flow into the core reaction stage.

Reactor System Part

The reactor system comprises three 10-L plexiglass columns (Reactors 1–3) arranged in series, with a combined capacity of 30 L. Each reactor is connected by downward-sloping tubes (Tube 03 and Tube 04) to facilitate gravity-fed flow, reducing energy consumption as shown in Fig. 14.3. A gas distribution system introduces a controlled flow of 10% CO_2 gas into each reactor via submerged tubes, promoting optimal gas-liquid contact time. A back-pressure regulator limits the internal pressure to 2 bar to avoid damaging the plexiglass structure. Alternative materials like stainless steel are recommended for future high-pressure testing to increase system durability.

Product Collection Part

The product collection unit captures the treated brine and separated solids in an effluent tank connected to the reactor system through Tube 05. Mixer 02 ensures homogeneous mixing of the slurry in the effluent tank, preventing solids from settling. The slurry is pumped through Tube 07 to a filter press equipped with 1 μm filter cloth, separating solids from the clarified brine. The filtered brine is stored in the product tank, ready for reuse or further treatment, while the solids can be

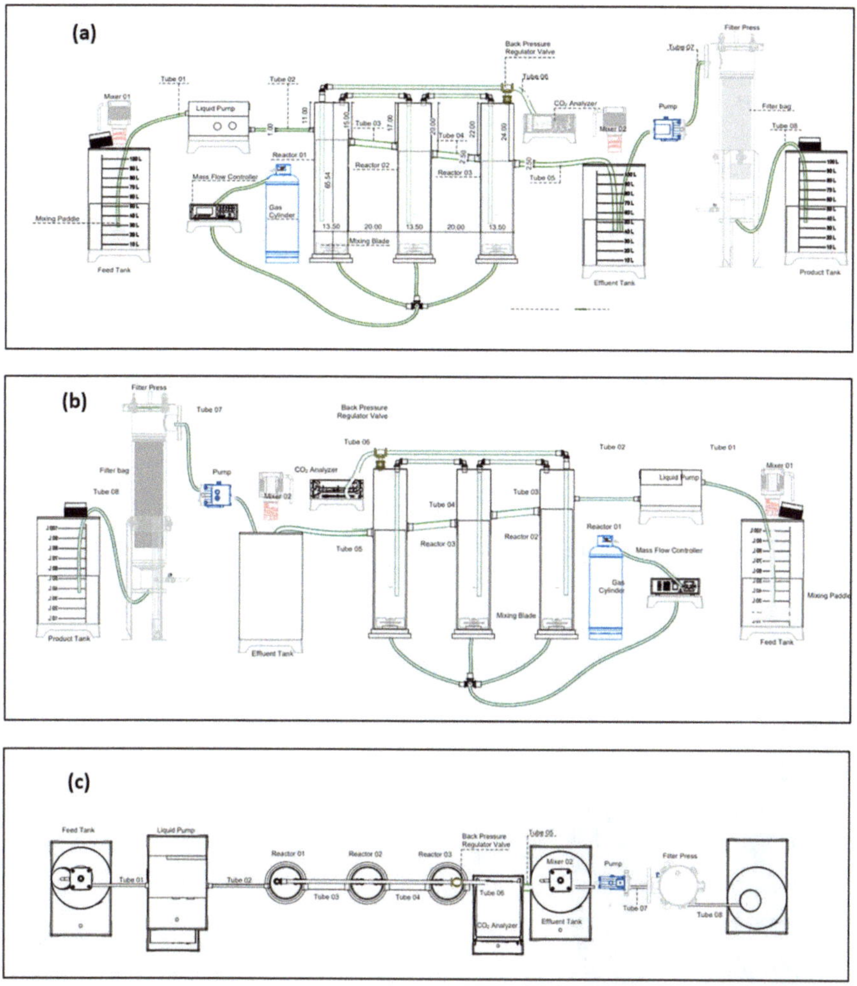

Fig. 14.2 The desalination and CO_2 capture system view from (a) front, (b) back, and (c) plan

analyzed for mineral recovery. Figure 14.4 shows a schematic diagram for the filter press unit.

14.3 Methods

Screening experiments were designed to optimize operational parameters, such as gas and liquid flow rates, and determine the ideal alkaline industrial solid waste concentration for effective CO_2 capture and brine desalination. Initial trials established a baseline for gas flow rate at 1800 mL/min and liquid flow rate at 110 mL/

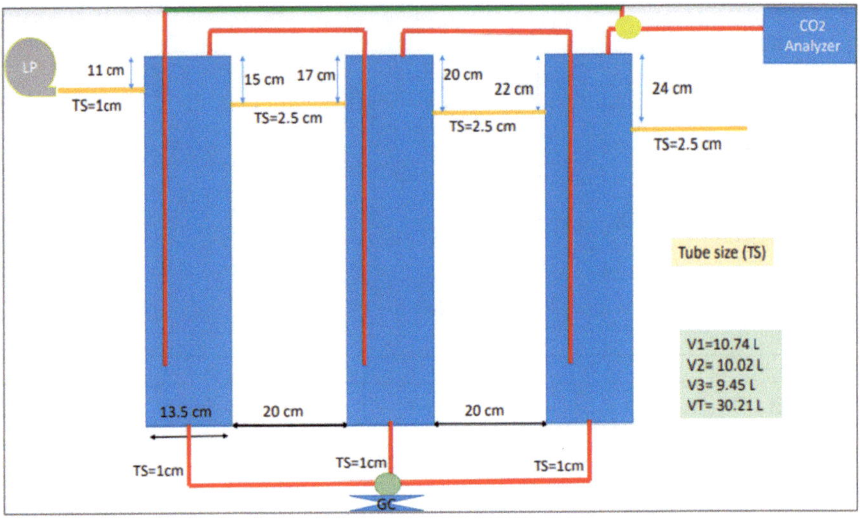

Fig. 14.3 The dimensions of the three-reactor system

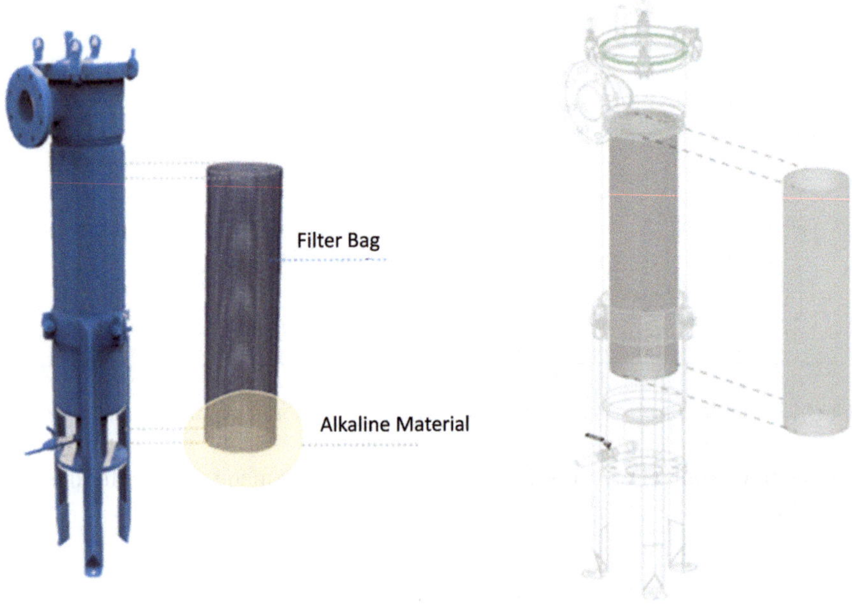

Fig. 14.4 A schematic diagram for the filter press unit

min, yielding a gas-to-liquid ratio of 16.4. The system's performance was evaluated based on the volume of treated brine and the extent of CO_2 capture over 24 h.

14.4 Results and Discussions

Optimization of Operational Parameters

Our trials revealed that the selected gas and liquid flow rates enabled effective CO_2 distribution within each reactor, ensuring a continuous bubbling state that maximized gas-liquid contact time. The 16.4 gas-to-liquid flow ratio was identified as the most efficient, producing an output of approximately 158 L of treated brine per 24-h cycle. Using alkaline industrial solid waste significantly enhanced CO_2 capture efficiency, likely due to the high alkalinity introduced by the waste material, which facilitated chemical binding with CO_2.

CO_2 Capture Efficiency

The CO_2 analyzer consistently detected a significant reduction in CO_2 content in the effluent gases. This reduction indicates effective capture and utilization of CO_2 within the reactors, which was further enhanced by recirculating unreacted CO_2 through the system. Adjusting the reactor's material and structure could optimize capture efficiency in future iterations.

Brine Desalination and Solids Analysis

Post-treatment analysis of the brine revealed a notable decrease in salinity, demonstrating the system's effectiveness in brine desalination. Solids collected in the filter press exhibited high carbonate minerals, suggesting they could be repurposed for industrial use, thereby adding economic value to the waste streams.

The optimized operational conditions for the reactor, including the gas-to-brine flow ratio, CO_2 capture efficiency, and brine desalination performance, are summarized in Table 14.1.

The reduction of brine salinity is a crucial challenge in desalination processes. Various methods have been explored to address this issue, with varying degrees of success. A modified Solvay process achieved a 19% overall salinity reduction in a single stage, capturing CO_2 and recovering valuable products [17]. An electrochemical precipitation system reduced total dissolved solids from 47,000 to 42,000 ppm, representing a modest improvement [19]. Moreover, solid sorbents have shown

Table 14.1 Optimized reactor operating conditions and corresponding CO_2 capture and brine desalination outcomes

Description	Measurement
Total reactor volume	30 L
Gas flow rate (10% CO_2 vol.%)	1800 mL/min
Liquid flow rate	110 mL/min
Brine volume treated in 24 h	158.4 L
Gas-to-liquid flow ratio	16.36
Total alkaline added to the treated brine	10.296 kg
Total collected solids after 24 h	14.167 kg
Total brine salinity reduction	39%
Total CO_2 captured after 24 h	0.5 g CO_2/g solid

promise for CO_2 capture, with various materials demonstrating different adsorption capacities. Zn-MOF outperformed activated carbon, achieving 0.1451 g/g CO_2 adsorption under optimal conditions [20]. Amine-modified montmorillonite nanoclay exhibited fast kinetics and capture capacities of 7.5 wt.% at atmospheric pressure and 17 wt.% at 2.07 MPa in the 75–85 °C range [21]. Hyperbranched aminosilica materials showed potential for direct air capture, maintaining adsorption capacity even at 400 ppm CO_2 concentration [22].

The structure and morphology of fine, dried solid samples after the reaction were analyzed using SEM. Figure 14.5a, b shows SEM images, while Fig. 14.5c, d display EDS analysis for each sample. The morphology primarily consisted of carbonated and bicarbonate solids, characterized by elongated blocks, plates, and needle-like structures. EDS analysis indicated a higher carbonation percentage in the collected solids than in the fresh alkalized material, with sodium in the collected solids after the reaction confirming desalination effects in the treated brine.

These optimized reactor conditions balance CO_2 capture and brine desalination, contributing to sustainable resource management. The results highlight the potential for integrating the design reactor for carbon capture with brine treatment for enhanced environmental benefits.

Energy Consumption and Saving

One of the key advantages of this system is its energy efficiency. The use of gravity-fed flow between reactors (as described in Sect. 14.2.2) significantly reduces the energy needed to pump liquids and gases, thereby lowering the overall system energy consumption. Additionally, the continuous recirculation of unreacted CO_2 enhances the efficiency of CO_2 capture, reducing the need for additional CO_2 input and energy costs.

A preliminary analysis indicates that the energy required for the CO_2 capture process is lower than that of conventional systems due to the reduced pumping energy and optimized gas-liquid contact. Future work will focus on quantifying the

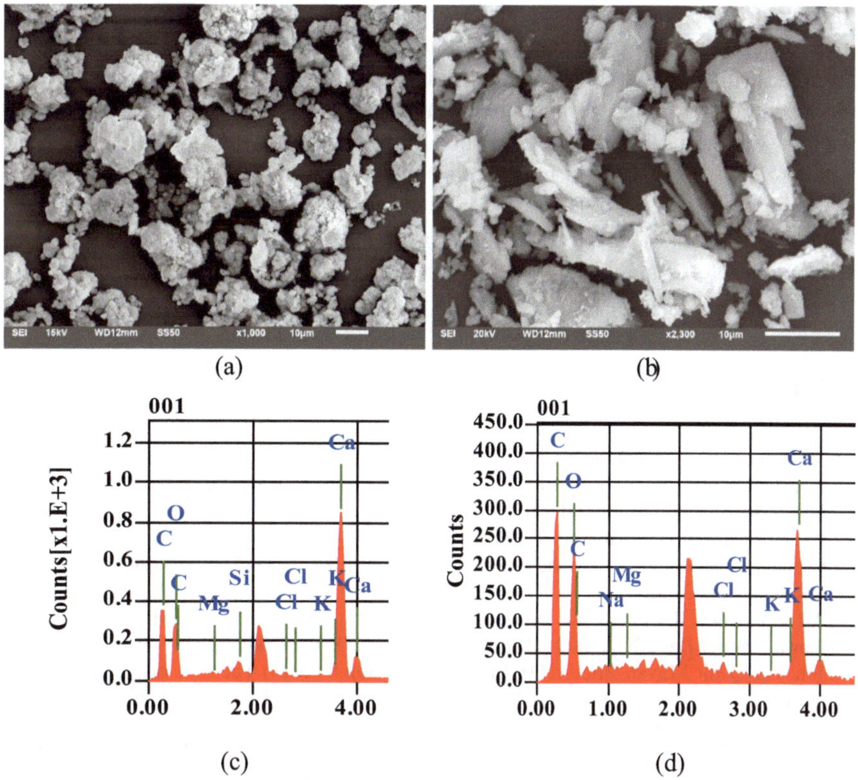

Fig. 14.5 SEM and EDS analyses of the collected solids (**a**, **c**) before and (**b**, **d**) after gas-brine contact

exact energy consumption, and we aim to compare it with other desalination technologies in the context of energy-saving potential.

14.5 Conclusions

The integrated system for brine desalination and CO_2 capture developed in this study significantly addresses two pressing environmental issues: industrial brine management and carbon dioxide emissions. Combining chemical precipitation and electrocoagulation treatments effectively reduced brine salinity and efficient CO_2 capture, transforming waste streams into valuable byproducts such as calcium and magnesium carbonates. These findings underscore the system's potential to reduce the environmental footprint of industries and contribute to a circular economy by converting waste materials into reusable resources. The successful optimization of operational parameters, including flow rates, brine concentration, and pH levels, enhances the feasibility of scaling this technology for larger industrial applications.

Furthermore, the purity and composition of the solid byproducts present opportunities for their reuse in various sectors, adding further economic value to the process. Ultimately, this integrated approach provides a sustainable and cost-effective solution for industries facing stringent environmental regulations on waste management and CO_2 emissions. Future work should focus on scaling the process to pilot and commercial levels and exploring the system's broader applicability across different industrial waste streams. This system's continued development and optimization will contribute to global efforts toward sustainable resource management and environmental protection.

References

1. Mavukkandy MO, Chabib CM, Mustafa I, Al Ghaferi A, AlMarzooqi F (2019) Brine management in desalination industry: from waste to resources generation. Desalination 472:114187. https://doi.org/10.1016/j.desal.2019.114187
2. Panagopoulos A, Haralambous K, Loizidou M (2019) Desalination brine disposal methods and treatment technologies—a review. Sci Total Environ 693:133545. https://doi.org/10.1016/j.scitotenv.2019.07.351
3. Ali MY, Riley JP (1989) The production of brominated methanes in desalination plants in Kuwait. Water Res 23(9):1099–1106. https://doi.org/10.1016/0043-1354(89)90153-X
4. Jones E, Qadir M, van Vliet MTH, Smakhtin V, Kang S (2019) The state of desalination and brine production: a global outlook. Sci Total Environ 657:1343–1356. https://doi.org/10.1016/j.scitotenv.2018.12.076
5. Panagopoulos A, Haralambous K (2020) Minimal liquid discharge (MLD) and zero liquid discharge (ZLD) strategies for wastewater management and resource recovery—analysis, challenges and prospects. J Environ Chem Eng 8(5):104418. https://doi.org/10.1016/j.jece.2020.104418
6. Semblante GU, Lee JZ, Lee LY, Ong SL, Ng HY (2018) Brine pre-treatment technologies for zero liquid discharge systems. Desalination 441:96–111. https://doi.org/10.1016/j.desal.2018.04.006
7. Panagopoulos A (2022) Study and evaluation of the characteristics of saline wastewater (brine) produced by desalination and industrial plants. Environ Sci Pollut Res 29(16):23736–23749
8. Jacobson Y, Bialik OM, Silverman J, Lazar B, Burd-Villanova D, Galilove E et al (2024) Desalination brines as a potential vector for CO2 sequestration in the deep sea. Desalination 574:117234. https://doi.org/10.1016/j.desal.2023.117234
9. Afgan NH, Darwish M, Carvalho MG (1999) Sustainability assessment of desalination plants for water production. Desalination 124(1):19–31. https://doi.org/10.1016/S0011-9164(99)00085-5
10. Mohammad AF, El-Naas MH, Al-Marzouqi AH, Suleiman MI, Al Musharfy M (2019) Optimization of magnesium recovery from reject brine for reuse in desalination post-treatment. J Water Process Eng 31:100810
11. El-Naas MH, Al-Marzouqi AH, Chaalal O (2010) A combined approach for the management of desalination reject brine and capture of CO2. Desalination 251(1):70–74. https://doi.org/10.1016/j.desal.2009.09.141
12. Mezher T, Fath H, Abbas Z, Khaled A (2011) Techno-economic assessment and environmental impacts of desalination technologies. Desalination 266(1):263–273. https://doi.org/10.1016/j.desal.2010.08.035

13. Mahmoudi A, Bostani M, Rashidi S, Valipour MS (2023) Challenges and opportunities of desalination with renewable energy resources in middle east countries. Renew Sust Energ Rev 184:113543. https://doi.org/10.1016/j.rser.2023.113543
14. Cornejo PK, Santana MV, Hokanson DR, Mihelcic JR, Zhang Q (2014) Carbon footprint of water reuse and desalination: a review of greenhouse gas emissions and estimation tools. J Water Reuse Desalin 4(4):238–252
15. Aya AM, Mohammad AF, Al-Marzouqi AH, El-Naas MH, Al-Marzouqi MH, Altarawneh M (2022) CO2 capture and ions removal through reaction with potassium hydroxide in desalination reject brine: statistical optimization. Chem Eng Process Process Intensif 170:108722
16. Mohammad AF, El-Naas MH, Suleiman MI, Al Musharfy M (2016) Optimization of a solvay-based approach for CO2 capture. Int J Chem Eng Appl 7(4):230
17. Mohammad AF, Mourad AAI, Al-Marzouqi AH (2023) Integrated brine salinity reduction, CO2 capture, and valuable product recovery via modified solvay process: an overview. Paper presented at the 2023 14th international renewable energy congress (IREC), pp 1–8
18. Lee C, Chen P, Chen W (2021) Recovery of alkaline earth metals from desalination brine for carbon capture and sodium removal. Water 13(23):3463
19. Lalia BS, Hashaikeh R (2021) Electrochemical precipitation to reduce waste brine salinity. Desalination 498:114796
20. Daud NK (2023) CO2 adsorption performance of AC and Zn-MOF for the use of carbon capture and sequestration (CCS). Mater Today Proc
21. Roth EA, Agarwal S, Gupta RK (2013) Nanoclay-based solid sorbents for CO2 capture. Energy Fuel 27(8):4129–4136
22. Choi S, Drese JH, Eisenberger PM, Jones CW (2011) Application of amine-tethered solid sorbents for direct CO2 capture from the ambient air. Environ Sci Technol 45(6):2420–2427

Chapter 15
Compositional Modeling of Impure CO_2 Injection into Depleted Gas Fields for Geological Storage

Richard T. J. Porter, Haroun Mahgerefteh, and Sergey Martynov

Abstract This study examines the impact of varying CO_2 stream qualities, particularly those containing nitrogen (N_2), on CO_2 injection strategies and storage site behavior using the MATLAB Reservoir Simulation Toolbox (MRST) compositional model. The research simulates the injection of impure CO_2 into a reservoir, representative of those in the North Sea, with a focus on plume migration and its associated reservoir flow processes. The study is structured in phases: in the first, relevant geophysical and geological parameters of the storage site are defined, and the most suitable Equation of State (EOS) is selected for accurate fluid behavior modelling. In the second phase, a sensitivity analysis is conducted, varying porosity and permeability to assess their impact on reservoir performance. The final phase evaluates how CO_2 compositions, including N_2, affect reservoir characteristics such as pressure, gas saturation, and fluid composition. Preliminary results indicate how the presence of N_2 in CO_2 streams influences plume migration and storage site stability. These findings will guide the optimization of CO_2 injection strategies for long-term storage efficacy and safety.

Keywords CO_2 storage · Depleted gas reservoir · Compositional simulation

15.1 Introduction

Geological CO_2 storage is recognised as a critical strategy for mitigating global climate change. The Intergovernmental Panel on Climate Change (IPCC) has estimated a technical storage capacity of around 1000 gigatons ($GtCO_2$) globally; a volume that could comfortably accommodate the expected CO_2 emissions through to the end of this century [1]. The International Energy Agency (IEA) further emphasises the importance of CO_2 storage in the Carbon Capture, Utilization, and

R. T. J. Porter (✉) · H. Mahgerefteh · S. Martynov
Department of Chemical Engineering, University College London, London, UK
e-mail: r.t.j.porter@ucl.ac.uk

L. Chen (ed.), *Clean Energy Research*, Green Energy and Technology,
https://doi.org/10.1007/978-3-031-99312-1_15

Storage (CCUS) chain, noting that its role will grow significantly in the coming decades [2]. In fact, without the widespread deployment of CO_2 storage, achieving global decarbonisation targets could increase costs by up to 70%.

Key geological formations identified for CO_2 storage include saline aquifers and depleted oil and gas reservoirs. Saline aquifers are abundant and widely distributed, while depleted oil and gas reservoirs benefit from the knowledge of their capacity and the availability of existing infrastructure [3]. However, the capacity for safe, long-term storage in these geological formations is largely contingent on the integrity of the reservoir and caprock, which together prevent CO_2 from migrating to the surface. While these geological formations present promising opportunities, challenges exist when transitioning from hydrocarbon extraction to CO_2 storage. The differences in pressure and temperature conditions between oil and gas production and CO_2 injection can introduce complexities, particularly around wellbores, where phenomena like Joule-Thomson cooling can occur and risk the integrity of well seals [4].

Impact of CO_2 Quality on Storage

The quality of CO_2 injected into geological formations plays a significant role in its behavior during injection and its long-term stability [5]. Impurities present in the CO_2 stream, such as nitrogen (N_2) and sulfur dioxide (SO_2), can alter the behavior of CO_2 in the subsurface, potentially compromising the integrity of both the well components and the geological storage site. SO_2 impurities in CO_2 streams can enhance brine acidity, leading to geochemical reactions that may alter reservoir properties, such as porosity and permeability, ultimately affecting storage capacity and stability [6]. Meanwhile, N_2 can reduce the overall density of CO_2, thereby altering the plume migration dynamics and potentially increasing the pressure required for injection. This is particularly relevant when considering depleted gas reservoirs where nitrogen may alter the migration and long-term containment of CO_2. Given that N_2 concentrations in CO_2 streams can vary significantly in industrial applications, understanding the effects of varying N_2 concentrations is crucial for optimizing CO_2 storage operations.

Research Objectives

The study aims to explore the impact of CO_2 impurities, particularly N_2, on the stability of geological storage sites. The focus of the research is on understanding how varying concentrations of N_2 influence CO_2 plume migration, pressure changes, and storage efficiency in depleted gas reservoirs. Through detailed modelling and simulation using the MATLAB Reservoir Simulation Toolbox (MRST) [7], the study will address the following key objectives:

- Investigate how N_2 impurities influence CO_2 plume migration and storage capacity in geological formations.
- Contribute to the development of strategies for the safe, efficient and long-term storage of CO_2 in the presence of impurities.

This work fills a gap in current research, which has primarily focused on pure CO_2 injection or the effects of other impurities such as SO_2. By modelling N_2 impurities in realistic reservoir conditions, this study aims to provide insights into the behavior of CO_2 plumes in the presence of N_2 and how this affects storage stability and operational strategies.

15.2 Methodology

Geological Model and Grid Development

A Cartesian grid was used to represent the P18-6 field reservoir [8], with a 1000 m × 1000 m × 100 m domain and a 30 × 30 × 10 cell configuration (9000 cells in total) covering. This grid design ensures a balance between computational efficiency and an accurate representation of the reservoir's lateral extent. The model includes five distinct layers, each represented by two cells to capture vertical variations in permeability and porosity:

Layer 1: Hardegsen 1 (10 m thickness, 0.09 porosity, 24 mD permeability).
Layer 2: Hardegsen High Perm (10 m thickness, 0.20 porosity, 550 mD permeability).
Layer 3: Hardegsen 2 (10 m thickness, 0.09 porosity, 24 mD permeability).
Layer 4: Upper Detfurth (35 m thickness, 0.075 porosity, 12.6 mD permeability).
Layer 5: Lower Detfurth (35 m thickness, 0.075 porosity, 0.29 mD permeability).

Geological heterogeneity was incorporated using a custom script to assign variations in permeability and porosity across the grid, while small-scale heterogeneities were omitted for computational efficiency. The P18-6 field was selected due to its comprehensive geological data, suitability for CO_2 storage, and well-characterised caprock and reservoir properties. A pressure range of 30–74 bar was applied to simulate CO_2 plume migration, with injection schedules reflecting field practices. In addition, a sensitivity analysis on the effect of increasing permeability and porosity by up to 40% percent was conducted, along with a study on the effect of varying nitrogen concentrations (50%, 80%, 95%, and pure CO_2) on CO_2 migration and retention.

Fluid Thermodynamics and Compositional Modeling

The fluid thermodynamics in this study were modelled using the Peng-Robinson Equation of State (EOS) within the MATLAB Reservoir Simulation Toolbox (MRST) [9]. This EOS simulated the phase behavior and density of CO_2-N_2 mixtures under reservoir conditions, focusing on how different nitrogen concentrations affect CO_2 injection dynamics. The injected gas was modelled as a mixture of CO_2 and N_2 in varying concentrations: 50%, 80%, 95%, and pure CO_2. The fluid model incorporated a six-component mixture to represent the relevant hydrocarbon groups along with CO_2 and N_2, following SPE 79691, Example 5 [10].

Initial reservoir conditions were set to 52 bar and 387.45 K. Binary interaction coefficients were used to model the interactions between CO_2 and N_2. The Peng-Robinson EOS was selected for its robustness in capturing the phase behavior of CO_2-N_2 mixtures, offering a balance between accuracy and computational efficiency. Thermal effects during CO_2 injection were approximated using an isothermal assumption, simplifying the treatment of thermal behavior while capturing key thermodynamic interactions relevant to plume migration and storage capacity.

Injection Well Configuration and Schedule

A vertical injection well, located in one corner of the reservoir, was used with a rate-controlled injection of CO_2 at 5.4 m^3/h. The well operates in a quarter-five-spot well configuration, with the injection rate ramping up gradually during the initial time steps to reflect operational practices. The simulation covers 2 years, with a focus on evaluating the reservoir's response to CO_2 injection.

Sensitivity Analysis of Porosity, Permeability, and CO_2/ N_2 Concentration

A sensitivity analysis was conducted to evaluate the impact of reservoir properties and CO_2 composition on plume migration and storage capacity. The initial analysis focused on the effect of porosity and permeability variations for pure CO_2 injection. Both parameters were systematically increased by 10%, 20%, 30%, and 40% relative to the base case values, within individual reservoir layers, to assess their influence on plume spread and overall storage efficiency. Permeability plays a crucial role in CO_2 flow and plume retention, while porosity affects storage capacity and fluid distribution.

Following this, the effect of injected CO_2/N_2 concentration was analysed to determine how variations in gas composition influence plume behavior. N_2 concentration affects fluid density, with higher N_2 content reducing the overall density, thereby altering buoyancy and plume migration rates [10]. The impact of varying N_2 concentration on plume spread and pressure changes was assessed to quantify its influence on storage efficiency and containment stability [11].

15.3 Results and Discussion

The simulations focused on the impact of both reservoir properties and N_2 impurities on CO_2 injection in depleted gas reservoirs. The primary goal was to understand how variations in porosity, permeability, and N_2 concentration influence plume migration, reservoir pressure, and CO_2 saturation, with an emphasis on optimising injection strategies and evaluating storage efficiency in heterogeneous geological formations. To achieve this, a range of key parameters were tested, including systematic increases in porosity and permeability (10–40%) and N_2 concentrations (0–50%).

Plume Migration and Behavior

The evolution of the CO_2 plume over a 2-year simulation period is illustrated in Fig. 15.1, which depicts the fractional molar composition of CO_2 at four selected timesteps: Day 33, Day 146, Day 365, and Day 730. At Day 33, the plume remains compact and localised near the injection well, with minimal lateral or vertical migration. By Day 146, the plume begins to spread radially, indicating the influence

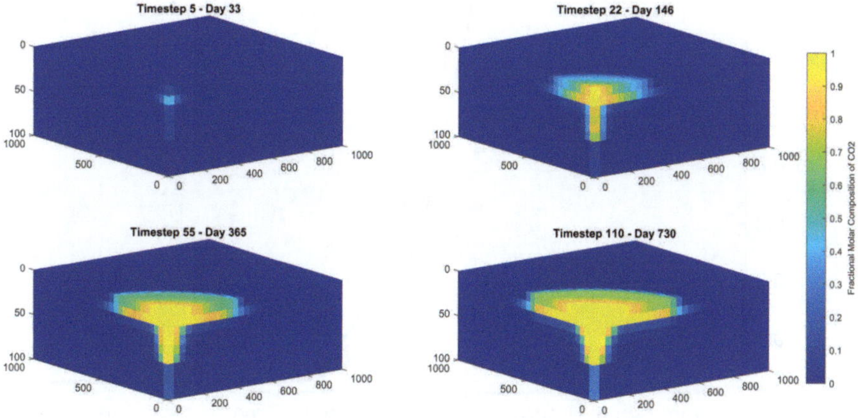

Fig. 15.1 3D plots showing the fractional molar composition of CO_2 at four selected timesteps (Days 5, 22, 55, and 110) during a 2-year simulation of pure CO_2 injection. The plots visualise the spatial distribution of CO_2 concentration, with the color bar representing the fractional molar composition. The results demonstrate the progression of CO_2 concentration across the simulation period

of reservoir properties such as permeability and pressure gradients. The concentration near the injection site remains high, suggesting efficient containment at this stage.

As the simulation progresses to Day 365, the plume exhibits significant radial and vertical expansion. High-concentration zones are observed further from the injection point, highlighting the plume's upward migration due to buoyancy effects. By the end of the simulation at Day 730, the plume stabilises, with a pronounced horizontal spread, and the high-concentration region near the injection well diminishes, indicating dispersion and pressure dissipation over time.

The results confirm that reservoir characteristics, particularly permeability variations, strongly influence plume migration. Effective management of injection rates and reservoir-specific properties is essential to control plume behavior.

Effect of Porosity and Permeability on Plume Migration

The impact of increasing porosity and permeability on lateral CO_2 plume migration was assessed in the uppermost, highly permeable section of the reservoir. The results, shown in Fig. 15.2, indicate a slight reduction in lateral spread as both

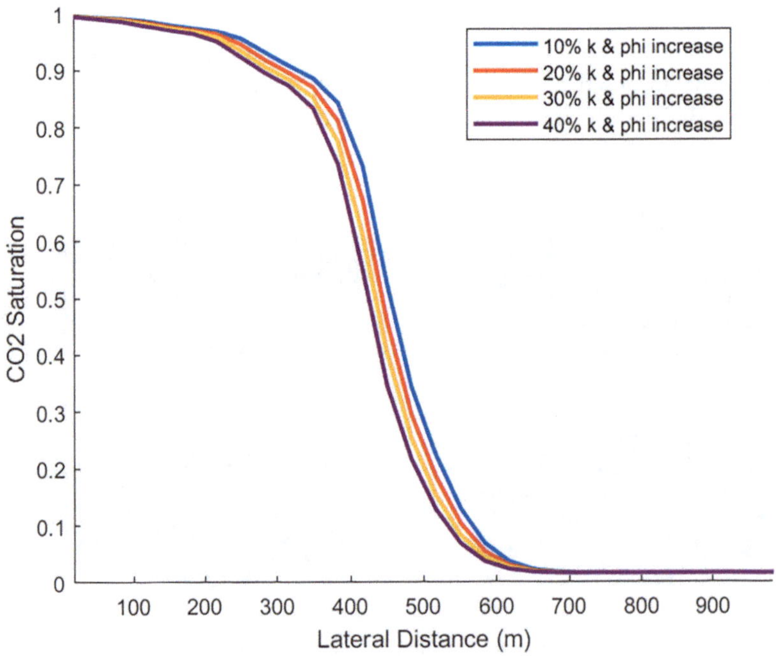

Fig. 15.2 Effect of increasing porosity and permeability on CO_2 saturation along the x-axis in the uppermost permeable reservoir section ($z = 15$ m)

parameters were simultaneously increased by 10–40%. However, the effect remains minimal, suggesting that the increase in permeability alone does not significantly influence plume migration within the tested range. This aligns with expectations, as permeability typically requires changes of several orders of magnitude to produce a notable difference in CO_2 mobility. Instead, the observed reduction in plume spread is primarily attributed to increased porosity, which enhances CO_2 retention within the pore space. A higher porosity allows for greater CO_2 storage within the rock matrix, thereby reducing the volume available for lateral migration and slowing plume expansion.

Effect of N_2 Dilution on Plume Migration

The evolution of the CO_2 plume under different injection compositions is illustrated in Fig. 15.3, where subplots (a–d) show the fractional molar composition of CO_2 at the final time step of the 2-year injection period for varying N_2 concentrations (pure CO_2, 95%, 80% and 50% CO_2). The dilution of CO_2 by increasing N_2 concentrations has a small but observable effect on the final plume spread, increasing it by 10–100 m in terms of lateral spread as the N_2 content rises from pure CO_2 to 50%.

This increase in plume spread, while modest, may have implications for long-term storage efficiency and containment over decadal timescales. Higher rates of N_2 dilution reduce the CO_2 fractional molar composition in the reservoir, as expected, and the impact of varying N_2 concentrations highlights how the buoyancy and migration behavior of CO_2 may be influenced by the presence of nitrogen.

In practical terms, the study's findings provide valuable insights for improving CO_2 injection strategies. By understanding how varying N_2 concentrations influence

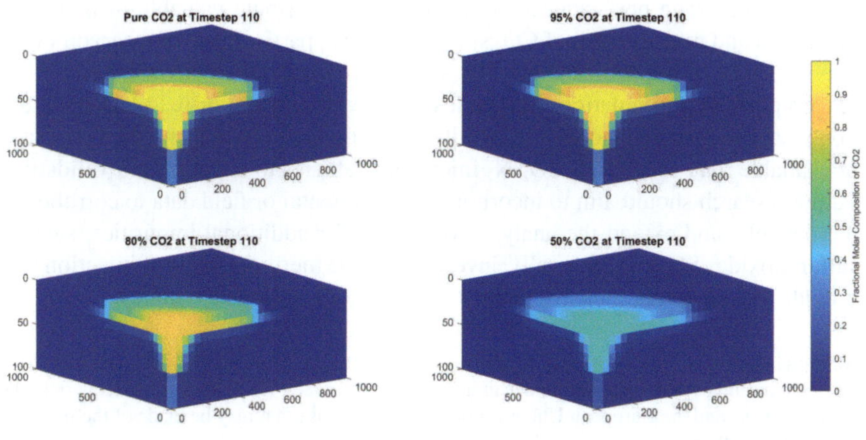

Fig. 15.3 Fractional molar composition of CO_2 in the reservoir at timestep 110 for varying initial CO_2 concentrations. Subplots correspond to (**a**) Pure CO_2, (**b**) 95% CO_2, (**c**) 80% CO_2, and (**d**) 50% CO_2 scenarios. The color scale represents the CO_2 mole fraction, ranging from 0 to 1

plume migration, operators can tailor injection rates to optimise plume containment and reduce risks associated with excessive CO_2 dispersion. This could lead to more efficient storage over extended periods, ensuring the integrity of geological reservoirs and minimising leakage potential.

The effect of N_2 dilution also contributes to enhancing long-term storage strategies by informing decision-making on injection compositions. As the plume spreads more widely with higher N_2 concentrations, it becomes essential to carefully manage reservoir pressure and CO_2 saturation levels to prevent over-pressurisation and enhance the stability of storage sites. This could potentially improve the overall effectiveness of CO_2 storage projects, contributing to the scaling up of CO_2 sequestration as a viable climate change mitigation strategy.

15.4 Conclusions

This study highlights the significant impact of CO_2 impurities, particularly nitrogen (N_2), on the behavior of CO_2 plumes during injection into depleted gas reservoirs. Using compositional modeling with the MATLAB Reservoir Simulation Toolbox (MRST), the results demonstrate that higher N_2 concentrations reduce CO_2 density, leading to increased plume migration and altered pressure distributions within the reservoir. These effects underline the importance of considering CO_2 composition and reservoir characteristics when designing injection strategies to optimise storage capacity and ensure reservoir stability.

The sensitivity analysis further identifies N_2 concentration as a critical factor influencing plume dynamics, with implications for operational costs and long-term storage efficiency. Permeability and porosity changes, considered in the sensitivity analysis, could influence fluid migration patterns and provide insight into potential impacts on storage performance. These findings contribute valuable insights into the design and management of CO_2 storage projects, particularly for reservoirs with varying levels of impurities in the CO_2 stream. Although this study primarily relies on computational simulations, the inclusion of experimental or field-based validation would further enhance the reliability of the findings. At present, there is a lack of available data specific to CO_2-N_2 injection strategies to support such validation. Future research should aim to incorporate experimental or field data to corroborate these results and extend the analysis to account for additional impurities, such as sulfur dioxide (SO_2). Additionally, investigating the thermal effects of injection will provide a more comprehensive understanding of CO_2 storage dynamics.

Acknowledgments This work has received funding from the European Union's Horizon 2020 research and innovation program under grant agreement no. 884418. The work reflects only the authors' views and the European Union is not liable for any use that may be made of the information contained therein.

References

1. IPCC (2005) Intergovernmental panel on climate change: special report on carbon dioxide capture and storage. In: Metz B, Davidson O, de Coninck H, Loos M, Meyer L (eds) IPCC special report. Cambridge University Press, Cambridge, pp 1–402
2. IEA (International Energy Agency) (2019) Exploring clean energy pathways: the role of CO_2 storage. IEA, Paris
3. IEAGHG (IEA Greenhouse Gas R&D Programme): managing the transition of depleted oil and gas fields to CO_2 storage: technical report 2024-05. IEA Greenhouse Gas R&D Programme, August 2024
4. Oldenburg CM (2006) Joule-Thomson cooling due to CO_2 injection into natural gas reservoirs. In: Proceedings, TOUGH symposium 2006, Lawrence Berkeley National Laboratory, Berkeley, California, May 15–17, 2006, pp 1–10
5. Razak AA, Saaid IM, Yusof MAM, Husein N, Zaidin MF, Sabil KM (2023) Physical and chemical effect of impurities in carbon capture, utilisation and storage. J Pet Explor Prod Technol 13:1235–1246
6. Hedayati M, Wigston A, Wolf JL, Rebscher D, Niemi A (2018) Impacts of SO_2 gas impurity within a CO_2 stream on reservoir rock of a CCS pilot site: experimental and modelling approach. Int J Greenhouse Gas Control 70:32–44
7. Lie K-A (2019) An introduction to reservoir simulation using MATLAB/GNU octave: user guide for the MATLAB reservoir simulation toolbox (MRST). Cambridge University Press, Cambridge
8. Neele F, Wildenborg T, Geel K, Loeve D, Peters L, Kahrobaei S, Candela T, Koenen M, Hopmans P, van der Valk K, Orlic B, Vandeweijer V (2020) CO_2 storage feasibility in the P18-6 depleted gas field. TNO report, TNO 2019 R11212, Utrecht, The Netherlands
9. Lie K-A, Møyner O (eds) (2021) Advanced modeling with the MATLAB reservoir simulation toolbox (MRST). Cambridge University Press, Cambridge
10. Yu Y, Li Y, Cheng F, Yang G, Ma X, Cao W (2021) Effects of impurities H_2S and N_2 on CO_2 migration and dissolution in sedimentary geothermal reservoirs. J Hydrol 603:126959
11. Zapata Y, Kristensen MR, Huerta N, Brown C, Kabir CS, Reza Z (2020) CO_2 geological storage: critical insights on plume dynamics and storage efficiency during long-term injection and post-injection periods. J Nat Gas Sci Eng 83:103542

Chapter 16
An Analysis of Using Flettner Rotors in the Black Sea Basin

Valerian Novac and Eugen-Victor-Cristian Rusu

Abstract The article briefly presents wind-assisted ship propulsion systems that underline the Flettner rotors. The authors analyzed the wind regime wind features in the Black Sea along the Constanta—Batumi route for 2024 using 26 selected waypoints. Then, based on a semi-fictitious scenario, the authors set conditions for a simulated voyage in the Black Sea basin with a ship using Flettner rotors in a realistic wind environment and found the values of fuel savings and emissions avoided. In conclusion, the paper deduces that using such devices can lead to a fair reduction in fuel consumption, contributing to the sustainability of maritime transport. The work reminds us of some other operational and technological measures to improve the efficiency of Flettner rotors and enhance fuel consumption reduction.

Keywords Flettner rotors · Black Sea · Sustainability · Green energy · Wind energy

16.1 Introduction

Freighting accounts for at least 90% of global trade, as traders consider it the most worthwhile means of conveying goods from one port to another.

Although shipping is one of the most important pillars of the world economy, and the demand for transportation is rising along with the economic trend, shipping and its adjacent infrastructure pollute the environment with emissions from ships and shore facilities [1].

The International Maritime Organization (IMO) acknowledged the rising trend and established a regulatory framework to reduce greenhouse gas (GHG) emissions. In 2011, IMO issued a regulation about GHG emissions by adopting the Energy Efficiency Design Index (EEDI), a measure of the volume of carbon dioxide

V. Novac (✉) · E. Rusu
'Dunarea de Jos' Galati University, Galați, Romania
e-mail: valerian.novac@ugal.ro

L. Chen (ed.), *Clean Energy Research*, Green Energy and Technology,
https://doi.org/10.1007/978-3-031-99312-1_16

dispatched into the atmosphere to produce the energy for the ship's propulsion and operational needs. Although EEDI covers just a fraction of the total CO_2 emissions related to the whole amount of GHG emitted from energy produced onboard, the EEDI regulation has sparked the interest of the seafarers in finding the most suitable technical, operational, constructive, or energetic measures to reduce the amount of GHG emitted from ships. Along the same line, in 2020, IMO's Marine Environment Protection Committee (MEPC 75), similar to EEDI, adopted the Energy Efficiency Existing Ship Index (EEXI), which enforced a certification requirement for existing ships, disregarding the year of build [2–4].

In the actual period, when the world is focusing on finding the best means to downsize energy consumption, wind-assisted ship propulsion appears as a sustainable solution to diminish emissions and curb the impact of anthropogenic activities on marine ecosystems [5].

On modern ships, the engines are the primary means of propulsion, assisted by sails or other devices that convert wind energy into propulsion force, usually controlled by computers (see Table 16.1) [5, 7].

Flettner rotors (FR) are wind-assisted devices that can augment the primary propulsion system studied and proposed in the 1920s by German researcher Anton Flettner.

Flettner rotor technology for ship propulsion has matured. A series of ships (see Table 16.2) were furnished with rotors to enhance propulsion and diminish fuel consumption.

The FRs are spinning cylinders installed vertically on the ship deck, which acquire the rotation motion from electric motors.

The Magnus effect forms the basis of FR's working principle. It occurs when the rotor cylinder is spinning in a wind flow. The pressure rises on the exposed-to-wind side and diminishes on the other side, resulting in a lift force perpendicular to the wind direction [8].

Table 16.1 Wind-assisted ship propulsion devices (WASP) [5, 6]

WASP	Description	Working principle
Towing kite	Similar to paragliders settled to the ship's bow	Use wind thrust to pull the ship—tail sector winds are favorable
Suction wings	Stationary wings with vents and fans	Achieve force due to the action of the frontier layers
Rigid sails	Foils whose position the crew can alter	Supply aerodynamic forces
Soft sails	Standard sails furnished with modern features	Supply aerodynamic forces
Wind turbines	Regular wind turbines adapted to ship usage	Generate electricity or thrust by the use of blades
Hull sail	The concept uses the ship hull as a sail	Create aerodynamic lift using the relative wind on the symmetrical foils

Table 16.2 Flettner rotor adoption [4–6]

Ship name	Ship type	Built year	Year of FR installation
Buckau	Schooner	1920	1925
Barbara	Freighter	1926	1926
E-Ship 1	General Cargo/Ro-Lo	2010	2010
Estraden	Ro-Ro	1999	2014
Viking Grace	Passenger	2013	2018
Copenhagen	Ferry	2012	2020

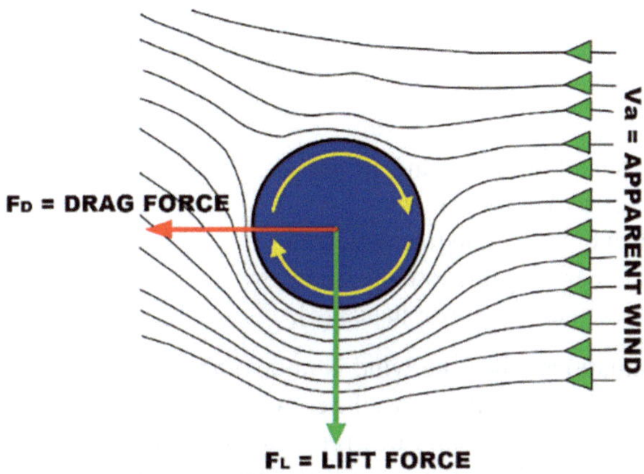

Fig. 16.1 Magnus effect and forces acting on FR [8]

The Magnus effect (shown in Fig. 16.1) is evident in sports that use a round ball, which deviates due to lift occurring in wind streams, or artillery projectiles, which alter the moving path due to incident wind.

The FR construction (shown in Fig. 16.2) consists of a vertical cylinder terminated with plates called Thom discs. Their name comes from the physicist with an identical name who studied increasing lift force by adding plates at the end of FR cylinders and even intermediate plates, which act like wind fences, improving the utile force [9].

Compared with other wind-assisted ship propulsion systems, FR has some merits: It can perform in almost all velocities or wind directions, has downsized dimensions, and can potentially increase efficiency with the number of FRs installed onboard.

The FR is a kind of wind-assisted ship propulsion (WASP) technology that is more efficient than kites in slower winds at lower altitudes (below 30 m), preferably side winds.

There are also some drawbacks when using FR, and they are related to their performance:

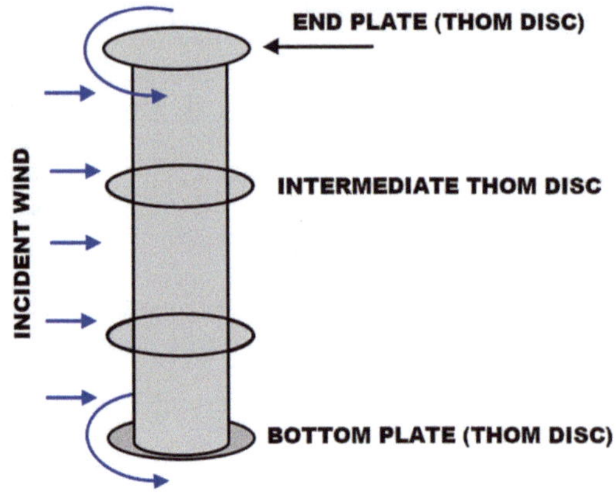

Fig. 16.2 FR construction and operation [9]

- FR is less efficient with fore and aft winds.
- Requires solid deck arrangement.
- Increased heel moment due to side forces.
- Additional rudder drag due to increased yaw.
- Vibration forces can reach resonance [3].

There are situations when the wind is unfavorable, and the rotors become pure energy consumers; in such situations, the crew can shut them off. During high winds and heavy seas, the bare existence of a rotor can increase the ship's heeling or, even worse, damage the ship's structure; in these cases, the ships have comprehensive emergency action plans, and the crew receives adequate training to manage the distress. Besides organizational measures, the vessels fitted with rotors have emergency stop buttons to switch off the rotors, and their design is tiltable, or they can move by rails to allow the crew to contain dangerous motions of the ship and enhance the stability [10].

The Black Sea is an almost landlocked body of water with a precarious exchange of waters, especially in the vertical plane. This results in a fragile ecosystem and sensitivity to pollution. Some studies revealed that shipping emissions have significant values, particularly in areas with important port facilities. Therefore, employing WASP devices in the Black Sea basin is paramount to cutting up to 20% of emissions [11].

16.2 Methods

First of all, it is mandatory to determine the wind speed and direction along the proposed path.

The authors chose to investigate the suitability of the use of FR during the year 2024 in the Black Sea in a voyage from Constanta (44.09°N, 28.7°E) to Batumi (41.67°N, 41.65°E), a heading constant 104.3°.

Table 16.3 Waypoint coordinates

Waypoint	Latitude	Longitude	Waypoint	Latitude	Longitude
WP1	44°N	29°E	WP14	42.75°N	35.5°E
WP2	44°N	29.5°E	WP15	42.75°N	36°E
WP3	44°N	30°E	WP16	42.5°N	36.5°E
WP4	43.75°N	30.5°E	WP17	42.5°N	37°E
WP5	43.75°N	31°E	WP18	42.5°N	37.5°E
WP6	43.5°N	31.5°E	WP19	42.25°N	38°E
WP7	43.5°N	32°E	WP20	42.25°N	38.5°E
WP8	43.25°N	32.5°E	WP21	42.25°N	39°E
WP9	43.25°N	33°E	WP22	42°N	39.5°E
WP10	43.25°N	33.5°E	WP23	42°N	40°E
WP11	43°N	34°E	WP24	41.75°N	40.5°E
WP12	43°N	34.5°E	WP25	41.75°N	41°E
WP13	43°N	35°E	WP26	41.75°N	41.5°E

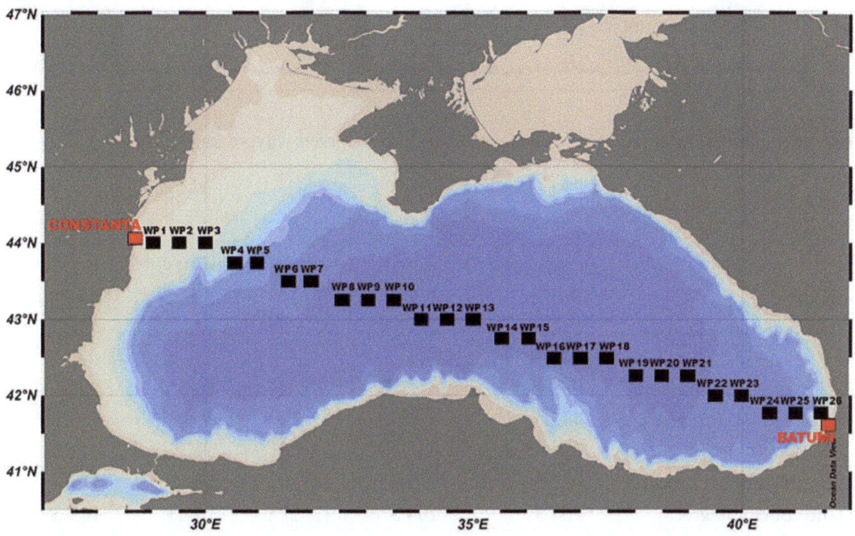

Fig. 16.3 Waypoints array [12]

The authors chose 26 convenient waypoints with the coordinates from Table 16.3 and the array in Fig. 16.3 to appraise the wind speed and direction. We chose only six waypoints (1, 6, 11, 16, 21, 26) and drew a wind direction diagram (Fig. 16.4). We also established five route segments depicted with their start, end, and midpoint coordinates and the length of segments in Table 16.4. In the same table, the authors state the real and apparent wind speed and direction used in calculus for each segment.

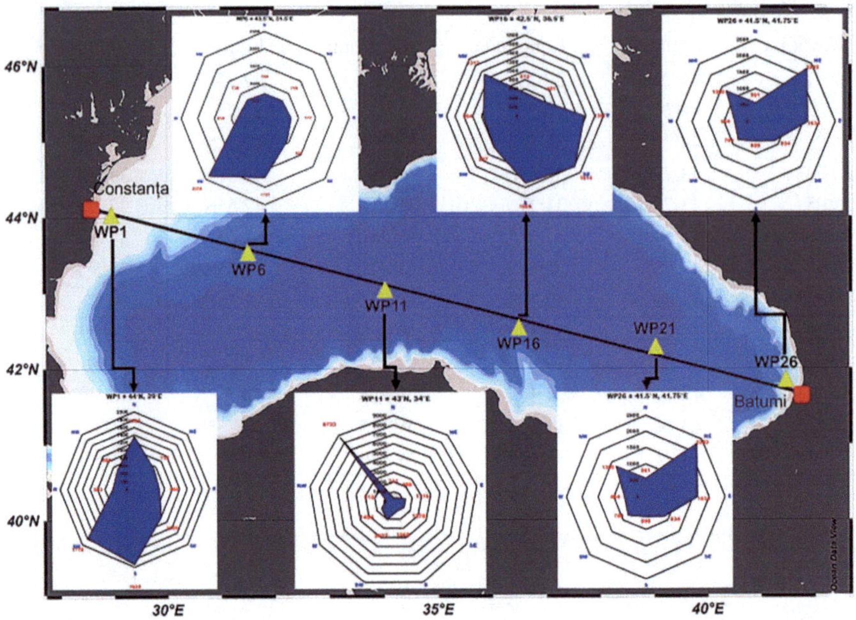

Fig. 16.4 Real wind directions for the year 2024 at considered waypoints [12, 13]

Table 16.4 Voyage data

Segment	1 Constanta/ WP 5	2 WP5/ WP10	3 WP10/ WP15	4 WP15/ WP20	5 WP20/ Batumi
Start point lat. [degrees]	Constanta 44.1°N	43.6°N	43.1°N	42.5°N	42°N
Start point long. [degrees]	28.7°E	31.3°E	33.9°E	36.5°E	39°E
End point lat. [degrees]	43.6°N	43.1°N	42.5°N	42°N	41.7°N
End point long. [degrees]	31.3°N	33.9°E	36.5°E	39°E	41.6 E
Length [nautical miles]	117.5	117.5	117.5	117.5	117.5
True course [degrees]	104.3°	104.3°	104.3°	104.3°	104.3°
Ship speed [knots]	15	15	15	15	15
Average V_R direction [degrees]	211.1°	217.6°	218.8°	157.9°	124.2°
Average V_R speed [m/s]	15.2	14	10.4	4.8	5.3
Date and time start [dd.mm hh:mm]	26.12 04.00	26.12 11.50	26.12 18.40	27.12 03.30	27.12 11.20
Date and time end [dd.mm hh:mm]	26.12 11.50	18.12 18.40	27.12 03.30	27.12 11.20	27.12 19.10

The actual wind speed and direction acting on FR is a vectorial composition of real wind V_R and ship wind V_S (same speed as ship speed but opposite direction), while the V_R comes from the U and V components of an ERA 5 dataset hourly values. The dataset has a regular latitude-longitude grid projection, a parallel resolution of 0.25° × 0.25°, and a gridded data type. Everything is in GRIB format [13–15].

The ship considered for the voyage is a bulk carrier with the following features:

Length overall: 229 m.
Width (beam): 32 m.
Draft: 14.47 m.
Deadweight: 82,048 tons.
Engine power: 9801 kW
Speed: 15 knots.
WASP: 3 × FRs with diameters = 5 m and height = 24 m.

The below figure depicts the array of waypoints and gives a spatial perspective of the voyage environment (shown in Fig. 16.3).

Formulas used to determine the fuel economy and subsequent emissions reduction are [10]:

$$V_A = \sqrt{V_R^2 + V_S^2 - 2V_R V_S \cdot \cos\,\theta} \tag{16.1}$$

V_A—apparent wind [m/s]
V_R—real wind [m/s]
V_S—ship wind [m/s]
θ—angle between the ship's true course and real wind [degrees] [10]

$$F_M = C_L \cdot \rho \cdot A \cdot V_A^2 \tag{16.2}$$

F_M—Magnus force [N]
C_L—lift coefficient (values from 5 to 10)—for this study the authors used 3.5
ρ—air density ~1.23 kg/m^3
A—rotor's surface = $n \bullet d \bullet h$ [m^2] [10]

$$R_t = C_t \cdot \frac{1}{2} \rho \cdot V^2 \cdot S \tag{16.3}$$

R_t—added resistance [kN]
ρ—water density ~1025 kg/m^3
V—ship speed [knots]
S—wetted ship body surface [m^2]

Magnus force will generate a propulsion component on the ship's course, reducing fuel consumption.

Besides fuel consumption, the emissions from ships decrease accordingly.

Concerning the quantities of air pollutants emitted with each ton of Heavy Fuel Oil burnt, the literature review gives the following values (see Table 16.5):

Table 16.5 Quantities of air pollutants emitted per ton of heavy fuel oil [1, 11, 16]

Pollutant	Quantity prescribed [tons]	Quantity used [tons]
CO_2 (carbon dioxide)	3.11	3.11
CO (carbon monoxide)	0.005–0.01	0.005
NO_x (nitrogen oxides)	0.08	0.08
SO_x (sulfur oxides)	0.05	0.05
PM (particulate matter)	0.002–0.008	0.002

Table 16.6 Real wind speed values at reference points considered in 2024

Month	No. of values	Minimum speed [m/s]	Maximum speed [m/s]	Average speed [m/s]
JAN	19,344	0.05	15.21	6.63
FEB	18,096	0.08	15.04	5.72
MAR	19,344	0.05	17.59	5.28
APR	18,270	0.05	16.32	4.75
MAY	19,344	0.05	12.44	4.48
JUN	18,270	0.02	14.17	4.32
JUL	19,344	0.04	11.08	4.42
AUG	19,344	0.01	10.13	3.75
SEP	18,270	0.02	14.44	4.42
OCT	19,344	0.04	16.69	5.14
NOV	18,270	0.04	16.31	5.63
DEC	19,344	0.06	16.75	6.54

16.3 Results and Discussion

Using the data from the previous section, we estimated wind availability by month. The results are in Table 16.6.

It is easily observable that December 2024 is the month with the highest average speed, and August has the lowest average speed values.

Concerning speed directions, the figure below presents its orientation for six chosen waypoints (shown in Fig. 16.4).

It is not difficult to notice that most wind blows from northern and southern sectors and rarely from transverse sectors.

Segments 1 and 2 are the most energetic due to their average wind speed and direction, leading to higher Magnus force values.

Table 16.7 shows that even when the Magnus force value is higher, the orientation of the apparent wind can diminish it.

The authors selected a period with wind speeds higher than the average for the voyage; this situation usually occurs in the winter when the wind is proper to enhance the lift force by the use of rotor sails for a prolonged period, leading to a consistent fuel economy and emissions reduction.

Table 16.7 Economy of fuel and emissions avoided

Segment	1	2	3	4	5	Total
Apparent wind speed [knots]	24.25	24.26	21.53	12.74	10.17	–
Apparent wind direction [degrees]	143.1	147.9	153.8	162.2	169.8	–
Magnus force [kN]	9.53	9.54	7.51	2.63	1.68	–
Propulsion power economy [kW]	2.25	2.23	1.86	1.10	0.80	–
Fuel consumption without FR [t]	32.8	32.8	32.8	32.8	32.8	**158**
FR fuel economy [t]/[%]	7.22 22%	7.21 22%	5.70 17.4%	1.95 5.9%	1.23 3.8%	**23.31** **14.75%**
CO_2 avoided [t]	22.47	22.45	17.75	6.07	3.83	**72.59**
CO avoided [t]	0.14	0.14	0.11	0.04	0.02	**0.46**
NO_x avoided [t]	0.50	0.50	0.39	0.14	0.09	**1.61**
SO_x avoided [t]	0.28	0.28	0.23	0.08	0.05	**0.92**
PM avoided [t]	0.07	0.07	0.06	0.02	0.01	**0.23**

16.4 Conclusions

Diminished fuel consumption of about 14% demonstrates that FR is a mature and effective technology that can contribute to maritime transport sustainability. The reduction in fuel consumption also led to a downgraded emission of atmospheric pollutants like GHG.

It is worth noting that Magnus force and fuel consumption economy are not always proportional due to the orientation of the apparent wind; on the segment where the wind direction was not favorable, the FR contribution was not so substantial.

Emissions reduction helps seafarers comply with IMO regulations related to GHG emission reduction and reduce taxes through the emissions certificates scheme enforced by the European Union.

The article certainly briefly presented the FR and its potential, but there are also other measures to increase voyage sustainability, like planning routes in accordance with wind features, optimal design of FR, combining FR with other green energy devices, or enhancing the ship resistance coefficient.

References

1. Novac V, Rusu E (2019) Air emissions from ships—western Black Sea case study. In: Proceedings of the 19th international multidisciplinary scientific geoconference SGEM, vol 19(4.1), pp 813–819. SGEM, Albena
2. Barreiro J, Zaragoza S, Diaz-Casas V (2022) Review of ship energy efficiency. Ocean Eng 257:1–15
3. Karnan C, Thanikachalam V, Rajesh M, Anbazhahan K (2018) A review of Flettner rotor sails. Int J Interdiscip Res Innov 6(4):561–565

4. De Marco A, Mancini S, Pensa C, Calise G, De Luca F (2016) Flettner rotor concept for marine applications: a systematic study. Int J Rotat Mach:1–12
5. Petković M, Zubčić M, Krcum M, Pavić I (2021) Wind assisted ship propulsion technologies—can they help in emissions reduction? Naše more 68(2):102–109
6. Chou T, Kosmas V, Acciaro M, Renken K (2021) A comeback of wind power in shipping: an economic and operational review on the wind-assisted ship propulsion technology. Sustainability 13(4):1–16
7. Pan P, Sun Y, Yuan C, Yan X, Tang X (2021) Research progress on ship power systems integrated with new energy sources: a review. Renew Sust Energ Rev 144:1–21
8. Nuttall P, Kaitu'u, J. (2016) The Magnus effect and the Flettner rotor: potential application for future oceanic shipping. J Pac Stud 36(2):162–183
9. Pradana MF, Noche B, Ichsan FA (2021) The pathway to decarbonization: Flettner rotor application in maritime logistics. Adv Eng Res 210:72–78
10. Li B, Zhang R, Li Y, Zhang B (2021) Study of a new type of Flettner rotor in merchant ships. Polish Maritime Res 28:28–41
11. Novac V, Rusu E (2021) On the abatement of shipping emissions—a potential approach for the Black Sea basin. In: Proceedings of the 21st international multidisciplinary scientific geoconference SGEM, vol 21(4.1), pp 1–8. SGEM, Albena
12. R. Schlitzer, Ocean data view. http://odv.awi.de. Last accessed 7 Feb 2025
13. Copernicus homepage. https://cds.climate.copernicus.eu/datasets/reanalysis-era5-single-levels?tab=overview. Last accessed 9 Feb 2025
14. Yildirir V, Rusu E, Onea F (2022) Wind variation near the Black Sea coastal areas reflected by the ERA5 dataset. Inventions 7(3):1–57
15. Rusu E (2023) An evaluation of the expected wind dynamics in the Black Sea in the context of the climate change. e-Prime Adv Electr Eng Electron Energy 4:100–154
16. Skjølsvik KO, Andersen AB, Corbett J, Skjelvik JM (2000) Study of greenhouse gas emissions from ships. Norwegian Marine Technology Research Institute—MARINTEK, MT report: MT00 A23-038, pp 1–169, Trondheim, Norway

Chapter 17
Eco-friendly Biodiesel Production and Emission Analysis for Pollution Reduction

Adeyanju A. Anthony

Abstract The efficient management of environmental pollution caused by fossil fuel combustion in internal combustion engines (ICE) remains a significant challenge. This study investigates the potential of soybean methyl ester, a biodiesel, as an alternative fuel for a four-stroke, four-cylinder diesel engine. Biodiesel was produced via the transesterification of soybean oil and blended with regular diesel fuel in volume ratios of 2% (B2), 5% (B5), and 10% (B10). The emission characteristics of the engine operating on these blends were analyzed and compared to those using regular diesel fuel. A portable Autoplus Automotive exhaust gas analyzer was employed to measure emissions of total nitrogen oxides (NOx), carbon dioxide (CO_2) and carbon monoxide (CO). Results indicated that CO_2 emissions were reduced by 8%, 8%, and 20%, and CO emissions decreased by 5.89%, 11.76%, and 35.3% for B2, B5, and B10 blends, respectively, compared to diesel. However, NOx emissions for B2, B5, and B10 were found to be 0.83%, 1.24%, and 1.84% higher than those of regular diesel. These findings demonstrate the potential of soybean biodiesel to mitigate CO and CO_2 emissions while highlighting the need to address increased NOx emissions.

Keywords Environmental Pollution · Soybean oil · Exhaust Emission · Biodiesel

17.1 Introduction

The reliance on fossil fuels in internal combustion engines (ICEs) contributes to significant health and environmental risks due to emissions of pollutants such as NOx, CO_2, SOx, CO, HC, and PM. These emissions cause respiratory diseases, climate change, smog, acid rain, and ozone pollution, particularly in urban areas of

A. A. Anthony (✉)
The University of the West Indies, Department of Mechanical and Manufacturing Engineering, St. Augustine, Trinidad
e-mail: anthony.adeyanju@sta.uwi.edu

L. Chen (ed.), *Clean Energy Research*, Green Energy and Technology,
https://doi.org/10.1007/978-3-031-99312-1_17

developing nations, where older vehicles and "super emitters" amplify the issue. Research indicates biodiesel's potential to reduce pollutants, with its higher oxygen content and lack of aromatic compounds lowering CO and PM emissions, as noted by Knothe et al. [1] and Lapuerta et al. [2]. Soybean oil biodiesel effectively reduces CO and CO_2 emissions, as reported by Demirbas [3], though it may increase NOx emissions due to higher combustion temperatures [4, 5].

Biodiesel, derived from renewable sources like vegetable oils or animal fats, is sulfur-free, biodegradable, and compatible with existing engines. It significantly reduces emissions of CO, CO_2, unburned hydrocarbons, and SOx, with studies showing up to 48% reduction in CO and 78.5% in CO_2 compared to diesel. This study focuses on soybean oil biodiesel blends (B2, B5, B10) and their emission profiles. Sustainable feedstocks like microalgae, with high lipid content and minimal agricultural resource demands, offer an alternative to traditional sources such as soybeans and palm oil [6]. Innovative transesterification methods, including enzymatic processes and heterogeneous catalysts, improve efficiency and reduce chemical waste compared to conventional sodium or potassium hydroxide catalysts [7]. Biodiesel reduces lifecycle CO_2 emissions by up to 78%, as feedstock plants partially offset emissions during growth [8]. It also lowers PM emissions but may increase NOx emissions, requiring mitigation strategies [9]. Higher oxygen content reduces HC and CO emissions through cleaner combustion [10].

Blends such as B20 reduce PM, CO, and HC emissions while minimally increasing NOx. Advanced options like hydrogenated vegetable oils further optimize emission profiles [11]. Electric-biodiesel hybrid systems integrate biodiesel's low emissions with electric power to minimize urban transport emissions [12]. Regulatory standards, including ASTM D-6751 and Ultra-Low Sulphur Diesel (ULSD) regulations by USEPA, have reduced sulfur content and promoted cleaner fuels [13–16]. Despite these advancements, diesel exhaust remains a health hazard, contributing to respiratory and cardiovascular diseases, with studies highlighting increased risks in polluted areas [17, 18].

17.2 Methodology

The methodology of this study encompasses the production of biodiesel, blending with conventional diesel fuel, and determining the physicochemical properties of the fuel samples. Emission testing protocols and the experimental setup are also comprehensively detailed.

Biodiesel Production

Biodiesel was produced via transesterification of soybean oil, resulting in soybean methyl ester (SME). This organic reaction, also known as alcoholysis, involves the conversion of one ester into another by exchanging alkyl groups. It occurs as an equilibrium process through the mixing of reactants, where triglycerides react with alcohol in the presence of an acid or base catalyst to yield fatty acid alkyl esters and glycerol. Base-catalyzed transesterification is notably faster than the acid-catalyzed method.

The general equation representing the transesterification reaction is as follows: RCOOR" + R" OH = RCOOR" + R"O.

As methanol was employed as the alcohol in this research, the process is specifically referred to as methanolysis. The reaction between triglycerides and methanol can be expressed by the following:

$$
\begin{array}{cccc}
 & CH_2OOR_1 & & CH_2OH & \\
 & | & & | & \\
3CH_3OH\ + & CHOOR_2 & \longleftrightarrow & CH_2OH & +\ 3CH_3OOR_x \\
 & | & & | & \\
 & CH_2OOR_3 & & CH_2OH & \\
\text{Methanol} & \text{Triglycerides} & & \text{Glycerol} & \text{Biodiesel}
\end{array}
$$

Materials Used The materials used are: methanol (CH_3OH), soybean oil, and sodium Hydroxide (NaOH)—Catalyst.

Preparation of Biodiesel Fuel Blends

Biodiesel fuel blending involves combining biodiesel with conventional diesel, with blends indicated by "B" followed by a number representing the biodiesel percentage. For instance, B5 contains 5% biodiesel and 95% diesel. In this study, B2, B5, and B10 blends were prepared. The apparatuses used are a beaker, a conical flask and a volumetric flask, as shown in Fig. 17.1.

Procedure Soybean methyl ester (SME) and diesel fuel were heated to approximately 40 °C. Using a 2000 mL volumetric flask, the required volumes of biodiesel and diesel were measured. For the B2 blend, 40 mL of biodiesel and 1960 mL of diesel were mixed. The B5 blend used 100 mL of biodiesel and 1900 mL of diesel. The B10 blend used 200 mL of biodiesel and 1800 mL of diesel. Each mixture was blended using a hand-held electric mixer and transferred to plastic containers, labeled as Samples A, B, and C.

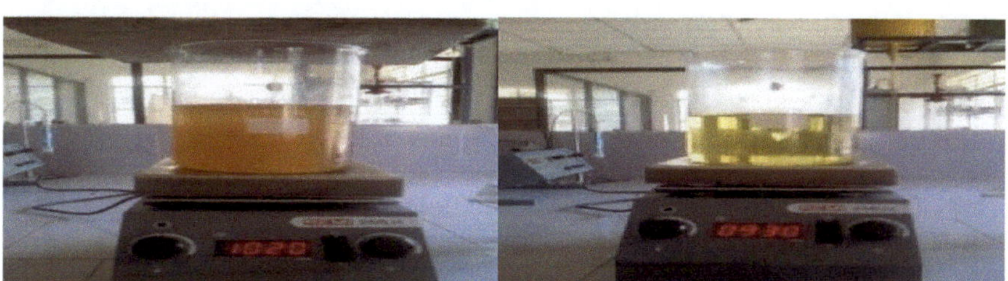

Fig. 17.1 Trans-esterification process of soybean oil

Physicochemical Property Test

The physicochemical properties of B0, B2, B5, and B10 samples were analyzed in the laboratory. The following properties were tested: ash content, specific gravity, refractive index, volatile matter content, fire point, cloud point, flash point, pour point, kinematic viscosity, heating value, fixed carbon content and moisture content.

(i) Volatile Matter Content Determination

The gravimetric method, following AOAC (1990) standards [19], was used. The mass of a clean, dried evaporating dish was measured (W1), then 10 g of the sample was added, and the combined mass was recorded (W2). The dish was dried at 70 °C for 8 h, then cooled in a desiccator and weighed again. This process was repeated until a constant weight (W3) was achieved. Volatile matter content was calculated as a percentage of the initial weight of the sample:

$$\%\text{Volatile matter content} = \frac{W_2 - W_3}{W_2 - W_1} \times \frac{100}{1} \tag{17.1}$$

where W_1 represents the weight of the empty evaporating dish; W_2 denotes the combined weight of the sample and the evaporating dish; W_3 refers to the combined weight of the sample and the evaporating dish after drying at 70 °C.

Determination of Ash Content

This was determined using the gravimetric method as per AOAC (1990) standards. The procedure involved measuring the weight of a washed and dried empty crucible (W1) using a Mettler balance. A 5 g sample was added to the crucible, and the combined weight was recorded (W2). The crucible was then placed in a muffle furnace at 550 °C for 4 h to ash the sample. After cooling in a desiccator, the final weight (W3) was recorded. The percentage ash content was calculated using the appropriate formula:

$$\%\text{Ash} = \frac{W_3 - W_1}{W_2 - W_1} \times \frac{100}{1} \tag{17.2}$$

where W_1 represents the weight of the empty crucible; W_2 denotes the weight of the crucible with the sample; W_3 refers to the weight of the crucible with the sample after ashing at 55 °C.

(ii) Specific Gravity (SG)

$$SG = \frac{\text{Weight of Samples}}{\text{Equivalent Weight of Water}} \tag{17.3}$$

(iii) Sample Refractive Index: This was determined at 25 °C using the Abbe refractometer equipment.

(iv) Sample Cloud Point: This is defined as the temperature at which a cloud of wax crystals first appears in a liquid form when the liquid is cooled under certain conditions.
(v) Sample Pour Point: This is defined as the lowest temperature at which a liquid remains capable of flowing. Each sample was homogenized and poured into a test jar up to the required level. The pour point for each sample was then determined in accordance with the ASTM D97 standard method [20].

Experimental Setup for Emission Test

The study assessed the effect of biodiesel blends on diesel engine exhaust emissions using a SIFANG S115 4-stroke, 4-cylinder diesel engine. Emissions were measured under no-load and steady-state conditions after the engine reached optimal temperature. The Autoplus Automotive Exhaust Gas Analyzer, with non-dispersive infrared sensors, was used to measure gases such as O_2, CO_2, CO, NO, HC, and NOx. Calibration of the analyzer was done before each test by purging with fresh air. The fuel samples were tested sequentially (B0, B2, B5, B10), with each tested twice, and average emission data was recorded as shown in Table 17.1.

Emission Test Procedure The engine's fuel hose was modified to prevent contamination. B0 was tested first, followed by B2, B5, and B10, with emission data collected after 30 minutes of engine operation. The test was repeated for each fuel type, and average data were used for analysis.

Challenges Faced During Biodiesel Production and Testing Feedstock Limitations: The availability and cost of traditional feedstocks like soybean, palm oil, and rapeseed pose challenges, with alternatives such as waste oils and jatropha being inconsistent and difficult to source at scale.

Production Efficiency and Energy Consumption: The transesterification process is energy-intensive, and conventional chemical catalysts create by-products that need disposal, while enzymatic catalysts are slower and more expensive.

Cost and Complexity of Emission Testing: Emission testing is costly and time-consuming, with results varying by engine type, load, and fuel blend, complicating standardization.

Table 17.1 Description of fuel samples tested

Fuel samples	Description
B0	100% pure diesel fuel
B2	98% diesel blend and 2% biodiesel
B5	95% diesel blend and 5% biodiesel
B10	90% diesel blend and 10% biodiesel

NOx Emission Concerns: Biodiesel tends to produce higher NOx emissions than conventional diesel, which contributes to air pollution and smog formation, requiring additional solutions like additives and EGR systems, adding to costs and operational complexity.

17.3 Results and Discussion

Physicochemical and Emission Test Results

The results of the physicochemical and emission tests for pure diesel (B0), pure biodiesel (B100), and biodiesel blends (B2, B5, B10) are presented in several tables. Table 17.2 outlines the physicochemical properties, while Tables 17.3 and 17.4 show emission characteristics from two testing periods, and Table 17.5 summarizes the average emission data with percentage changes in regulated pollutants for each fuel blend compared to pure diesel.

Emission Test Results Emission results for B0, B2, B5, and B10 are compared in Tables 17.3, 17.4, and 17.5, showing differences in pollutant levels across fuel samples.

Physicochemical Properties of Fuel Samples Pure biodiesel (B100) has higher values for flash point, ash content, fire point, moisture content, and fixed carbon content than pure diesel (B0). As biodiesel content increases in blends (B2, B5, B10), these properties tend to rise, except for fire point and fixed carbon, which decrease but remain higher than in pure diesel. Refractive index, specific gravity, heating value, volatile matter, pour point, and kinematic viscosity are higher in pure diesel and decrease as biodiesel content increases.

Table 17.2 Physicochemical properties of fuel samples

S/N	Property	Unit	B0	B2	B5	B10	B100
1.	Refractive index (@30 °C)		1.460 ± 0.000	1.474 ± 0.002	1.472 ± 0.001	1.453 ± 0.002	1.460 ± 0.000
2.	Specific gravity	g/cm^3	0.882 ± 0.002	0.8552 ± 0.001	0.8561 ± 0.002	0.8435 ± 0.001	0.882 ± 0.002
3.	Kinematic Viscositv	Pas/sec	5.890×10^{-3}	5.044×10^{-3}	5.066×10^{-3}	5.188×10^{-3}	5.516×10^{-3}
4.	Ash content	%	0.101 ± 0.001	0.245 ± 0.010	0.402 ± 0.003	0.586 ± 0.005	0.738 ± 0.002
5.	Flash point	°C	105.00 ± 0.00	130.00 ± 0.20	133.00 ± 0.10	140.00 ± 0.00	170 ± 0.200
6.	Volatile matter	%	22.404 ± 0.012	1.861 ± 0.021	2.217 ± 0.010	3.165 ± 0.004	0.749 ± 0.017
7.	Cloud point	°C	−4.000 ± 0.000	1.000 ± 0.000	−2.000 ± 0.000	−4.000 ± 0.010	−4.000 ± 0.001
8.	Pour point	°C	14.000 ± 0.000	11.000 ± 0.000	10.000 ± 0.000	7.000 ± 0.000	4.000 ± 0.000
9.	Fire point	°C	118.00 ± 0.05	178.00 ± 0.20	158.00 ± 0.10	144.00 ± 0.05	242.00 ± 0.10
10.	Moisture content	%	0.560 ± 0.010	0.380 ± 0.003	0.425 ± 0.003	0.568 ± 0.002	0.633 ± 0.002
11.	Heating value	KJ/kg	33917.225 ± 0.220	34101.603 ± 0.302	34029.272 ± 0.202	33909.069 ± 0.156	29401.920 ± 0.033
12.	Fixed carbon	%	76.935 ± 0.110	97.517 ± 0.210	96.956 ± 0.058	95.699 ± 0.100	98.880 ± 0.021

Table 17.3 Results of the emission test for B0, B2 and B5 and fuel samples after the first testing period

Constituent	Unit	B0	B2	B5	B10
CO_2	Percentage by volume	1.3	1.2	1.2	1.0
CO	Percentage by volume	0.09	0.08	0.08	0.06
O_2	Percentage by volume	18.66	17. 39	19.34	19.38
NO_X	PPM	605	607	610	615

Table 17.4 Results of the emission test for B0, B2 and B5 and B10 fuel samples after the second testing period

Constituent	Unit	B0	B2	B5	B10
CO_2	%Vol.	1.2	1.1	1.1	1.0
CO	%Vol.	0.08	0.08	0.07	0.05
O_2	%Vol.	19.39	19.51	19.50	19.10
NO_X	ppm	600	608	610	612

Table 17.5 Mean values and percentage variations of engine emissions for regulated pollutants

Constituent	Unit	B0	B2	B5	B10
CO_2	Percentage by volume	1.25	1.15	1.1	1.0
Change in CO_2	%		+8	+8	+20
CO	%Vol.	0.085	0.080	0.075	0.055
Change in CO	%		+5.89	+11.76	+35.3
O_2	%Vol.	19.03	18.45	19.42	19.24
NOx	ppm	602.5	607.5	610	613.5
Change in NOx	%		−0.83	−1.24	−1.84

PPM—parts per million; %Vol.—concentration in percentage by volume; (+)—sign implies decrement and (−)—sign implies increment

Emission Characteristics CO_2 Emissions: All biodiesel blends showed reduced CO_2 emissions compared to pure diesel, with B10 showing the largest reduction (20%), as shown in Fig. 17.2.

CO Emissions: CO emissions decreased with increasing biodiesel content. B10 showed the greatest reduction (11.7%), as shown in Fig. 17.3.

NOx Emissions: Biodiesel blends produced higher NOx emissions than pure diesel, with increases of 0.83%, 1.24%, and 1.84% for B2, B5, and B10, respectively, as shown in Fig. 17.4.

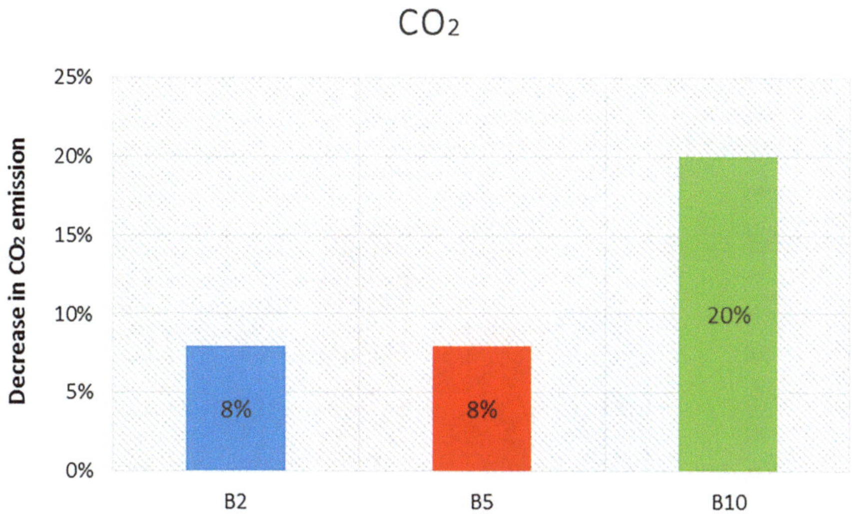

Fig. 17.2 % reduction in CO_2 emissions for B2, B5, and B10 fuel blends compared to pure diesel fuel (B0)

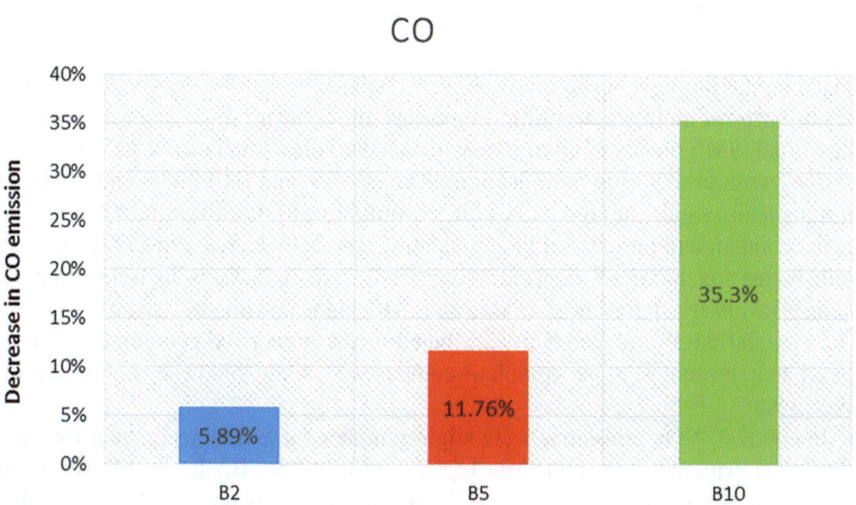

Fig. 17.3 % reduction in CO emissions for B2, B5, and B10 fuel blends compared to pure diesel fuel (B0)

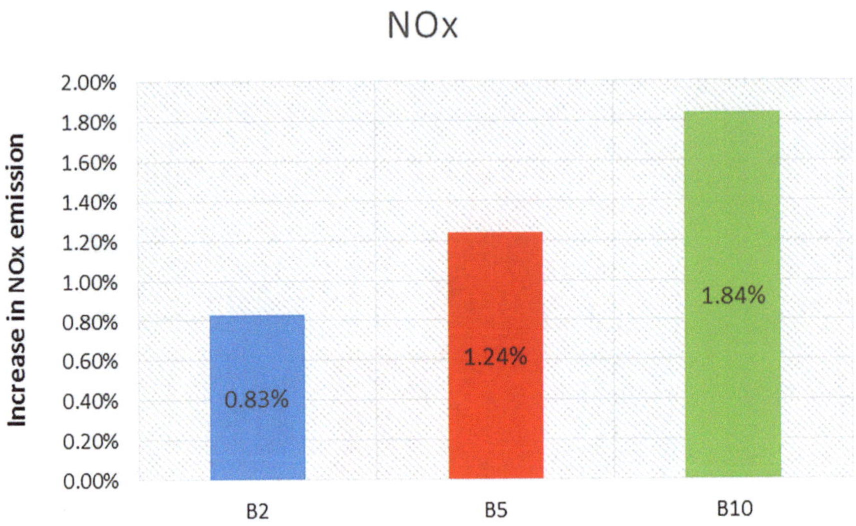

Fig. 17.4 % reduction in NO_x emissions for B2, B5, and B10 fuel blends compared to pure diesel fuel (B0)

17.4 Conclusion

Based on the findings of this study, several conclusions can be drawn. The properties of soybean methyl ester (SME) biodiesel are comparable to those of standard diesel fuel. SME blends of up to 10% with standard diesel fuel can be used in diesel engines without requiring modifications. The B2, B5, and B10 fuel blends resulted in statistically significant reductions in CO_2 emissions by 8%, 8%, and 20%, respectively, compared to pure diesel fuel, indicating a notable decrease in CO_2 emissions with increasing biodiesel concentration. Significant reductions in CO emissions were also observed for the B2, B5, and B10 blends, with decreases of 5.89%, 11.76%, and 35.3%, respectively, attributed to the higher oxygen content in biodiesel that promotes more complete combustion as the biodiesel concentration increases.

In contrast, NOx emissions were slightly higher for SME blends than for pure diesel fuel, with increases of 0.83%, 1.24%, and 1.84% for B2, B5, and B10 blends, respectively. However, the increase in NOx emissions was not statistically significant, particularly for the B2 blend. Overall, SME blends of up to 10% provide superior environmental performance compared to pure diesel fuel, particularly in reducing CO and CO_2 emissions. These results align with findings from Lapuerta et al. [2] for CO emissions reductions, Demirbas [3] for CO_2 emissions reductions, and Ramadhas et al. [5] for the slight increase in NOx emissions observed with SME blends.

References

1. Knothe G, Krahl J, Van Gerpe J (2005) The biodiesel handbook. American Oil Chemist's society (AOCS) Press, Champagn
2. Lapuerta M, Armas O, Rodríguez-Fernández J (2008) Effect of biodiesel fuels on diesel engine emissions. Prog Energy Combust Sci 34:198–223
3. Demirbas A (2009) Progress and recent trends in biodiesel fuels. Energy Convers Manag 50(1):14–34. https://doi.org/10.1016/j.enconman.2008.09.001
4. Graboski MS, McCormick RL (1998) Combustion of fat and vegetable oil derived fuels in diesel engines. Prog Energy Combust Sci 24(2):125–164. https://doi.org/10.1016/S0360-1285(97)00034-8
5. Ramadhas AS, Jayaraj S, Muraleedharan C (2005) Characterization and effect of using rubber seed oil as fuel in the compression ignition engines. Renew Energy 30(5):795–803. https://doi.org/10.1016/j.renene.2004.07.003
6. Chisti Y (2023) Microalgae as sustainable feedstock for biodiesel production. J Renew Energy 14(3):215–232
7. Madhavan A, Singh AK, Gupta R (2022) Enzymatic catalysis in biodiesel production: green chemistry approaches. Biotechnol Adv 40:107906
8. Zhang Z, Liu Y, Chen W (2023) Comparative life cycle analysis of biodiesel and fossil diesel: GHG emissions and environmental impact. Renew Energy Rev 162:1267–1281
9. Verma P, Sharma N, Singh A (2023) Advances in emission control Technologies for Biodiesel Engines. J Clean Prod 284:124617
10. Dwivedi G, Sharma MP (2021) A comprehensive review on the impact of biodiesel on engine emissions. Energy Rep 7:2345–2359
11. Pérez F, Morales M, Rodríguez J (2023) Hybrid biofuels: optimizing emission reduction from combustion engines. Fuel Process Technol 251:106716
12. Kumar A, Singh R, Verma P (2024) Future prospects of electric-biodiesel hybrid vehicles for emission control. Transp Res Part D: Transp Environ 98:102323
13. United States Environmental Protection Agency (USEPA) (2013) Human health and environmental effects of emissions from power generation: Focus on sulfur dioxide (SO2), nitrogen oxides (NOx) and mercury (Hg). www.epa.gov/airmarkets. Retrieved on 31-06-2016
14. Shida T, Frank B, Lanni T, Rideout G, Meyer N, Beregszaszy C (2007) Unregulated emissions from a heavy-duty diesel engine with various fuels and emission control systems. Environ Sci Technol 41(14):5037–5043
15. Laroo C, Mason R, Buckingham J (2005) Ultra low sulfur diesel (ULSD) sulfur test method variability: a statistical analysis of reproducibility. USEPA ULSD Round-Robin Test Program, SAE International
16. ASTM International (2002) ASTM International standard test method for oxidation onset temperature of hydrocarbons by differential scanning calorimetry (E-2009-02). Annual book of standards, Section 12. 10. ASTM International, West Conshohocken, pp 734–738
17. Unites States Environmental Protection Agency (USEPA). (2002a). National-scale air toxics assessment for 2002
18. Burnett C, Richard T, Pope A III, Michael JT, Eugenia EC, Krewski D, Kazuhiko I, Thurston GD (2002) Lung cancer, cardiopulmonary mortality, and long-term exposure to fine particulate air pollution. JAMA 287:1132–1141
19. AOAC (1990) Official methods of analysis, 15th edn. Association of Official Analytical Chemists, Arlington
20. ASTM International (2017) ASTM D97-17: standard test method for pour point of petroleum products. ASTM International, West Conshohocken. https://doi.org/10.1520/D0097-17

Chapter 18
Simulation of Thermal Performance of Microalgae Photobioreactor Façade: A Review

Bernadetha Grace Wisdayanti, Napassawan Wongmongkol, and Chatchawan Chaichana

Abstract Microalgae offer substantial potential for reducing CO_2 emissions associated with buildings across their entire lifecycle by improving energy efficiency and actively capturing CO_2. Integrating microalgae photobioreactor (PBR) façades into building systems provides multiple benefits, including acting as an effective insulation system, offering shading, harnessing solar energy, and converting light into biomass. Various methods have been employed to simulate the thermal performance of microalgae PBR, aiming to evaluate their effectiveness in reducing heat transfer into buildings. These methods include the finite volume approach, self-developed Computational Fluid Dynamics (CFD) codes, and integral analytical models. These simulations are often supported by advanced software tools such as ANSYS, Design Builder, TRNSYS, COMSOL Multiphysics, EnergyPlus, as well as plugins like Ladybug and Honeybee, which assist in analyzing energy performance and thermal behavior. This work reviews the state-of-the-art knowledge of the simulation of thermal performance of microalgae PBRs. It highlights the potential of simulation methods such as CFD in improving energy efficiency, reducing building energy demand, and supporting sustainable architectural practices. Furthermore, current research gaps, including limited exploration of multi-scale modeling and cost-effective computational frameworks, are discussed.

Keywords Microalgae · Building façade · Photobioreactor · Thermal performance · CO_2 fixation

B. G. Wisdayanti
Faculty of Engineering, Chiang Mai University, Chiang Mai, Thailand

N. Wongmongkol (✉) · C. Chaichana
Department of Mechanical Engineering, Faculty of Engineering, Chiang Mai University, Chiang Mai, Thailand
e-mail: napassawan.k@cmu.ac.th

L. Chen (ed.), *Clean Energy Research*, Green Energy and Technology,
https://doi.org/10.1007/978-3-031-99312-1_18

18.1 Introduction

Photobioreactors (PBRs) are innovative systems integrating microalgae into the built environment to enhance energy efficiency and thermal management. By utilizing passive cooling, they lower indoor temperatures and reduce air conditioning demands, making them effective in mitigating the urban heat island effect [1]. Algae panels on façades can decrease water temperatures by 6 °C, minimizing reliance on mechanical cooling [2]. Computational Fluid Dynamics (CFD) simulations optimize PBR designs by analyzing heat transfer, light distribution, and biomass productivity [3]. Compared to open-pond systems, PBRs offer greater thermal stability and energy efficiency [4, 5].

Beyond cooling, PBRs contribute to sustainability by capturing CO_2 from building emissions and improving indoor air quality. Through photosynthesis, microalgae convert CO_2 into biomass for biofuels or wastewater treatment, aligning with carbon reduction strategies [6]. Their microbial activity aids in bioremediation, enhancing urban ecosystems. Architecturally, PBRs integrate seamlessly into façades, complementing green building technologies. Advances such as 3D printing enable lightweight, modular components with minimal waste [7]. Although installation costs are high, long-term benefits such as energy savings, biomass revenue, and government incentives ensure financial viability [8].

PBRs contribute to energy-efficient urban planning by supporting urban agriculture, vertical farming, and sustainable building design [9]. Innovations in mixing efficiency and nutrient delivery further enhance algae growth and energy output, making PBRs valuable climate-responsive solutions. This study reviews the thermal performance of microalgae PBR façades, evaluating simulation methods such as CFD and energy modeling tools to assess their effectiveness in temperature regulation and energy efficiency. The research explores the environmental benefits of PBR façades, including CO_2 sequestration and air quality improvement, while addressing key challenges like overheating risks, cost constraints, and scalability. The study aims to identify research gaps and highlight future directions for optimizing PBR technology in sustainable urban development. Abbreviations and definitions used in this paper are summarized in Table 18.1.

Table 18.1 List of abbreviations and acronyms used in this paper

Abbreviation	Definition	Abbreviation	Definition
PBR	Photobioreactor	HVAC	Heating, ventilation, and air-conditioning
CFD	Computational fluid dynamic	CCM	CO_2 concentrating mechanism
TRNSYS	Transient system simulation tool	AI	Artificial intelligence
EES	Engineering equation solver	IoT	Internet of things

18.2 Thermal Simulation of Photobioreactor

Recent studies highlight the thermal performance of microalgae PBR façades, emphasizing their benefits and limitations. Efficient heat transfer modeling is essential for optimizing energy use and reducing carbon footprints. Various simulation methods, including CFD, finite volume approaches, and analytical models using software like Ansys, TRNSYS, and EnergyPlus, assess PBR thermal efficiency [10]. CFD studies, such as Yaman et al.'s work, show that adjusting biomass density influences light transmittance, improving insulation and energy savings in office buildings [11].

PBRs also support sustainability by aiding wastewater treatment and biomass production. Díaz et al. demonstrate their role in nutrient removal, reducing buildings' ecological footprints [12]. Additionally, multi-objective optimization studies reveal that dynamic PBR shading systems enhance daylighting, leading to energy savings and improved indoor comfort [13]. Further research is needed to maximize productivity across different climates and operational conditions.

Yu [14] used ANSYS CFX5 and CFD to optimize PBR structure, improving thermal efficiency and uniform temperature distribution for microalgae growth but emphasized the need for experimental validation. Huang [15] combined numerical and experimental methods to assess a split flat-plate airlift reactor using ANSYS ICEM CFD 12.1 and CFX 12.1, demonstrating stable thermal regulation but lacking comparative analysis with other designs. A COMSOL Multiphysics study analyzed PBR hydrodynamics and heat transfer, identifying poor mixing efficiency in certain configurations and highlighting the need for design modifications, though it focused on specific geometric parameters without broader experimental validation [16].

CFD simulations and energy balance equations were used to analyze PBR thermal performance in urban environments, revealing overheating issues in high-exposure areas. To enhance thermal stability, the study recommends integrating shading and cooling mechanisms. However, balancing energy inputs and outputs remains a challenge, requiring further research on practical thermal management solutions. While simulations provide valuable insights into heat transfer, experimental validation is essential for real-world applicability. Table 18.2 summarizes the key simulation tools used in PBR thermal performance studies, highlighting their applications and limitations.

18.3 Effects of PBR on the Thermal Environment of Building

Integrating PBR façades into buildings enhances thermal performance and fosters a favorable microclimate. Microalgae use sunlight for photosynthesis that absorbs CO_2, releases oxygen and improves indoor air quality while reducing carbon emissions. PBRs also act as thermal insulators, stabilizing indoor temperatures and

Table 18.2 Summary of simulation tools for thermal performance of PBR facades

Simulation tools	Finding results	Strengths	Limitations	Ref.
ANSYS ICEM CFD 12.1; ANSYS CFX 12.1	The split flat-plate airlift reactor had the highest biomass productivity (0.018 g L^{-1} h^{-1}), surpassing other designs.	The model integrated fluid dynamics and irradiation transfer for efficient PBR optimization and evaluation.	The study lacked a prior comparison of PBR types, highlighting a research gap.	[15]
ANSYS fluent MATLAB co-modelling EES MATLAB	PBRs overheat, hindering biomass growth (optimal: 28 °C). Adjusting inlet temperature to 22–24 °C maintains optimal conditions, boosting productivity.	PBRs generate bioenergy and provide heating, requiring no arable land and offering high CO_2 sequestration for sustainability.	Temperature control in PBRs is challenging, increasing costs. Overheating in summer and cold in winter reduce biomass productivity.	[17]
CFD simulation: ANSYS R17.2	Effective simulation enhances hydrodynamic understanding, optimizing energy use and biomass yield in commercial PBRs.	Efficient mixing and low energy input requirements,	Low mixing efficiency due to design constraints highlights the need for improved reactor configuration.	[3]
CFD, ANSYS	Reduced energy costs in tubular photobioreactors through improved light yield	Enhanced productivity, prevention of contamination, cost-effective resources	Higher operation costs compared to open raceway ponds	[18]
Open studio-energy plus software	Algae PBR façades cool interiors, improve air quality, and cut cooling energy. Simulations show significant ΔT, reducing solar heat gain, with secondary skins boosting efficiency.	Algae façades boost energy efficiency and air quality, guiding future sustainable architecture and urban energy studies.	Reliance on simulations may introduce uncertainties. Further research is needed to validate results post-construction.	[19]

lowering energy consumption, particularly in regions with seasonal variations [20]. A case study on a retrofitted office building in a Mediterranean climate showed improved thermal comfort and reduced energy demand [11]. By reducing reliance on HVAC systems, PBRs lower operational costs and extend system lifespans.

Studies also indicate that PBRs effectively minimize heat gain, reducing cooling loads during warmer months [21]. PBR façades buffer external conditions, reducing thermal stress while enhancing structural integrity and aesthetics. The integration of microalgae systems into architecture reduces carbon footprints, reinforcing their role in sustainable construction. This synergy between biological and architectural systems aligns with broader energy efficiency goals and promotes climate resilience in urban environments [11].

Beyond thermal benefits, PBR façades enhance sustainability by mitigating urban heat island effects through optimized heat exchange and shading. Microalgae absorb CO_2, improving air quality and reducing greenhouse gas emissions [1]. Their dynamic nature allows year-round energy efficiency, providing insulation in winter and shading in summer to reduce solar heat gain. This dual function makes PBRs a cutting-edge solution for sustainable architecture.

Retrofitting buildings with PBRs enhances their applicability, with multi-criteria optimization showing superior thermal comfort, energy efficiency, and visual comfort over traditional systems [22]. By harnessing solar energy for microalgae growth, PBRs enable biofuel production and support the circular economy. They also improve acoustic insulation, enhancing indoor quality and occupant well-being. As a climate-responsive solution, PBRs aid carbon reduction and climate resilience while promoting urban sustainability through carbon sequestration and biomass generation. To maximize their potential, future research should focus on material optimization, system integration, long-term performance, lifecycle analysis, and cost-benefit assessments for wider adoption in modern architecture.

18.4 CO_2 Fixation by Microalgae

Microalgae play a vital role in CO_2 fixation, helping mitigate climate change and promote sustainability. Through photosynthesis, they convert CO_2 into biomass, reducing atmospheric levels while producing valuable compounds. The efficiency of this process depends on factors like carbon sources, culture conditions, and microalgal strains. This section examines CO_2 fixation mechanisms, influencing factors, and applications in sustainable building materials.

Microalgae utilize CO_2 as their primary carbon source, assimilating it through photosynthesis into organic matter. This process occurs in the chloroplasts, where microalgae absorb light energy to drive the conversion of CO_2 and water into glucose and oxygen. The fundamental biochemical pathway responsible for this fixation is the Calvin cycle, which is catalyzed by the enzyme ribulose-1,5-bisphosphate carboxylase/oxygenase (RuBisCO). The effectiveness of CO_2 fixation is enhanced by the CO_2 Concentrating Mechanism (CCM), a specialized adaptation that actively transports bicarbonate ions (HCO_3^-) and generates localized CO_2 gradients around RuBisCO to optimize carbon assimilation [23]. The principle of microalgae photosynthesis and the processes of CO_2 sequestration by the microalgae are explained in detail in Fig. 18.1.

The efficiency of CO_2 uptake in microalgae depends on external factors like CO_2 concentration, light intensity, pH, and temperature. Optimizing these conditions significantly boosts biomass production and carbon sequestration. Higher CO_2 levels in photobioreactors enhance biomass and carbohydrate accumulation [26], while red and blue light spectra improve photosynthesis efficiency [27]. Recent research explores microalgae in sustainable construction, integrating them into facades and materials for CO_2 sequestration and thermal performance. Hybrid films with corn

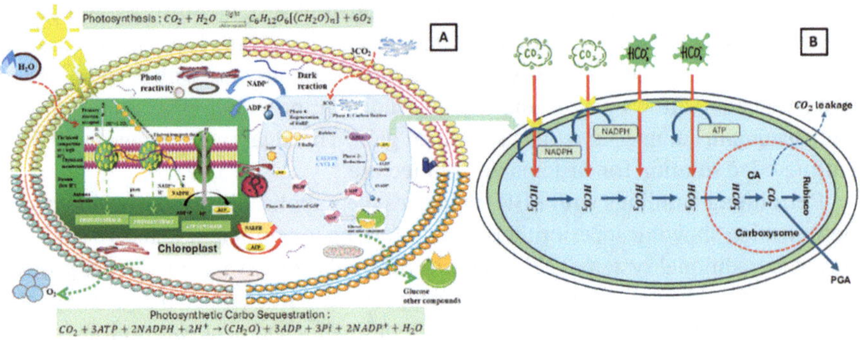

Fig. 18.1. (**a**) Microalgae photosynthetic carbon fixation [24]; (**b**) Calvin cycle: CO_2 sequestration in microalgae [25]

starch and microalgae offer strong mechanical properties [28], while photosynthetic textile biocomposites enhance indoor air quality and structural durability.

Microalgae are also being integrated into porous building materials that facilitate algal colonization. Studies on porous bricks and sandstone have shown that these materials can support microalgal growth, thereby promoting continuous CO_2 sequestration throughout the building's lifespan [29]. Additionally, living building facades utilizing microalgal photobioreactors can improve energy efficiency by providing shading and thermal insulation, reducing the need for artificial cooling and heating systems [30].

Despite the promising applications of microalgae in CO_2 fixation and sustainable architecture, several challenges remain. The scalability of microalgal cultivation for large-scale CO_2 sequestration is hindered by high production costs, the need for precise environmental control, and limitations in mass transfer efficiency. Advances in photobioreactor design and automation have sought to address these challenges by optimizing growth conditions and improving CO_2 capture efficiency [31].

Genetic engineering represents another avenue for enhancing CO_2 fixation efficiency in microalgae. By modifying CCM pathways and improving RuBisCO activity, researchers aim to develop microalgal strains with superior carbon sequestration capabilities, particularly in high-CO_2 environments [32]. Furthermore, integrating microalgal systems into waste treatment processes offers a dual benefit of CO_2 sequestration and wastewater remediation, enhancing the feasibility of large-scale applications [23].

18.5 Summary

Microalgae Photobioreactor (PBR) façades offer a sustainable solution for improving energy efficiency, reducing CO_2 emissions, and enhancing indoor environmental quality. These living façades provide shading, thermal insulation, and carbon sequestration. Various simulation methods, including CFD, finite volume approaches, and energy modeling tools like ANSYS, TRNSYS, and EnergyPlus, have been used to assess their thermal performance. Key findings show that PBRs contribute to passive cooling, lower energy demand, and improve air quality by capturing CO_2 and generating biomass for biofuels or wastewater treatment.

Despite their advantages, several challenges remain. Many studies rely on computational simulations without sufficient experimental validation, leading to uncertainties in environmental, fluid, and light interaction assumptions. Most research is limited to small-scale prototypes, restricting real-world applicability. Thermal overheating can also affect microalgae growth and building efficiency, necessitating better shading, ventilation, and cooling strategies. High costs, long payback periods (16–24 years), and manufacturing challenges hinder commercialization. Additionally, current studies often separate biological and architectural modeling, lacking an integrated approach.

Future research should combine experiments with simulations to enhance reliability. Optimizing reactor designs—improving mixing, nutrient delivery, and CO_2 transfer—can boost microalgae productivity and thermal performance. Smart technologies like AI, IoT, and automation can further optimize PBR operations. Beyond façades, PBRs have potential in urban agriculture, wastewater treatment, and carbon-negative building materials. Financial models, subsidies, and policy incentives are also essential for large-scale adoption in the construction industry.

Acknowledgments This research was supported by CMU Junior Research Fellowship Program, Chiang Mai University, Thailand, and was partially supported by the Murata Science Foundation.

References

1. Sedighi M, Pourmoghaddam Qhazvini P, Amidpour M (2023) Algae-powered buildings: a review of an innovative, sustainable approach in the built environment. Sustainability 15:3729
2. El-Mahdy D, Youssef AK (2021) Wastewater matter: from algae to bio-algae plastic 3D printed façade element, pp 153–165
3. Aslanbay Guler B, Deniz I, Demirel Z, Imamoglu E (2020) Evaluation of scale-up methodologies and computational fluid dynamics simulation for fucoxanthin production in airlift photobioareactor. Asia-Pacific J Chem Eng 15:15
4. Mata TM, Oliveira GM, Monteiro H, Silva GV, Caetano NS, Martins AA (2021) Indoor air quality improvement using nature-based solutions: design proposals to greener cities. Int J Environ Res Public Health 18:8472
5. Shokri A, Talebi S, Salami-Kalajahi M (2020) Polybutadiene rubber/graphene nanocomposites prepared via in situ coordination polymerization using the neodymium-based ziegler–natta catalyst. Ind Eng Chem Res 59:15202–15213
6. Salguero-Rodríguez Y, Gómez-Pérez CA, Arango-Restrepo JP, Espinosa J (2021) Static mixer proposal for tubular photobioreactors to reduce mixing energy consumption and enhance light–dark cycles. J Chem Technol Biotechnol 96:113–124
7. Oudot E, Gholmane K, Fakra DAH, Benelmir R (2024) Energetic valorization of the innovative building envelope: an overview of electric production system optimization. Sustainability 16:2305

8. Arora R, Sudhakar K, Rana RS (2024) Photobioreactors for building integration: a overview of designs and architectural potential. Heliyon 10:e35168
9. Milani I, Bongioanni C, Colone F, Lombardo P (2021) Fusing measurements from Wi-Fi emission-based and passive radar sensors for short-range surveillance. Remote Sens 13:3556
10. Seyrek CI, Widera B, Woźniczka A (2021) Sustainability-related parameters and decision support tools for kinetic green Façades. Sustainability 13:10313
11. Yaman Y, Altunacar N, Tokuç A, Köktürk G, Deniz İ, Ezan MA (2022) Effects of photobioreactor façades on thermal and visual performance of an office in izmir. Eskişehir Tech Univ J Sci Technol A—Appl Sci Eng 23:68–75
12. Díaz V, Antiñolo L, Poyatos Capilla JM, Almécija MC, Muñío M d M, Martín-Pascual J (2022) Nutrient removal and membrane performance of an algae membrane Photobioreactor in urban wastewater regeneration. Membranes (Basel) 12:982
13. Lo Verso VRM, Javadi MHS, Pagliolico S, Carbonaro C, Sassi G (2019) Photobioreactors as a dynamic shading system conceived for an outdoor workspace of the State Library of Queensland in Brisbane: study of daylighting performances. J Daylighting 6:148–168
14. Yu G, Li Y, Shen G, Wang W, Lin C, Wu H, Chen Z (2009) A novel method using CFD to optimize the inner structure parameters of flat photobioreactors. J Appl Phycol 21:719–727
15. Huang J, Kang S, Wan M, Li Y, Qu X, Feng F, Wang J, Wang W, Shen G, Li W (2015) Numerical and experimental study on the performance of flat-plate photobioreactors with different inner structures for microalgae cultivation. J Appl Phycol 27:49–58
16. Loomba V, Huber G, von Lieres E (2018) Single-cell computational analysis of light harvesting in a flat-panel photo-bioreactor. Biotechnol Biofuels 11:149
17. Duarte M de A, Oliveira RD and Lima FRS 2023 Thermal energy generation of PBR integrated in a commercial building for water heating under humid subtropical climate. Int J Sustain Build Technol Urban Dev 14 153–168
18. Gómez-Pérez CA, Espinosa J, Montenegro Ruiz LC, van Boxtel AJB (2015) CFD simulation for reduced energy costs in tubular photobioreactors using wall turbulence promoters. Algal Res 12:1–9
19. Martokusumo W, Koerniawan MD, Poerbo HW, Ardiani NA, Krisanti SH (2017) Algae and building façade revisited. A study of façade system for infill design. J Archit Urban 41:296–304
20. Yaman Y, Tokuç A (2024) An optimization of thermal, visual and energy indicators for retrofit with photobioreactors in the mediterranean climate ed P Mendonça. E3S Web Conf 546:03006
21. Mohammadivahidi N (2023) To reduce the annoying light with microalgae window. Innov Sci Technol 2:12–18
22. Todisco E, Louveau J, Thobie C, Dechandol E, Hervé L, Durécu S, Titica M, Pruvost J (2022) A dynamic model for temperature prediction in a façade-integrated photobioreactor. Chem Eng Res Des 181:371–383
23. Handayani T, Priyanto FE, Sulistia S, Oktaviani AN, Sopiah N, Santoso AD, Putra AS, Djarot IN, Widyastuti N, Tjokrokusumo D, Apriyanto H, Rifai A, Aziz A, Wijayanti SP, Bahua H, Nuha N, Isharyadi F, Paminto AK, Ariyani NR (2023) Integrated microalgae cultivation for sustainable wastewater treatment and carbon dioxide bio-fixation in milk factories. Int J Sustain Dev Plan 18:2941–2952
24. Dahai H, Zhihong Y, Lin Q, Yuhong L, Lei T, Jiang L, Liandong Z (2024) The application of magical microalgae in carbon sequestration and emission reduction: removal mechanisms and potential analysis. Renew Sustain Energy Rev 197:114417
25. Ighalo JO, Dulta K, Kurniawan SB, Omoarukhe FO, Ewuzie U, Eshiemogie SO, Ojo AU, Abdullah SRS (2022) Progress in microalgae application for CO2 sequestration. Clean Chem Eng 3:100044
26. Mahmoudi H, Kossalbayev B, Sarsekeyeva F, Zaletova D, Lyaguta M (2024) The influence of varying concentrations of CO 2 on the buildup of carbohydrates in microalgae biomass ed Y Alibek. BIO Web Conf 100:02013

27. Cecchin M, Paloschi M, Busnardo G, Cazzaniga S, Cuine S, Li-Beisson Y, Wobbe L and Ballottari M 2021 CO 2 supply modulates lipid remodelling, photosynthetic and respiratory activities in chlorella species
28. Fabra MJ, Martínez-Sanz M, Gómez-Mascaraque LG, Coll-Marqués JM, Martínez JC, López-Rubio A (2017) Development and characterization of hybrid corn starch-microalgae films: effect of ultrasound pre-treatment on structural, barrier and mechanical performance. Algal Res 28:80–87
29. Quagliarini E, Gregorini B, D'Orazio M (2022) Modelling microalgae biofouling on porous buildings materials: a novel approach. Mater Struct 55:158
30. Hariyanto L, Arifin LS, Damayanti R (2022) Microalgae as a sustainable facade for occupants' health: a review. Adv Civ Eng Sustain Archit 4:26–36
31. Swilam O, El-Belely E, El-Gamal A (2022) Modified closed flat-plate photobioreactor for optimizing CO2 bio-fixation by the cyanobacterium Synechococcus elongatus. Egypt J Phycol 23:89–113
32. Burlacot A, Dao O, Auroy P, Cuiné S, Li-Beisson Y, Peltier G (2022) Alternative photosynthesis pathways drive the algal CO2-concentrating mechanism. Nature 605:366–371

Chapter 19
Thermal Energy Storage Solutions for Off-Grid Solar Thermal Power Plant Application: A Review

Rohitha Weerasinghe, Praveen Moonasinghe, Vladan Velisaljevic, Mina Mortazavi, Kirsty Hinchliffe, and Jon Fenton

Abstract The role of thermal energy storage (TES) is critical in addressing the intermittency of solar energy, enabling continuous power generation in solar thermal power plants. The present paper evaluates TES options for a 20 kWe Organic Rankine Cycle (ORC) based thermal power plant that is to be commissioned in Sri Lanka, developed under the Innovate UK project RESORCS (Off-Grid Renewable Energy Production and Storage). The study explores sensible heat, latent heat, and thermochemical storage technologies, with a focus on commercially available options such as molten salts, water-based systems, and phase change materials (PCMs). HiTEC salt and Na-Li nitrate eutectic salts are optimal TES solutions for low- to medium-temperature applications, offering low melting points, high specific heat capacities, and robust energy densities. Performance analysis at 200 °C and 220 °C confirms their thermal stability and efficiency. Preliminary calculations support 24-hour plant operation but emphasize the need for further modeling to account for system thermal losses. This analysis provides a framework for selecting an efficient, cost-effective TES medium to ensure reliable off-grid energy production with an ORC system.

Keywords Thermal Energy Storage · Organic Rankine Cycle · Solar Thermal power plant

R. Weerasinghe (✉) · P. Moonasinghe · V. Velisaljevic · M. Mortazavi
School of Computer Science and Technology, University of Bedfordshire, Luton, UK
e-mail: rohitha.weerasinghe@beds.ac.uk

K. Hinchliffe · J. Fenton
FeTu Limited, Elland, UK

L. Chen (ed.), *Clean Energy Research*, Green Energy and Technology,
https://doi.org/10.1007/978-3-031-99312-1_19

19.1 Introduction

Thermal energy storage (TES) is a crucial component in solar thermal power plants, as it addresses the intermittency of solar energy and enables continuous power generation even when sunlight is not available. TES systems allow solar power plants to store excess thermal energy produced during peak sunlight hours, which can then be used during low or no sunlight periods [1]. In a small-scale organic Rankine cycle-based solar thermal power plant with a capacity of 20 kWe, efficient and cost-effective TES options are essential to maximize energy output and ensure reliability. This paper evaluates various TES options suitable for a 20 kWe solar thermal power plant, highlighting commercially available technologies and their respective advantages and limitations. A 20 kWe solar thermal plant is to be commissioned in Sri Lanka through Innovate UK project 10079411: Off-Grid Renewable Energy Production and Storage with Organic Rankine Cycle, Solar and Waste (RESORCS). This study lays the foundation for selecting a suitable storage option that will supply continuous heat energy to run a Rankine cycle thermal expander, a FeTu engine.

Thermal Energy Storage Options

TES systems for solar thermal power plants are generally classified into three types: (1) sensible heat storage, (2) latent heat storage, and (3) thermochemical storage. Each category has unique characteristics, making them suitable for different operational requirements and thermal management strategies [2].

Sensible Heat Storage (SHS) Sensible heat storage systems store thermal energy by changing the temperature of a storage medium, such as water, oil, or molten salts. The energy is stored in the form of temperature difference within the medium, as the heat capacity is a product of temperature difference times the specific heat capacity. This is a widely used and commercially available TES option because of its simplicity, reliability, and relatively low cost [3]. However, the storage capacity of SHS is limited by the specific heat of the material and the temperature range that the material can handle.

Latent Heat Storage (LHS) Latent heat storage involves storing energy through phase changes of materials, typically from solid to liquid or liquid to gas. Materials used in latent heat storage are known as phase change materials (PCMs). LHS systems offer higher energy density than SHS because they exploit the latent heat associated with the phase transition of the material [4]. The constant temperature during phase transition makes PCMs ideal for applications requiring stable thermal output. However, LHS systems can be more expensive due to the cost of PCM materials and the complex thermal management systems required.

Thermochemical Storage (TCS) Thermochemical storage involves storing energy through reversible chemical reactions. When heat is applied, a chemical reaction occurs, and the products can later recombine to release heat when energy is required. TCS offers very high energy density and can store energy for extended periods without significant loss [5]. However, TCS systems are still in the research and development stage and are not widely available commercially due to complexities in reaction control and material stability.

Comparison of Commercially Available TES Systems

Commercial TES systems are available for different scales and applications, ranging from small systems suitable for residential or small commercial use to large, grid-scale systems. For a 20 kWe ORC solar thermal power plant, several commercially viable options are suitable based on capacity, temperature range, and economic feasibility (Table 19.1).

Molten Salt Systems Molten salt TES systems are among the most commonly used TES options in solar thermal power plants. These systems use a combination of sodium nitrate and potassium nitrate salts to store energy as sensible heat. Molten salt systems operate at temperatures between 142 and 565 °C, making them ideal for applications requiring medium to high temperatures. Examples include the SolarReserve's Crescent Dunes project, which utilizes molten salt storage [6].

HiTEC Salt HiTEC salt is a ternary eutectic salt mixture commonly used in thermal energy storage applications due to its favorable thermal properties. It consists primarily of sodium nitrate ($NaNO_3$), potassium nitrate (KNO_3), and sodium nitrite

Table 19.1 Commercially available TES systems

	Molten salt	Water	Paraffin PCM
Storage type	Sensible heat	Sensible heat	Latent heat
Operation temperature (°C)	123–565	100	40–80
Storage medium	Na/K/LI nitrates & Nitrites	Water	Paraffin wax
Energy density (kWh/m^3)	100–150	50–80	200–300
Cost per kWh (USD)	30–50	5–10	100–200
Advantages	High energy density, scalable	Low cost, simple design	Stable output temperature, compact design
Disadvantages	High initial cost, requires high temperatures	Limited to low-temperature applications	Expensive PCM materials, lower heat transfer rate

($NaNO_2$). The typical composition of HiTEC salt is approximately 53%, 40% $NaNO_2$, and 7% $NaNO_3$ [7]. It has a melting temperature around 142 °C, making it suitable for low-temperature heat storage applications [7]. It also has a specific heat capacity of about 1.56 J/g·K at temperatures near its melting point, with thermal stability in the range of 142–450 °C [8].

Li-Na nitrate salts It is a binary eutectic mixture of lithium nitrate ($LiNO_3$) and Sodium nitrate ($NaNO_3$). The typical composition of Li-Na nitrate salts is approximately 53% $NaNO_3$ and 47% $LiNO_3$.The mixture has a melting temperature of 123 °C with a specific heat capacity of 1.50 J/g·K, making it viable for low temperature applications [7, 8].

Fluoride and Chloride Salts Fluoride and chloride salts, such as mixtures of LiF-NaF-KF and NaCl-KCl-$MgCl_2$, have high melting points, often above 400 °C, with some even exceeding 500 °C. This high melting point poses a significant challenge for concentrating CSP applications, which typically need storage materials that remain in a liquid state at lower temperatures to simplify handling and reduce operational costs. The need to keep these salts above their high melting points requires complex heating systems to prevent solidification, adding complexity and cost to CSP operations. This makes fluoride and chloride salts less practical as thermal storage media in CSP, where lower-melting-point materials are preferred to enhance system efficiency and reliability [9].

Water-Based Storage Systems Water-based storage systems are simple and cost-effective TES solutions suitable for small-scale organic Rankine cycle (ORC) solar thermal plants. Water has a high specific heat capacity of 4.187 kJ/kg·K, which makes it effective for storing thermal energy at low temperatures.

However, a significant disadvantage of this system in concentrated solar power (CSP) applications is that storing energy at temperatures higher than 100 °C requires maintaining water under pressure. This is necessary to prevent the formation of steam, which significantly reduces the specific heat capacity to 1.996 kJ/kg·K. To achieve a temperature of at least 200 °C in liquid form, a pressure of about 16.5 bars is needed. This necessitates the use of specially designed storage tanks capable of withstanding such pressure [10].

Paraffin-Based Phase Change Materials (PCM) Paraffin-based phase change materials (PCMs) possess high latent heat capacity and chemical stability, making them initially attractive for thermal energy storage [4]. However, they are not ideal for ORC-based CSP applications due to several limitations. Their low thermal conductivity (~0.2 W/m·K) impedes efficient heat transfer, necessitating additional measures to enhance conductivity, which increases system complexity and cost [11]. Moreover, their melting temperatures are generally below 100 °C, incompatible with the high operating temperatures between 150x and 220c required in ORC

CSP systems. Safety concerns arise from their flammability and potential degradation at elevated temperatures, further limiting their suitability for CSP [4, 12].

Comparison of the Storage Technologies

Among the thermal storage solutions for a 20 kWe solar thermal power plant, HiTEC and Na-Li nitrate salts stand out as optimal choices, particularly when compared to alternatives such as fluoride/chloride salts, water-based systems, and paraffin-based PCMs.

HiTEC salt, a mixture primarily comprising sodium nitrate, potassium nitrate, and sodium nitrite, offers several advantages due to its relatively low melting point (around 142 °C) and high specific heat capacity of about 1.56 J/g·K. This combination makes it highly suitable for low-temperature storage needs, supporting stable operation without the complexity of excessive heat management systems. In a similar vein, the Na-Li nitrate salts (lithium nitrate and potassium nitrate) blend also serves low to medium-temperature applications effectively. With a melting point of 123 °C and a specific heat capacity of 1.50 J/g·K, this mixture provides the necessary thermal efficiency while maintaining cost-effectiveness. This low melting temperature minimizes the risk of solidification, a major benefit compared to higher melting point options like fluoride and chloride salts, which necessitate additional heating systems due to their melting points above 400 °C.

Furthermore, HiTEC and Li-Na nitrate salts offer robust energy storage density, an essential factor for scaling energy output in small solar power applications. Their high energy density (100–150 kWh/m^3) is especially advantageous when compared to water-based systems that operate at lower temperatures and require pressurization to prevent steam formation at elevated temperatures. Additionally, these salts maintain operational stability at temperatures compatible with organic Rankine cycle (ORC) systems, an asset over paraffin-based PCMs, which suffer from low thermal conductivity and instability at the required temperatures.

19.2 Molten Salt Storage Types

Direct Molten Salt Systems

In direct molten salt systems, the molten salt serves both as the heat transfer fluid (HTF) in the solar field and as the thermal energy storage medium [13]. This configuration eliminates the need for an intermediate heat exchanger, reducing thermal losses and system complexity (Fig. 19.1). The direct use of molten salts enhances the overall efficiency of the CSP plant due to reduced thermal resistance and lower capital costs associated with fewer components [1].

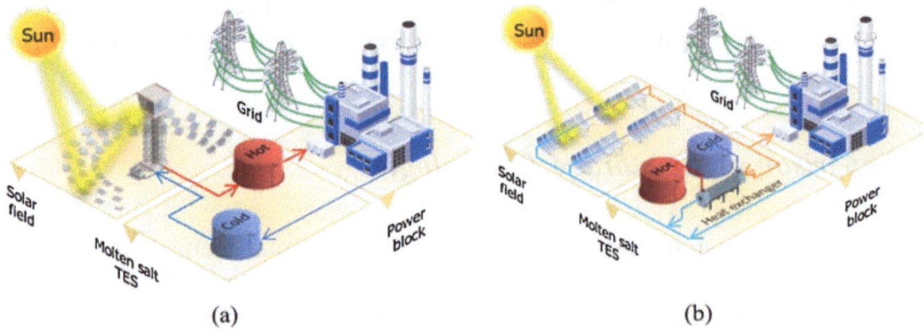

Fig. 19.1 Direct (**a**) and indirect (**b**) molten salt system [19]

Indirect Molten Salt Systems

Indirect molten salt systems employ a different HTF in the solar field, such as synthetic oils or steam, and transfer the absorbed heat to the molten salt storage medium via heat exchangers [14] (Fig. 19.1). While this configuration allows for the use of HTFs with lower freezing points and established operational experience, it introduces additional thermal losses due to the heat exchange process. Indirect systems may be advantageous when the HTF used in the solar field is incompatible with storage requirements or when retrofitting existing plants [13].

Dual-Tank Storage Systems

The dual-tank system is the most widely used configuration in commercial CSP plants [1]. It consists of two separate tanks: a hot tank and a cold tank (Fig. 19.2). During the charging phase, molten salt heated by the solar receiver is stored in the hot tank at high temperatures (around 565 °C). During discharging, the hot molten salt flows from the hot tank to a turbine to produce electricity and is then returned to the cold tank at a lower temperature (approximately 290 °C) [14].

This system offers operational simplicity and reliability, effectively separating the hot and cold fluids, minimizing thermal losses, and allowing precise temperature control. However, the need for two large tanks increases the initial capital investment and requires more physical space [15].

Single-Tank Thermocline Storage Systems

Single-tank thermocline systems aim to reduce costs by using a single tank to store both hot and cold molten salt, relying on thermal stratification to maintain a temperature gradient within the tank [16]. The tank is often filled with a filler material,

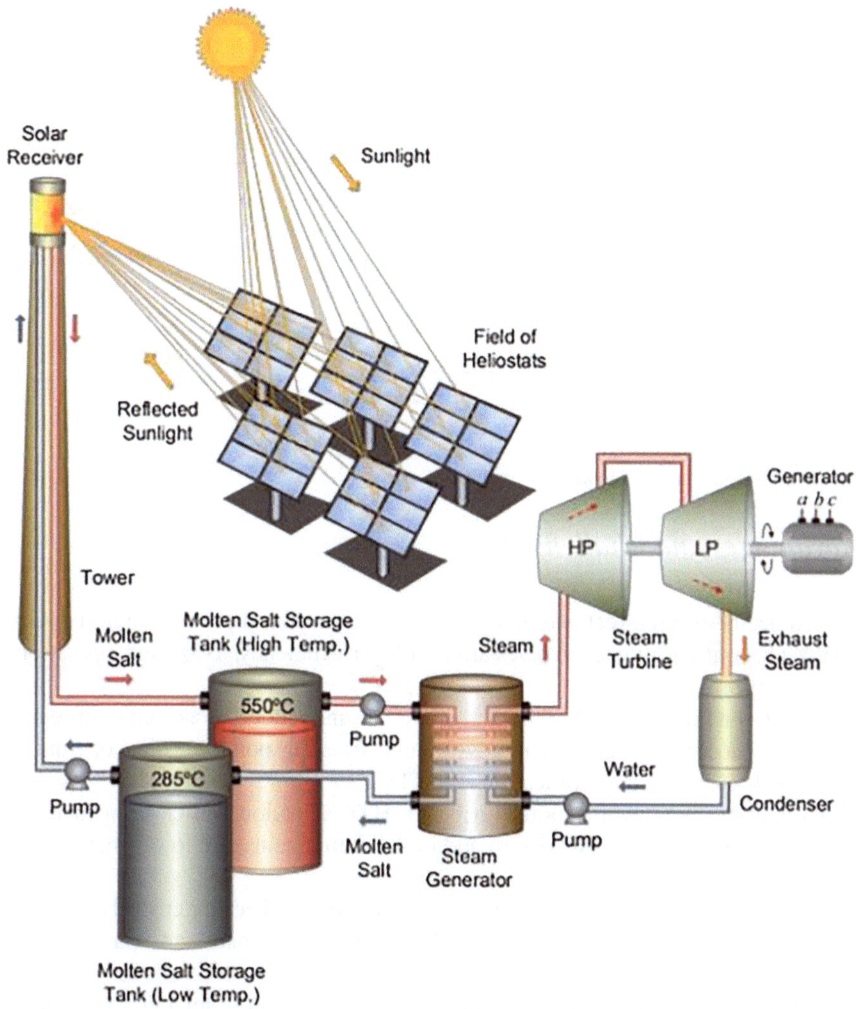

Fig. 19.2 Dual tank molten salt system [20]

such as quartzite rock or ceramic spheres, which enhances thermal stratification and reduces the required volume of molten salt (Fig. 19.3) [17].

During operation, hot molten salt is introduced at the top of the tank during charging, while cold salt is extracted from the bottom. During discharging, the flow is reversed. While single-tank systems are cost-effective due to reduced material requirements and fewer components, they face challenges in maintaining thermal stratification over time, which can lead to decreased efficiency [1].

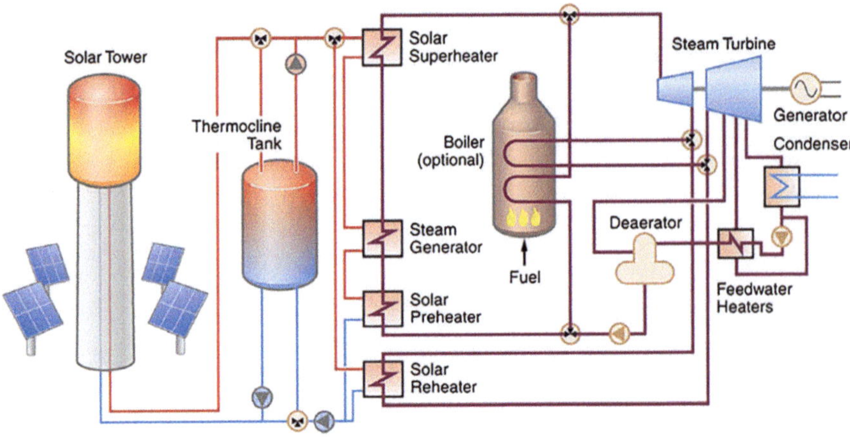

Fig. 19.3 Single tank molten salt system [20]

Advantages and Challenges

The choice between direct and indirect systems, as well as between single-tank and dual-tank configurations, involves trade-offs in efficiency, cost, and operational complexity.

Direct systems offer higher thermal efficiency and reduced costs but require molten salts that remain liquid at relatively high freezing points, posing operational challenges such as salt solidification risks [18].

Indirect systems, while more flexible in HTF selection, suffer from increased capital and operational costs due to additional components and lower thermal efficiency [1].

Dual-tank systems provide higher reliability and have been extensively proven in commercial applications but come with higher initial costs and space requirements [15].

Single-tank systems offer significant cost savings and a smaller footprint but require careful design to maintain the thermocline and prevent thermal degradation, impacting performance [16].

Application of Molten Salt Storage in the FeTu ORC

Determining the Ideal Temperature for the FeTu ORC

Using data provided by FETU and the following standard thermodynamics equations:

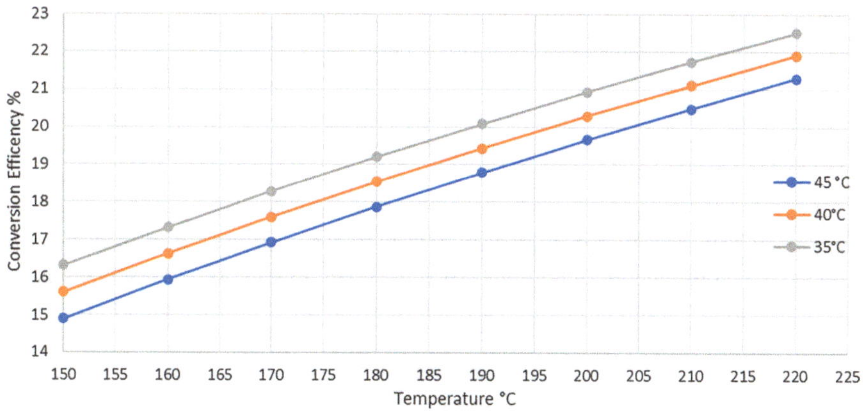

Fig. 19.4 Conversion Efficiency vs temperature graph

$$\eta_{(\text{carnot})} = 1 - \frac{T_C}{T_H}$$

$$\eta_{(\text{electrical})} = \eta_{(\text{carnot})} \times 0.6$$

A plot was generated to illustrate thermal to electrical conversion efficiency with fixed cold source temperatures and varying hot source temperatures (Fig. 19.4). The cold source temperatures were selected based on the typical ambient conditions in Sri Lanka, where temperatures rarely fall below 28 °C. As a result, cold source temperatures of 35 °C, 40 °C, and 45 °C were chosen. The hot source temperatures tested ranged from 150 °C to 220 °C, with an upper limit of 220 °C imposed by FETU due to lubrication oil limitations.

As per the plot, it is clearly visible that hot source temperatures above 200 °C are the most ideal, as they achieve a conversion efficiency of approximately 20% or higher, depending on the cold source temperature. This is consistent with the observation that Carnot Efficiency rises with increasing hot source temperature.

The resulting plot is shown below.

Assumption in the Determination of the Temperature The plot is solely based on the projected efficiencies listed on the H2P website. It does not include information about thermal losses at the evaporator or the generator's efficiency. Additionally, losses in the condenser and thermal losses in the piping were not accounted for.

Thermal Storage Capacity Required for the FeTu 20 KWe ORC The thermal storage is designed to operate for 24 h, and the total energy storage requirement is determined using the H2P efficiencies and the formula:

$$\text{Input Energy} = \frac{\text{Output Energy}}{\text{Efficiency}}$$

In this case, the output energy is specified as 20 kWe. Table 19.2 below outlines the total input energy and Total Energy Storage required, depending on the temperatures of the hot and cold sources.

Based on literature reviews and solar irradiation data in Sri Lanka, it was estimated that parabolic trough collectors (PTCs) could achieve temperatures exceeding 500 °C. Consequently, a nominal storage temperature of 500 °C was established as the baseline. A straightforward calculation using the following equation

$$E = mC\Delta T$$

was carried out to model the mass required for both HiTech salt and Li-Na Eutectic Salt, as shown in Table 19.3.

Using the above mass calculations, thermal depletion was determined relative to the final temperature and the remaining energy. Plots were generated for both salts, covering cold source temperatures ranging from 35 to 45 °C and hot source temperatures of 200 and 220 °C. Figures 19.5 and 19.6 illustrate the trend for HiTech salt with a hot source temperature of 220 °C and a cold source temperature of 45 °C.

Assumptions in the Thermal Storage Capacity Calculations Since the specifications of the evaporator and the molten salt tanks are currently unknown, it has

Table 19.2 Energy calculation

Cold source temperature (°C)	Hot source temperature (°C)	Efficency	Total run time on storage (h)	Input energy (J)	Total energy storage required (KJ)
45	200	0.1966	24	101729	8789420
45	220	0.2129	24	93941	8116487
40	200	0.2029	24	98571	8516511
40	220	0.219	24	91324	7890411
35	200	0.2092	24	95602	8260038
35	220	0.2251	24	88849	7676588

Table 19.3 Table of mass calculations

Cold source temperature (°C)	Hot source temperature (°C)	Total energy storage required (kJ)	Mass of Hitech (kg)	Mass of Li-NA (kg)
45	200	8789420	18781	19532
45	220	8116487	18582	19325
40	200	8516511	18198	18926
40	220	7890411	18064	18787
35	200	8260038	17650	18356
35	220	7676588	17575	18278

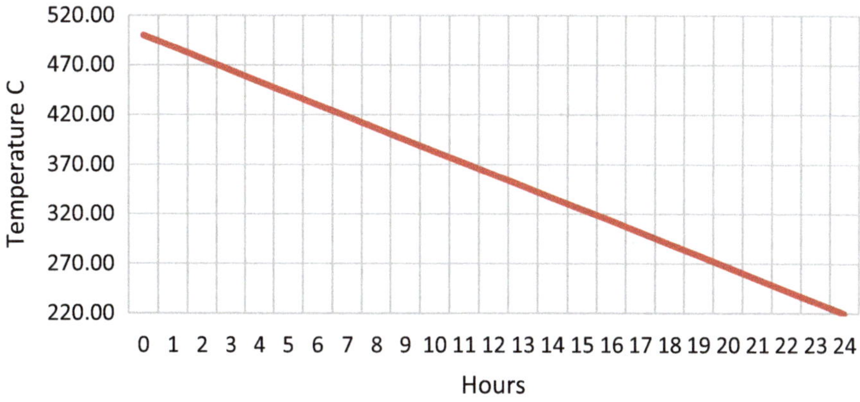

Fig. 19.5 Temperature depletion graph

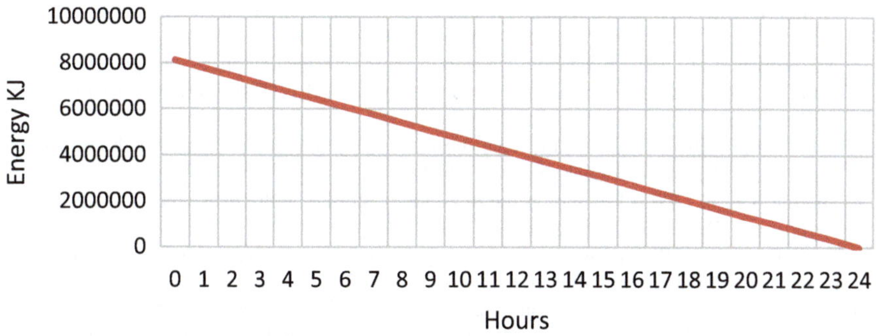

Fig. 19.6 Energy depletion graph

been assumed that heat transfer occurs with near-perfect efficiency and no losses. Furthermore, the power consumption of the molten salt pumps and the losses within the circulation loop have not been taken into consideration.

19.3 Conclusion

This study highlights the suitability of HiTEC and Na-Li nitrate salts as thermal energy storage options for a 20 kWe organic Rankine cycle-based solar thermal power plant. Both salts demonstrate favorable thermal properties, including low melting points and high specific heat capacities, making them well-suited for low- to medium-temperature applications. However, the choice between the two remains contingent upon the final determination of the plant's operating input temperature, which is yet to be confirmed as either 200 or 220 °C. This decision will significantly influence the thermal performance, stability, and cost-effectiveness of the chosen TES medium.

To ensure an informed selection, further modeling and analysis are imperative. This will involve incorporating critical loss parameters, including thermal losses at the evaporator, piping, and condenser, as well as inefficiencies in the generator and the molten salt circulation systems. These factors, currently unaccounted for in preliminary calculations, are essential for accurately evaluating the overall system performance. The incorporation of these parameters into detailed simulations will not only refine the mass and energy balance calculations but also provide insights into operational efficiency and long-term reliability under real-world conditions.

References

1. Gil A, Medrano M, Martorell I, Lázaro A, Dolado P, Zalba B, Cabeza LF (2010) State of the art on high-temperature thermal energy storage for power generation. Part 1—concepts, materials and modelization. Renew Sust Energ Rev 14(1):31–55
2. Agyenim F et al (2010) A review of materials, heat transfer and phase change problem formulation for latent heat thermal energy storage systems (LHTESS). Renew Sust Energ Rev 14(2):615–628
3. Solé A et al (2015) Review of the solar thermal energy storage options for high temperature applications. Renew Sust Energ Rev 47:386–398
4. Sharma A et al (2009) Review on thermal energy storage with phase change materials and applications. Renew Sust Energ Rev 13(2):318–345
5. Dutta R et al (2020) Thermochemical energy storage systems for concentrated solar power plants: state of the art. Int J Energy Res 44(9):7350–7377
6. Jenne S et al (2015) Techno-economic comparison of solar-only vs. hybrid CSP-biomass power plants with up to 20 hours of thermal energy storage. Sol Energy 122:92–104
7. Bradshaw RW (2010) Molten nitrate/nitrite salt mixtures for solar thermal energy systems. Sol Energy Mater Sol Cells 95(8):2437–2442
8. Brosseau DA, Hlava PF, Kelly MJ (2005) Testing of thermocline filler materials and molten-salt heat transfer fluids for thermal energy storage systems in parabolic trough power plants. Sandia National Laboratories, Report SAND2005-2492. Available at: https://www.osti.gov/servlets/purl/876232
9. Calderón M, Martínez A, Fernández AG (2019) Molten salts for sensible thermal energy storage: a review and global benchmark of molten salt mixtures. Renew Sust Energ Rev 103:24–37
10. Kuravi S et al (2013) Thermal energy storage technologies and systems for concentrating solar power plants. Prog Energy Combust Sci 39(4):285–319
11. Farid MM, Khudhair AM, Razack SAK, Al-Hallaj S (2004) A review on phase change energy storage: materials and applications. Energy Convers Manag 45(9–10):1597–1615
12. Mehling H, Cabeza LF (2008) Heat and cold storage with PCM. Springer, Berlin
13. Fernández AG, Galione A, Pérez FJ (2012) Molten salt corrosion of stainless stelels and low-Cr steel in CSP plants. Oxid Met 78(5–6):329–348
14. Pacheco JE, Showalter SK, Kolb WJ (2002) Development of a molten-salt thermocline thermal storage system for parabolic trough plants. J Solar Energy Eng 124(2):153–159
15. Laing D, Bahl C, Bauer T, Fiss M, Breidenbach N, Hempel M (2011) High-temperature solid-media thermal energy storage for solar thermal power plants. Proc IEEE 100(2):516–524
16. Flueckiger SM, Iverson BD, Garimella SV (2011) System-level simulation of a latent heat thermal energy storage system for concentrating solar power plants. Sol Energy 85(5):2468–2480

17. Xu C, Wang Z, He Y, Li X, Bai F (2012) Sensitivity analysis of the numerical study on thermal performance of a packed-bed molten salt thermocline thermal storage system. Appl Energy 92:65–75
18. Li G, Wang Z, Xiao B, Xu N (2014) A review of supercritical carbon dioxide technologies and applications in solar energy and natural gas combined cycle. Appl Energy 131:495–505
19. Prieto C, Tagle-Salazar PD, Patiño D, Schallenberg-Rodriguez J, Lyons P, Cabeza LF (2024) Use of molten salts tanks for seasonal thermal energy storage for high penetration of renewable energies in the grid. J Energy Storage 86(Part A):111203
20. Libby C (2010) Solar thermocline storage systems: preliminary design study. EPRI, Palo Alto, p 1019581

Chapter 20
Environmental, Economic and Regulatory Framework Assessment of Renewable Hydrogen Production

Hannah Hyunah Cho, Vladimir Strezov, and Tim Evans

Abstract Hydrogen is one of the promising energy carriers for global decarbonisation. Hydrogen can be produced by natural, renewable and fossil sources with or without integrating carbon capture technologies. This study provides an overview of hydrogen production, application, economic and environmental assessment, and international regulatory frameworks. The hydrogen applications include the use of hydrogen blends with natural gas in existing gas networks, electricity production, and green steelmaking. Higher production cost of renewable hydrogen compared to fossil hydrogen is expected to decrease along with lower cost of renewable electricity, while transport of renewable hydrogen using hydrogen carriers provides potential for a sustainable hydrogen supply chain. For the scale-up of the clean hydrogen market, policy support will be important.

Keywords Renewable hydrogen · Hydrogen production cost · Environmental impact

20.1 Introduction

The global energy supply has increased from 415.1 EJ in 2000 to 631 EJ in 2022, with 81.3% of the total energy supplied by fossil fuels in 2022 (Fig. 20.1), while 14.1% of the total is renewable energy [1]. While global production of wind and solar power had the most significant increase between 2010 and 2023, driven by lower production cost [2], hydrogen production has also increased to the global production of 97 Mt. in 2023 [2]. Hydrogen is a promising energy carrier that can alleviate intermittent issues in wind and solar electricity by converting the electricity into hydrogen [2]. Hydrogen can help to reduce 6.2 tonnes of CO_2 emissions by

H. H. Cho · V. Strezov (✉) · T. Evans
School of Natural Sciences, Faculty of Science and Engineering, Macquarie University, Sydney, Australia
e-mail: vladimir.strezov@mq.edu.au

L. Chen (ed.), *Clean Energy Research*, Green Energy and Technology,
https://doi.org/10.1007/978-3-031-99312-1_20

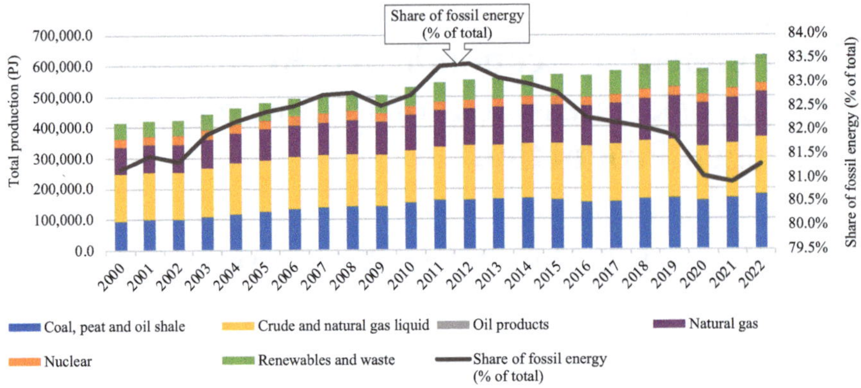

Fig. 20.1 World energy production by source between 2000–2022. (Produced using data from [1])

replacing 1 tonne of natural gas with hydrogen for combustion [3]. Hydrogen has about 2.4 times higher heating value (120 MJ/kg) than natural gas [3], while only water is emitted when the hydrogen is combusted for energy generation [4]. Higher efficiency and production scale of wind and solar photovoltaic (PV) power generation have promoted hydrogen production [4]. However, a lack of policy support and higher production costs are some of the obstacles to renewable hydrogen. This study discussed fossil and renewable pathways of hydrogen production with their applications, as well as the cost and environmental impacts of the production. Examples of international hydrogen frameworks are also presented.

20.2 Overview of the Hydrogen Industry

Current and Future Projection of Hydrogen Production

Hydrogen can be produced from natural, fossil and renewable sources, although the majority of current hydrogen production relies on fossil sources. In 2023 about 63% and 20% of the global hydrogen was produced from natural gas reforming and coal gasification, respectively, with less than 1% of low-carbon hydrogen production [2]. The natural gas-produced hydrogen is referred to as grey hydrogen (Table 20.1), while the natural gas-produced hydrogen with carbon capture and storage (CCS) is assigned a blue colour [4]. There are currently four plants in Canada, France, Japan and the USA that have adopted CCS to hydrogen production by natural gas steam reforming [5], with 163 global hydrogen projects planned with fossil-produced hydrogen with CCS [6]. Hydrogen production by renewable-powered electrolysis is termed as green hydrogen [4], with about 57% of the total 2440 global hydrogen projects planned with electrolysis [6].

Table 20.1 Colour spectrum of hydrogen

Colour	Feedstock	Process
White	N/A	Natural deposits
Orange	N/A	Stimulation of iron-rich geological Formations to obtain hydrogen-saturated water
Brown	Brown coal (lignite)	Steam reforming or gasification
Black	Black coal	Steam reforming or gasification
Grey	Natural gas	Steam reforming
Blue	Natural gas	Steam reforming followed by carbon capture and storage
Cyan	Water and renewable energy	Electrolysis of water with carbon capture, storage and utilisation
Turquoise	Natural gas	Pyrolysis of natural gas to hydrogen and solid carbon
Pink	Water and nuclear power	Electrolysis of water
Purple	Water, heat and nuclear power	Chemo-thermal electrolysis of water
Red	Nuclear power	High-temperature catalytic splitting of water
Green	Water and renewable energy	Electrolysis of water
Yellow	Solar power	Electrolysis of water

Source: [4]

Hydrogen Applications

Hydrogen applications include energy generation and storage, production of chemicals and use in the ironmaking process. The current use of hydrogen is in oil refining (33%) for hydrothermal cracking of long chain hydrocarbons, but also to reduce the sulphur content of the diesel fuels, followed by ammonia (27%), methanol (11%) and steel production (3%) [7]. Hydrogen can be directly combusted or used in fuel cells to produce electricity, while blending hydrogen with natural gas can be an immediate application option. The current threshold of hydrogen blending varies depending on countries, with 20% of hydrogen by volume being operated in Portugal and Romania (Table 20.2), while 30% blending was demonstrated in Germany [8]. 30–60% by volume of hydrogen can be blended with natural gas or ammonia to produce power in gas turbines, but thermal NOx emissions are a challenge in the combustion [9]. Less than 9 ppm of NOx emissions were found in the demonstration plant of the South Korean company, Hanwha, from the combustion of 100% hydrogen, indicating the potential of low emissions from the combustion. Electricity generation by hydrogen fuel cells presents a high efficiency of up to 60% [10], with 205 fuel cell power generators currently in operation in the US [11].

Hydrogen can also be used as a transportation fuel. A wide range of flammability, low ignition energy and high ignition temperature of hydrogen compared to petroleum fuels, as well as no direct greenhouse gas emissions when converted to energy, make hydrogen a promising transportation fuel. The advantages of using hydrogen were discussed by Sharma and Strezov [12], who conducted an

Table 20.2 Hydrogen blending limits with gas blending projects

Blending limit	
Limit	Countries
10%	Germany
6%	France
5%	Spain
4%	Austria
2%	Belgium, Canada, Czech Republic, Lithuania, Slovakia, Switzerland

Blending projects	
Hydrogen concentration (%)	Countries
5%	Australia, USA
8%	Indonesia
10%	Australia, Italy
20%	Portugal, Romania
30%	Germany

Adapted from [9]

environmental and economic assessment of hydrogen as a transportation fuel and operating hydrogen fuel cell vehicles. The study found the best environmental performance per unit of energy content in hydrogen among different types of fuels, while the second-best economic performance was found in the hydrogen-fuelled vehicle among the vehicles operated by other types of fuels. Despite the rapid increase by 55% of the hydrogen use in the transportation sector in 2023, hydrogen demand in road transport was less than 0.1% of the global hydrogen demand, with 60 kt in 2023 [2].

Natural gas or coal can be used in conventional iron reduction and replaced by hydrogen to produce carbon-neutral iron. 21.4% and 37.4% lower CO_2 emissions were estimated by using hydrogen in blast furnaces for ironmaking with basic oxygen furnaces for the steelmaking process [13], and in the gas-based direct reduced iron to electric arc furnace process [14], respectively.

Environmental Impacts of Hydrogen Production

Current natural gas-produced hydrogen possesses environmental burdens due to the use of the fossil fuel and consequent emissions. A study by Cho et al. [5] estimated the environmental impacts of the current natural gas-based hydrogen production based on 33 hydrogen production facilities in the USA using different US EPA pollutant inventories. The study found the average global warming impacts of 9.35 kg CO_2 eq/kg H_2 based on direct emissions from the facilities, while the impacts were increased to 11.22 kg CO_2 eq/kg H_2 when natural gas extraction with 1.87 kg CO_2 eq/kg H_2 of emissions from the gas extraction were included [15]. The global

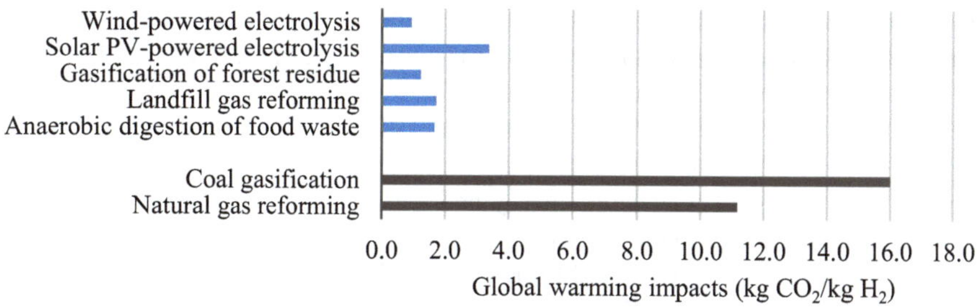

Fig. 20.2 Global warming impacts of fossil- and renewable-produced hydrogen. (Source: [7, 16])

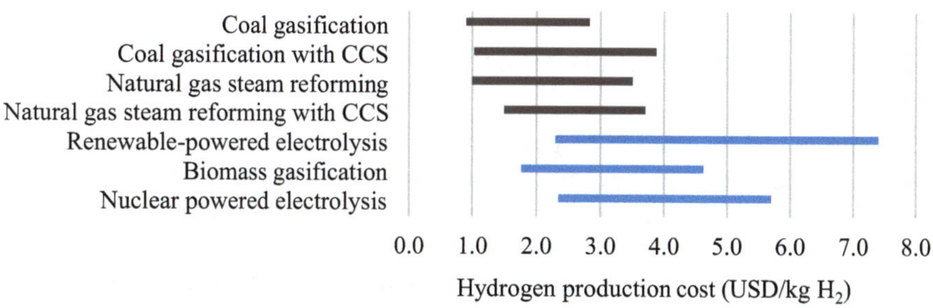

Fig. 20.3 Cost of hydrogen production. (Source: [7, 18, 20])

warming impact of hydrogen production by coal gasification is 16 kg CO_2 eq/kg H_2, which is about 43% higher than natural gas steam reforming (Fig. 20.2). Replacing natural gas with landfill gas significantly decreases the impact to 2.27 kg CO_2 eq/kg H_2 [16].

Hydrogen produced by electrolysis presents varying global warming impacts depending on the type of process energy, where wind-powered electrolysis has the lowest impact of 0.93 kg CO_2 eq/kg H_2, while different types of electrolysis (e.g., proton exchange membrane and alkaline) presented insignificant differences in the impacts [8]. The higher impact of solar-powered electrolysis than wind is due to greater energy consumption for manufacturing solar panels and differences in energy conversion efficiencies [8].

Cost of Hydrogen Production

Increasing the supply of renewable hydrogen is hindered by the higher production cost of renewable hydrogen compared to fossil-produced hydrogen. As shown in Fig. 20.3, coal gasification is currently the cheapest production option, with the cost ranging between \$0.9 and 2.8 USD kg H_2, which increases to \$1.0–3.9 USD kg H_2 when integrated with carbon capture and storage (CCS). The second lowest cost is found in natural gas reforming, with 40–68% of the cost attributed to the natural gas supply [17]. Renewable-powered electrolysis presents the highest ranges of production cost among the low-carbon hydrogen pathways, with wide cost variations because the cost is largely affected by the cost of electricity used for the electrolysis [8]. Cost reduction is expected by 2030 for the electrolysis [18], achieved by decreasing the cost of renewable energy and electrolysers with higher capacities. Current average cost of wind and solar electricity, $50 USD/MWh and $78 USD/MWh, respectively [19] also need to be decreased to the cost ranges of $14 USD/MWh by 2030 and $11 USD/MWh by 2050 [18] for the renewable-powered hydrogen to be cost-competitive.

Hydrogen Storage and Transport

Compressed and liquefied hydrogen facilitate long-term energy storage and transport of hydrogen to other locations, but energy-intensive compression and liquefaction processes are disadvantageous for the storage options [8]. The compression of hydrogen consumes about 1.2–3.8 kWh/kg H_2 electricity [8], while the low energy density of hydrogen necessitates more efficient types of hydrogen storage [21]. Ammonia can be used as a hydrogen storage where it can be transported to other locations, while hydrogen can be separated by ammonia cracking when necessary [16]. Transport of ammonia has already been practiced commercially, with about 35% of the announced hydrogen projects by 2030 selected ammonia as a hydrogen carrier [22]. Transport of hydrogen as ammonia can be an environmentally sustainable option, as estimated by Cho et al. [16], who found a global warming impact of 2.9 kg CO_2 eq/kg H_2 from 5000 km marine transport of hydrogen produced by wind-powered electrolysis. Another hydrogen carrier, liquid organic hydrogen carriers (LOHC), has advantages that include high volumetric storage density, no boil-off loss during storage and easy handling at atmospheric temperature and pressure. Recyclability of LOHC up to 1000 transport cycles also makes the LOHC economic [21]. Toluene and dibenzyltoluene are the most discussed LOHC, with toluene commercially applied by the Japanese company, Chiyoda, since 2022 at a 210 t/year scale [21].

National Strategies and Regulations on Renewable Hydrogen

Global hydrogen project developers spent USD 3.5 billion in 2023 for currently constructing hydrogen supply projects, with about 80% of the investment used for constructing electrolysis facilities, while the rest was spent on the hydrogen projects integrated with carbon capture, storage and utilisation [2]. In order to promote hydrogen production, policy support can be important, particularly in acknowledging the production of renewable hydrogen. The recent regulatory acts by the

European Commission set the definition of renewable hydrogen from non-biological origin and the methods that should be used for calculating the greenhouse gas emissions of renewable hydrogen [23]. The European Commission defined the targets to achieve full renewable hydrogen maturity by 2050 by establishing a three-phase gradual roadmap where the first phase would be completed by 2024 in which 6 GW of renewable hydrogen electrolysers will be installed across the EU states with the increased capacity of 40 GW during the second phase between 2025 to 2030. The target of full maturity of the renewable hydrogen during the third phase, between 2030 and 2050, aims to expand the use of the hydrogen to hard-to-decarbonise sectors, including use in buildings, aviation and shipping, and renewable gas production. The US also announced prioritised pathways of hydrogen production by publishing the U.S. National Clean Hydrogen Strategy and Roadmap in 2023. The production pathways include renewable- and nuclear-produced hydrogen, and integrated fossil- and waste-produced hydrogen with carbon capture and storage. The US strategy plans to increase production of clean hydrogen to 50 Mt./y by 2050.

Australia has updated the National Hydrogen Strategy in 2024 since its implementation in 2019. The strategy aims to increase hydrogen production to 15 Mt. and up to 30 Mt. per year by 2050 with financial support of $2 per kg of hydrogen through the Hydrogen Production Tax Incentive [3]. Future Made in Australia Innovation Fund will provide funds for the deployment of renewable hydrogen facilities [3], while the South Australian Government and Port of Rotterdam Authority in the Netherlands signed a Memorandum of Understanding (MoU) to support the development of large-scale export of renewable hydrogen from South Australia to Europe [24]. Germany also announced plans for hydrogen distribution infrastructure by 2028 through constructing 1800 km of repurposed and newly built hydrogen pipelines, which will be expanded to connect all major hydrogen producers, importers and consumers by 2030 [25].

20.3 Conclusions

Hydrogen can play a critical role in decarbonising the economy by 2050, due to its versatile applications including blending with natural gas, electricity generation, as well as use in the steel industry, and production of chemicals and transport fuels. While the prevalent method of hydrogen production involves natural gas, technological advancements enable its generation from diverse energy sources, including renewables and waste materials, rendering it a viable clean energy carrier. Production of renewable hydrogen is estimated to have 65–91% lower emissions than current hydrogen production by natural gas reforming, although the production cost of renewable hydrogen is higher than that of natural gas-produced hydrogen. The use of hydrogen not only offers flexibility to the electricity sector but can also be a storage medium of renewable energy. Hydrogen transport can be a viable option for satisfying the hydrogen demand, while achieving sustainability by using renewable hydrogen carriers, such as liquid organic hydrogen carriers and ammonia.

Hydrogen is identified in the recent roadmaps and strategies by countries worldwide, including Australia, the European Union, Germany and the USA, with significant committed investments to scale production of renewable and clean hydrogen. Given the rapid progress and hydrogen projects worldwide, hydrogen will play an important role in decarbonising the economy, where policy support and production infrastructure with increasing capacity will be important.

References

1. IEA (2024) World Energy Balances, IEA, Paris https://www.iea.org/data-and-statistics/data-product/world-energy-balances, Licence: Terms of Use for Non-CC Material. Last accessed 26 Feb 2025
2. IEA (2024) Global Hydrogen Review 2024. IEA, Paris https://www.iea.org/reports/world-energy-outlook-2024, Licence: CC BY 4.0 (report); CC BY NC SA 4.0 (Annex A)
3. DCCEEW (2024) National Hydrogen Strategy 2024, Australian Government Department of Climate Change, Energy, the Environment and Water, Canberra
4. Zainal BS, Ker PJ, Mohamed H, Ong HC, Fattah I, Rahman SA, Nghiem LD, Mahlia TI (2024) Recent advancement and assessment of green hydrogen production technologies. Renew Sust Energ Rev 189:113941
5. Cho HH, Strezov V, Evans TJ (2022) Environmental impact assessment of hydrogen production via steam methane reforming based on emissions data. Energy Rep 8:13585–13595
6. IEA (2024) Hydrogen Projects Database. IEA, Paris. https://www.iea.org/reports/hydrogen-projects-database. Last accessed 02 Dec 2024
7. Cho HH, Strezov V, Evans TJ (2023) A review on global warming potential, challenges and opportunities of renewable hydrogen production technologies. Sustain Mater Technol 35:e00567
8. IEA (2024) Hydrogen Infrastructure Database. IEA, Paris. https://www.iea.org/reports/hydrogen-projects-database. Last accessed 02 Dec 2024
9. Cecere D, Giacomazzi E, Di Nardo A, Calchetti G (2023) Gas turbine combustion technologies for hydrogen blends. Energies 16(19):6829
10. US DOE (2015) Fuel cells fact sheet, US Department of Energy, Washington, DC
11. EIA (2023) Hydrogen explained, Use of hydrogen, US Energy Information Administration. https://www.eia.gov/energyexplained/hydrogen/use-of-hydrogen.php. Last accessed 12 Mar 2024
12. Sharma A, Strezov V (2017) Life cycle environmental and economic impact assessment of alternative transport fuels and power-train technologies. Energy 133:1132–1141
13. Yilmaz C, Wendelstorf J, Turek T (2017) Modeling and simulation of hydrogen injection into a blast furnace to reduce carbon dioxide emissions. J Clean Prod 154:488–501
14. Fan Z, Friedmann SJ (2021) Low-carbon production of iron and steel: Technology options, economic assessment, and policy. Joule 5(4):829–862
15. NETL (2014) Life cycle analysis of natural gas extraction and power generation. National Energy Technology Laboratory, US Department of Energy, Washington, DC
16. Cho HH, Strezov V, Evans TJ (2024) Life cycle assessment of renewable hydrogen transport by ammonia. Int J Hydrog Energy 94:1018–1035
17. Wulf C, Kaltschmitt M (2013) Life cycle assessment of biohydrogen production as a transportation fuel in Germany. Bioresour Technol 150:466–475
18. IEA (2022) Global Hydrogen Review 2022. IEA, Paris
19. Evans A, Strezov V, Evans T (2022) Sustainability in power generation technologies. In: Reference module in earth systems and environmental sciences. Elsevier

20. Al-Qahtani A, Parkinson B, Hellgardt K, Shah N, Guillen-Gosalbez G (2021) Uncovering the true cost of hydrogen production routes using life cycle monetisation. Appl Energy 281:115958
21. Cho HH, Strezov V, Evans TJ (2024) Life cycle assessment of renewable hydrogen transport by liquid organic hydrogen carriers. J Clean Prod 469:143130
22. Hydrogen Council (2022) Hydrogen insights 2022: an updated perspective on hydrogen market development and actions required to unlock hydrogen at scale, Hydrogen Council
23. European Commission (2023) Renewable hydrogen production: new rules formally adopted. European Commission, Brussel. https://energy.ec.europa.eu/news/renewable-hydrogen-production-new-rules-formally-adopted-2023-06-20_en. Last accessed 13 Mar 2024
24. Department for Energy and Mining (2021) South Australia-Port of Rotterdam. Hydrogen supply chain: pre-feasibility study. Department for Energy and Mining
25. BMWK (2023) National Hydrogen Strategy Update 2023, Federal Ministry for Economic Affairs and Climate Action, Berlin. https://www.bmwk.de/Redaktion/EN/Hydrogen/Downloads/national-hydrogen-strategy-update.pdf?__blob=publicationFile&v=4. Last accessed 7 Sep 2023

Chapter 21
Change of Attitudes About Nuclear Energy in the European Union Countries Between 2009 and 2024

Guna Jakobsone-Snepste, Guillaume Deffuant, Ieva Vitolina, Janis Kapenieks, and Valdis Vitolins

Abstract With the willingness of the European Union to drastically reduce its dependence on Russian gas, the issues of nuclear energy production came to the fore in the European Union. Some European countries show a renewed interest in nuclear energy and argue for its development while others do not support such initiatives.

This issue is important in the context of the European climate goals. Each member state chooses its own scenario to achieve the goals. "The European Green Deal strategy targets accelerated sustainability transitions, yet the pathways and processes thereto differ markedly across EU member-states, particularly in the energy sector. Politics and institutional contexts largely determine the transition dynamics." (Zepa I. Energy Res Social Sci 93, 2022. https://doi.org/10.1016/j.erss.2022.102809).

This article uses the literature on nuclear energy policy and its impact on EU member states and public opinion. The main data sources are Eurostat, European Commission reports, International Energy Agency data, World Bank data, World Resources Institute data, and others to analyze and compare data from 2009 and 2024.

Keywords Nuclear energy · European Union · Energy policy · Public meaning

G. Jakobsone-Snepste (✉) · I. Vitolina · J. Kapenieks
Riga Technical University, Riga, Latvia
e-mail: guna.jakobsone-snepste@rtu.lv

G. Deffuant
INRAE, Paris, France

V. Vitolins
University of Latvia, Riga, Latvia

L. Chen (ed.), *Clean Energy Research*, Green Energy and Technology,
https://doi.org/10.1007/978-3-031-99312-1_21

21.1 Introduction

In the context of recent geopolitical shifts, particularly Europe's decision to reduce Russian energy imports, the support for nuclear energy increased. Surveys show growing positive perceptions of nuclear energy in the EU (from 46.02% to 60.51% overall), with opposition decreasing. Notably, Germany's rejection of nuclear power has dropped significantly.

Despite opposition in some EU countries, such as Portugal and Austria, nuclear energy is gaining support globally due to the energy crisis and climate security concerns.

This issue is important in the context of the European climate goals. The article examines the shift in positive attitudes towards nuclear energy in European Union countries by comparing data from 2009 and 2024. In general, most of the European Union (EU) countries currently demonstrate a positive attitude towards nuclear energy.

21.2 Public Attitudes Towards Nuclear Energy in Historical Cycles

The Chernobyl accident in April 1986 had a significant impact on public attitudes about nuclear power. This can be seen in the abrupt change in public opinion in Finland in 1986." [2] Not only in Finland, but also in other countries, the public support for nuclear power decreased. One of the reasons was that Chernobyl was in the Soviet Union and information about the disaster was hidden. Public opinion polls show that after negative events related to nuclear power, public attitudes tend to decrease, but over time, if the public is informed of the risks and benefits, especially if the benefits outweigh the potential risks, public positive attitudes tend to increase.

An interesting study is that it is not so important what information is presented to the population, but what is important is who delivers this information to the population and through which media. "The public gains most of its information on energy and nuclear power from the media but does not trust it. Scientists and environmental protection or consumer organizations are the most trusted groups. National governments are, in general, even less trusted on these issues than the media. This presents a clear problem to governments who wish to educate and influence their publics." [2] This finding is important because it indicates that the opinion expressed by scientists can be crucial in the development of nuclear power.

After the accident in Fukushima, public support for nuclear power plummeted. However, studies determined that the causes of death were not radiation (except for one case), but the evacuation plan (the residents died in the evacuation process, leaving Fukushima). It is important to remember that other types of disasters with human casualties have not occurred with nuclear reactors, as happened in Chernobyl and Fukushima. Considering this fact, it is interesting to look at society's attitude, its change from positive to negative and back to positive again.

The willingness of most European governments and of the European Commission to decrease Russian energy imports as much as possible has caused the opposite change of attitude—support for nuclear energy as a solution for ensuring the stability of the electricity supply.

The OECD Nuclear Energy Agency (NEA), Organization for Economic Co-operation and Development, published a report on "Public Attitudes to Nuclear Power" in 2010, which collected survey data on nuclear energy and public attitudes. It is interesting that "in those countries where nuclear already forms a part of the energy mix, the public tends to show distinctly more supportive attitudes." [2]. The report mentions that public awareness of mining can reduce public anxiety about the possible risks of nuclear energy. "In the absence of dramatic events, opinion changes slowly and, in several countries surveyed, it has become more supportive of nuclear energy. However, a large minority of respondents hold no firm views. The attitude of this middle ground will be critical for any future developments in the role of nuclear energy." [2] "A recent survey published by the Hungarian Századvég Foundation shows a growing positive perception of nuclear energy in all EU Member States. According to the results of the Project Europe Research conducted by Századvég, in the last year, the share of EU citizens opposing the use of nuclear energy has fallen from 26% (Autumn 2021) to 15% (Autumn 2022). Among the main factors, the Foundation underlines the current energy crisis, which has accelerated a trend already ongoing since 2015. The percentage of those in favor of nuclear power—who say the technology should produce very much or much energy—has increased from 26% to 40%. Moderate opinions (not too much or little) remain stable at 35%." [3] "Some of the member states with renowned negative perceptions of nuclear energy continue to maintain their relative position in country rankings; however, the degree of rejection in these countries has also significantly decreased over the last 12 months. The most instructive example is Germany, where the proportion of those rejecting nuclear technology has dropped from 65% to 20% in 6 years so today only a fifth of the population represents the previously majority position." [3] "This is the most significant moment in the relationship between society and energy in 100 years. In the decade, which is decisive in terms of collaborative energy actions worldwide, societies everywhere are reeling from multiple energy price crises and grappling with the interconnected challenges of energy and climate security." [4] "With Italian Foreign Minister Antonio Tajani in attendance at the 2024 Brussels summit, where Italy was one of 32 countries to sign a declaration stating," [we] reaffirm our strong commitment to nuclear energy," Italy's nuclear future remains up for debate once again.

There are also barriers to developing nuclear in several other EU member states, including Portugal, Denmark and Austria—another strong opposing voice, which in 2022 filed a legal challenge against the EU, claiming its categorization of nuclear energy as green investment was "greenwashing." [5].

21.3 Public Opinion on Nuclear Energy in the EU Before the Fukushima Disaster

The 2010 OECD survey (for the year 2009) on nuclear energy found that "Across Europe, people expect the share of nuclear energy to stay approximately the same in the future as it is today, with 14 EU states expecting it to be in the top three future energy sources in their country." [2] "To explore perceptions of energy dependency, citizens were asked to what extent their country and the EU as a whole is dependent on energy coming from abroad. Spanish citizens stand out as thinking their country is more energy independent than it truly is. Only 42% believe Spain is dependent on imported energy whereas the country actually imports 81% of its needs." [2] "Countries where citizens appear to be more aware than the average about new energy technologies are France, Germany, Finland, Luxembourg and Sweden." [2] "On question about nuclear security, it was found that 69% of people agree that nuclear power makes their country less dependent on energy imports and thus increases security of supply, 50% agree that it ensures lower and more stable energy prices and 46% agree that nuclear power helps to limit global warming. In all cases, respondents who agree with these three statements are in the largest group." [2] "The risks of nuclear power as an energy source were judged to outweigh its advantages by 53% of respondents overall, whilst only 33% judge that the advantages outweigh the risks it poses." [2] "However, in only six countries do respondents who consider that the advantages of nuclear energy are greater than the risks it poses outnumber those who are of the opposite view. These are Sweden, where 61% of respondents think this (despite their government's policy at the time to phase out nuclear energy), Bulgaria, the Czech Republic, Estonia, Finland and Slovakia. Estonia is unusual in this regard, as it does not have nuclear power." [2] "People who feel informed about nuclear safety tend to perceive less risk than those who feel uninformed. This is particularly the case in Sweden. The feeling of being informed appears to play a more significant role than does personal experience when people form opinions on the advantages and risks of nuclear energy." [2] "In addition, where nuclear plants operate, the majority of people trust the companies that run them—except in Germany." [2] "In France, 65% of citizens do not believe that radioactive waste can be managed safely even though the French Government has made firm decisions on the matter and 80% of French electricity comes from nuclear plants. Although the Eurobarometer poll did not directly address views on the cost of nuclear energy, it seems likely that people would be less opposed if it were clear that nuclear would reduce the price they paid for electricity" [2] "Support for building new nuclear plants is one third or more in the United Kingdom. In France, a country with a significant fleet of nuclear power plants, only 25% support building new nuclear plants. However, 50% respectively want to keep existing plants running. Support for closing all nuclear plants is highest in all countries without nuclear power." [2] "Europeans are less well disposed to nuclear energy than are their counterparts in the 18 countries sampled by Globescan." [2] "In the EU, Bulgaria had the highest level of public acceptance before the accident. Greece had the lowest public acceptance." [6]

21.4 Attitudes of the EU Countries Towards Nuclear Energy After the Fukushima Disaster

"A survey conducted shortly after the accident with more than 23,000 respondents from 41 countries has shown that geographical distance from the accident mattered: Contradicting a previous study, the decrease in support for nuclear energy was stronger in countries closer to Fukushima. In addition, support for nuclear energy decreased more in countries where new nuclear reactors were under construction." [7] "A decrease in support was observed not only in Japan, but also in countries such as Switzerland, Belgium and Italy. Yet in other countries, such as the UK, no drastic change occurred." [7] "The Fukushima nuclear accident reduced public support for nuclear energy in all countries studied in this paper, with the exception of Spain."[7] "The changes in the public acceptance of nuclear energy as an electricity source in 42 countries after the Fukushima nuclear accident. Of the 24,556 respondents, 52.7 percent favored the use of nuclear energy before the accident, and only 45.4 percent favored it after the accident." [6] "The general citizens and the doctors outside of Fukushima were anxious about health problems and food and soil pollution. We concluded that a high level of education about radiation decreased the anxiety about the radiation effects. It is important to spread knowledge about radiation." [8] Germany, Spain and Switzerland opposed nuclear power. "The Belgian twin-units Doel 1 and Doel 2 were initially planned to be phased out in 2015, reaching their maximum life time of operation of 40 years as per their initial design." [9] "However, for energy security, the Government decided in the same year to extend the life time of operation of the two units, which raised the need to resume their operation" [9] "Almost linear relationship existed between age and support for nuclear power, as the level of support for nuclear power increased with age" [10]

Support for nuclear energy in the EU in 2024, compared to data from 2009, is illustrated in the graph above, which shows the changes in positive attitudes across each country (see Fig. 21.1). Increase in the overall positive attitude of EU countries towards nuclear energy from 2024, compared to 2009, is illustrated in the graph below (see Fig. 21.2). The desire to move to 100% clean energy consumption could be more utopian than it is planned. Currently, without the use of nuclear energy, countries may not be able to achieve their set climate goals within the specified deadlines. "World simulation 100% renewable energy against global temperature scenarios—target year 2100." [11].

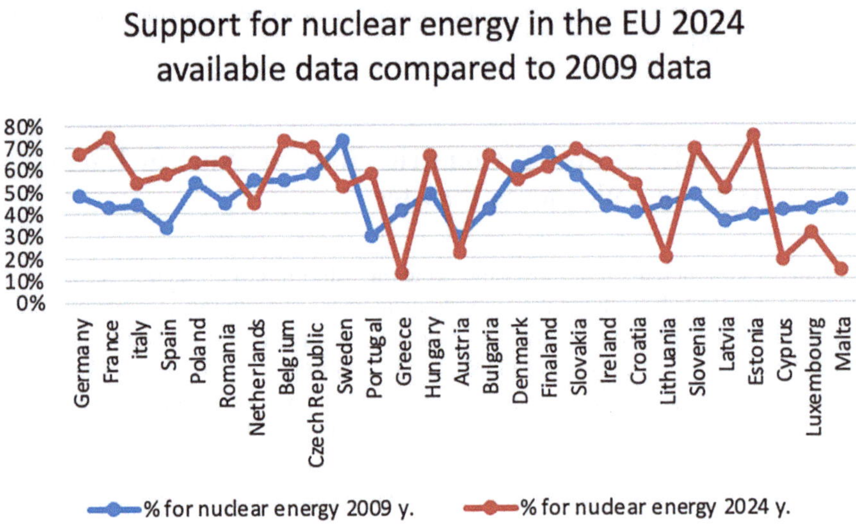

Fig. 21.1 The increase in the positive attitude of EU countries towards nuclear energy from 2024 compared to 2009

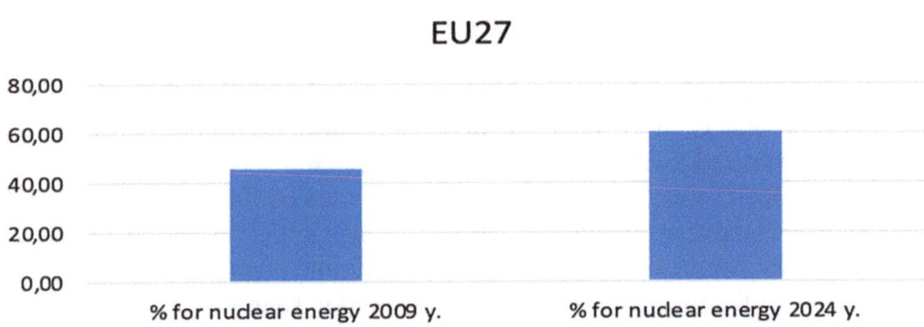

Fig. 21.2 Increase in the overall positive attitude of EU countries towards nuclear energy from 2024 compared to 2009

21.5 Policies and Regulations: Changes in Nuclear and Renewable Energy Policies from 2009 to 2024

The decision to promote nuclear energy is a national competence. The support should remain necessary and proportionate and not adversely affect trading conditions to an extent contrary to the common interest." [12] "Following the entry into force of the new electricity market design rules in July 2024, the Commission also assesses compliance with the CfD design principles set out in Regulation 2024/1747." [12] "EU taxonomy for sustainable activities entered into force on 12 July 2020. It establishes the basis for the EU taxonomy by setting out 4 overarching conditions that an economic activity must meet in order to qualify as environmentally sustainable." [13] It helps direct investments to the economic activities most needed for the transition, in line with the European Green Deal objectives. The taxonomy is a classification system that defines criteria for economic activities that are aligned with a net-zero trajectory by 2050 and the broader environmental goals other than climate." [13] Green Deal—"Firstly, the Green Deal has become a robust instrument to the extent that it is now part of the May 2022 REPowerEU plan. This means, on the one hand, that renewable energies (RE) are clearly presented as one of the solutions to be developed; on the other, that the issue of energy demand is once again gradually becoming a priority, with objectives for efficiency and the reduction of consumption" [14] Regulation (EU) 2020/852 states that "by providing a stable baseload of energy supply, nuclear energy facilitates the deployment of intermittent renewable sources and does not hamper their *development*." [15].

In 2024, the EU strengthened nuclear safety standards, especially for the life-cycle management of nuclear installations, from construction to decommissioning. In 2010, these standards were less strict [16]. More stringent regulations have been introduced for the safe and responsible management of radioactive waste and spent nuclear fuel. These rules of 2010 were less detailed [16]. In 2024, the EU actively promotes the use of small modular reactors (SMRs) and highly enriched uranium (HALEU) to improve the efficiency and safety of nuclear energy. [17] In 2010, these technologies were not yet so developed and widely discussed. In December 2023, the European Commission proposed and established the existing Euratom security framework, which was established in 2005. In 2010, this regulation was still valid without significant changes [17].

21.6 Conclusions

The research uses a multidisciplinary approach that combines energy, sociology and political science. This combination of research methods aims at a wider and deeper understanding of the synergy of political events with the change of public opinion, technological development in the context and energy as a common cornerstone of national security. It focuses on how public attitudes about nuclear energy evolve.

The main conclusions from the data analysis are that nuclear disasters affect the public's attitude towards nuclear energy negatively, but after some time the attitude tends to change to the positive side. Moreover, due to concerns about energy security, rising electricity prices and the transition to green energy, society is moving more positively towards nuclear energy. Countries that do not produce nuclear energy are more skeptical about nuclear energy, while countries that produce nuclear energy are more positive. More educated and informed people are more positive about nuclear energy. The public trusts the opinion of scientists the most, but often questions what politicians say. Political parties and the global political background influence countries' decisions regarding nuclear power generation. "When the climate change benefits of nuclear energy are explained, the support for nuclear energy amongst respondents increases significantly." [2].

"In summary, the Eurobarometer poll shows that respondents are rather more concerned about the "by-products" of nuclear energy than about the operation of the nuclear plants themselves." [2] "It is immediately clear that opposition to nuclear energy would reduce considerably if the matter of waste disposal were resolved." [2] Nuclear power has a positive impact on climate goals as it helps the EU achieve its decarbonization targets. The changes in public opinion triggered by the war in Ukraine and its consequences in the European and national public debates do not show any marked differences between the left and the right: overall, support for nuclear energy is increasing everywhere [18].

Acknowledgments This work has been supported by the European Union HORIZON programme, Project number: 101079206, Project acronym: TED4LAT, Call: HORIZON-WIDERA-2021-ACCESS-03.

References

1. Zepa I (2022, 2022) From energy islands to energy highlands? Political barriers to sustainability transitions in the Baltic region. Energy Res Soc Sci 93. https://doi.org/10.1016/j.erss.2022.102809
2. The OECD Nuclear Energy Agency (NEA) (2010) Organisation for economic co-operation and development, report on Public Attitudes to Nuclear Power, NEA Nr. 6859. OECD Publications, France., ISBN 978-92-64-99111-8
3. European Nuclear Society, Support for nuclear energy is increasing in all EU member states, 2022., www.euronuclear.org (2022)
4. Towards climate neutrality: economic impacts, opportunities and risks, Monograph. (2023.) University of Latvia Press, 272 pages, 9 pp., ISBN 978-9934-18-947-0. https://doi.org/10.22364/tcn.23
5. International news channel "Euro News", Frost R., Europe is divided on nuclear power: Which countries are for and against it? 01.04.2024, www.euronews.com (2024)
6. Younghwan K, Minki K, Wonjoon K (2013) Effect of the Fukushima nuclear disaster on global public acceptance of nuclear energy. Energy Policy 61:822–828. https://doi.org/10.1016/j.enpol.2013.06.107
7. Latré E, Perko T, Thijssen P (2017) Public opinion change after the Fukushima nuclear accident: the role of national context revisited. Energy Policy 104:124–133. https://doi.org/10.1016/j.enpol.2017.01.027
8. Okazaki R, Ootsuyamaa A, Kubo T, Abe T (2012) Questionnaire survey about public's image of radiation after the Fukushima Daiichi nuclear power plant accident, J-Stage. J UOEH 34(1):91–105. https://doi.org/10.7888/juoeh.34.91
9. Van Parys R, Haedens M, Schneidesch C, Zhang J, Meert A (2024) Enhanced reactor core design and safety evaluation for the final cycles of Belgian nuclear power plants. Nuc Eng Design 421. https://doi.org/10.1016/j.nucengdes.2024.113123
10. Yamagata H (2024) Public opinion on nuclear power plants in Japan, the United Kingdom, and the United States of America: a prescription for peculiar Japan. Energy Policy 185. https://doi.org/10.1016/j.enpol.2023.113939
11. Akpan J, Olanrewju O (2024) The transition to 100% renewable energy versus the global temperature scenarious: a perspective analysis. In: Green energy and technology: trends in clean energy research, selected papers from the 9th International Conference on Advances on Clean Energy Research (ICACER 2024). Springer, pp 153–154

12. European Commission (2024) Commission opens in-depth State aid investigation into Belgian support for lifetime extension of two nuclear reactors, Business, Economy, Euro., 22 July 2024, Europa.eu
13. European Commission. EU taxonomy for sustainable activities, finance.ec.europa.eu
14. McWilliams B, Tagliapietra S, Zachmann G, Deschuyteneer T (2023) Preparing for the next winter: Europe's gas outlook for 2023. Bruegel Institute
15. Official Journal of the European Union (2022) COMMISSION DELEGATED REGULATION (EU) 2022/1214, Document 32022R1214, EUR-Lex., 9 March 2022, eur-lex.europa.eu
16. European Commission (2022) In focus: EU nuclear energy policy—why it matters to us all, Energy, 13 March 2024, energy.ec.europa.eu
17. European Parliament (2024) Strategic autonomy and the future of nuclear energy in the EU, Briefing, February 2024, www.europarl.europa.eu
18. Brugidou M, Bouillet J (2023) A return to grace for nuclear power in European public opinion? Some elements of a rapid paradigm shift, Fondation Robert Schuman. Schuman Papers n°662

Chapter 22
Enhancing Wind Speed Prediction: A Comparative Analysis of Data Preprocessing Techniques

Eya Aloui, Mohsen Moomkesh, Imed Khabbouchi, and Uwe Ritschel

Abstract Wind energy holds immense promise for sustainable development, yet its integration into power grids remains challenging due to the complex and variable nature of wind. These inherent complexities make accurate wind speed prediction a critical area of research, aimed at enhancing the reliability and efficiency of wind energy systems. Among various techniques applied, data preprocessing has played a crucial role in improving the accuracy of wind speed prediction models. This paper focuses on evaluating data preprocessing algorithms, distinguishing itself by adopting a data-driven approach. By systematically analyzing numerical data from existing scientific literature, this study enables a comparative assessment of diverse data preprocessing methods used for wind speed prediction. The findings offer valuable insights, ultimately guiding practitioners in selecting or reconsidering specific preprocessing techniques based on their impact on the prediction accuracy. This research contributes to advancing predictive reliability, supporting the broader adoption of wind energy within sustainable power grids.

Keywords Wind speed forecasting · Data preprocessing · Comparative analysis

22.1 Introduction

The global environmental crisis, driven by excessive fossil fuel use, has led to significant ecological and climatic issues, pushing policymakers to encourage a shift towards renewable energy. As part of this shift, wind energy has gained attention as

E. Aloui (✉) · M. Moomkesh · U. Ritschel
University of Rostock, Faculty of Mechanical Engineering and Marine Technology, Chair of Wind Energy Technology, Rostock, Germany
e-mail: eya.aloui@uni-rostock.de

I. Khabbouchi
Energy and Thermal Systems Study, Laboratory (LESTE), National Engineering School of Monastir, University of Monastir, Monastir, Tunisia

L. Chen (ed.), *Clean Energy Research*, Green Energy and Technology,
https://doi.org/10.1007/978-3-031-99312-1_22

a promising solution for carbon neutrality, aligning with the Paris Agreement's mid-century targets. Wind power now accounts for 26.28% of the world's renewable capacity in 2023, showcasing its growth potential [1]. However, due to the complex dynamics of wind, influenced by fluid mechanics, geography, and microclimates, the efficient integration of this energy source has become challenging, complicating grid dispatching and load balancing across multiple time scales [2]. These issues highlight the need for accurate wind speed and power prediction models, prompting researchers to explore and refine numerous forecasting methods and algorithms.

Driven by this demand, numerous algorithms have been developed to accurately predict wind speed using diverse methodologies, motivating us, among others, to explore these approaches and evaluate their performance in enhancing the forecasting accuracy. In this context, [3] provided a thorough state-of-the-art review, evaluating various approaches and highlighting emerging trends such as spatial correlation and regional forecasting. Authors in [4] classified different methods in hybrid forecasting, including data preprocessing and parameter optimization. In [5], the authors explored artificial intelligence applications in wind farms, discussing various prediction algorithms for wind speed and power. While [6] examined data processing techniques, reviewing several methods such as feature selection and outlier detection for improving forecasting models. Lastly, [7] discussed the role of multi-objective optimization in enhancing the accuracy and stability of forecasting models.

In this paper, we aim to examine various aspects of recent advancements in wind speed and power prediction within the field of wind energy, specifically focusing on the application of data preprocessing tools. By providing a comprehensive comparative analysis, we will offer a guided reference that recommends the use or abandonment of certain algorithms based on their accuracy. Our work will rely on meticulously extracted numerical data from a selection of research papers, allowing us to identify and classify the diverse prediction tools and techniques available in the literature. The novelty of this work lies in the use of the collected numerical data to directly compare the performance of different algorithms, in contrast to other studies that rely solely on reported conclusions without utilizing the results in their raw form. The layout of this paper is as follows: the first section presents the state of the art in wind speed prediction and data preprocessing techniques. Following this, the results obtained through our analysis will be outlined. Finally, we conclude this paper with our findings and potential directions for future work.

22.2 State of the Art

General Overview of Wind Speed Prediction

In the literature, numerous studies have explored wind speed and power prediction, employing various algorithms and methodologies to address the complexity of forecasting. For this study, we conducted a comprehensive review of research from

Elsevier, Springer, IEEE Xplore, and MDPI, using keywords such as wind speed prediction and wind power prediction across fields such as engineering, computer science, environmental science, and sustainability. Given space limitations, we narrowed our focus to a subset of papers that apply data preprocessing algorithms, providing a targeted view of these methods' impact on predictive accuracy. Before presenting these findings, we first outline the fundamental principles of wind speed and power forecasting.

Forecasting wind power and wind speed is commonly categorized as either deterministic or probabilistic. Deterministic models predict a single future value accompanied by an accuracy metric, whereas probabilistic models incorporate uncertainty through interval forecasts or probability distributions, offering a broader range of potential outcomes [8].

Beyond this primary distinction, wind prediction methods are also classified based on spatial and temporal correlations. Spatial models, such as those focusing on geographic variation, address how location-specific features influence wind behavior [9]. Temporal models leverage time-series data, which are commonly segmented by time horizon ranging from ultra-short, short, medium, to long-term. These algorithms can also be classified according to their forecasting approach, including physical, statistical, intelligent, and hybrid models.

While categorization by temporal correlation is more common, there is no accurate description of the classification of wind power and speed prediction algorithms based on a predefined time scale [6]. Different categorizations reflect researchers' unique perspectives and goals. Short-term forecasts are more accurate and useful for immediate grid adjustments, whereas longer-term forecasts support load management and maintenance planning.

Prediction approaches differ substantially in methodology. Physical models use fundamental physical principles based on fluid dynamics to predict future wind speed values with the help of static information including the site and wind farm conditions, and dynamic information consisting of numerical weather prediction [10]. The main limitations for this method are the heavy computational requirements and the need for frequent and regular updates [11], which makes it inefficient for short-term predictions.

The statistical approach, conversely, uses the historical data to build a model able to mine the time-varying relationships within the time series and predict the value of wind speed one step ahead in the future [12]. However due to the nonlinear characteristics of wind power, the use of statistical methods alone hasn't provided satisfactory results that reflects accurately wind speed forecasts [13].

Intelligent methods, utilizing Machine Learning (ML), Deep Learning (DL), and Ensemble Learning (EL) techniques, have emerged to address these limitations, offering tools well-suited to the chaotic nature of wind power data. These tools were first introduced to compensate for the lack of statistical and physical approaches in terms of forecasting performance but were quickly re-exploited in hybrid methods.

Hybrid methods, as the name suggests, combine different approaches with the objective of enhancing the accuracy of the overall model [14]. In this regard, many algorithms have emerged exploiting the advantages of multiple tools like preprocessing, post-processing, and parameter optimization, so that the model can better capture the complex behavior of wind speed and consequently provide performant results.

Data Preprocessing

Data preprocessing involves transforming data before it is processed by forecasting algorithms or other data science tools. As illustrated in Fig. 22.1, three main methods have been enumerated, including data decomposition, dimensionality reduction and noise reduction.

Data decomposition is a popular preprocessing method in the field of wind energy prediction that breaks down data sequences into multiple components in order to reduce the influence of the non-stationarity of wind speed on the predicted result [15]. There exist multiple approaches in the literature including empirical mode decomposition (EMD)-based methods, wavelet transform (WT)-based methods, etc. A brief summary of these algorithms is summarized in Table 22.1.

Noise reduction, or denoising, is a preprocessing technique that improves data quality by isolating and removing unwanted signals [16]. Common denoising methods in wind speed and energy forecasting include Denoising Autoencoder (DAE), Stacked Denoising Autoencoder (SDAE), Maximal Overlap Discrete Wavelet Transform (MODWT), Weighted Standard Deviation (WSTD), Low-Pass Filter (LPF), and Wavelet Threshold Denoising (WHTD).

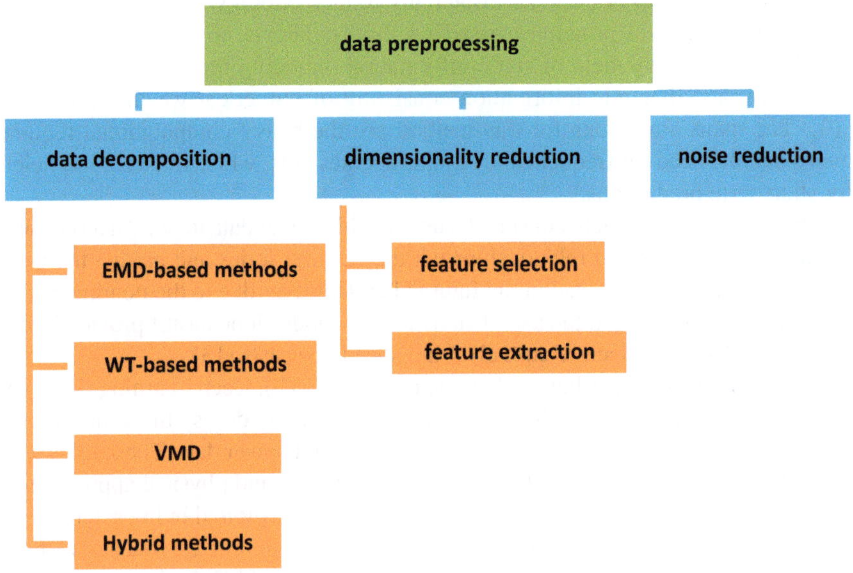

Fig. 22.1 Graphical summary to data preprocessing

Table 22.1 Data decomposition methods

Method	Description	Key innovations/advantages
EMD-based methods	Decomposes data by extracting Intrinsic Mode Functions (IMFs) through a process known as sifting.	To tackle the issue of modal aliasing associated with EMD, Ensemble Empirical Mode Decomposition (EEMD) and Complete Ensemble Empirical Mode Decomposition (CEEMD) were introduced by incorporating white noise, which have been shown to enhance the performance of various prediction algorithms.
WT-based methods	This method involves capturing different frequency components of the processed signal at various scales [17].	Continuous Wavelet Transform (CWT) and Discrete Wavelet Transform (DWT) allow for localized analysis for both time and frequency domains simultaneously. Empirical mode decomposition (EWT)'s adaptability and lack of preset parameters make it suitable for wind speed forecasting.
Variational Mode Decomposition (VMD)	VMD addresses EMD's limitations by utilizing techniques like the Hilbert transform, enabling robust subsequence generation for prediction models.	Provide several improvements to the forecasting accuracy
Hybrid decomposition methods	Hybrid methods combine multiple decomposition techniques or integrate additional processes like denoising and optimization for enhanced performance.	Outperform traditional methods in accuracy and computational efficiency.

Dimensionality reduction enhances forecasting by selecting relevant features (feature selection) or transforming data for a meaningful representation (feature extraction). This process removes irrelevant and non-informative features, reducing overfitting and improving model accuracy, efficiency, and robustness. Various methods, including correlation, clustering, Principal Component Analysis (PCA), Robust ReliefF (RReliefF), Recursive Feature Elimination (RFE), Mutual Information (MI) and AI-based approaches, are widely used in wind speed and power forecasting.

22.3 Results and Discussions

In this paper, we analyzed multiple scientific publications to gather accuracy metrics for various prediction algorithms, each incorporating different preprocessing techniques, to compare their effectiveness in handling wind speed data. To enhance transparency and reproducibility, a link to the compiled dataset is provided here: https://figshare.com/s/a5402d79daa91d0efa5c.

The accuracy metrics considered in this analysis include MAE, MAPE, and RMSE as detailed in Eqs. 22.1, 22.2 and 22.3.

$$\text{MAE} = \frac{1}{\text{n}}\sum_{i=1}^{n}\left|\left(\hat{y}_i - y_i\right)\right| \tag{22.1}$$

$$\text{MAPE} = \frac{1}{n}\sum_{i=1}^{n}\left|\frac{\hat{y}_i - y_i}{y_i}\right| x100\% \tag{22.2}$$

$$\text{RMSE} = \sqrt{\frac{1}{n}\sum_{i=1}^{n}\left(\hat{y}_i - y_i\right)^2} \tag{22.3}$$

where y_i and $\hat{y}_i$ represent, respectively, the true and predicted values of the wind speed. To quantify and compare performance improvements of each data preprocessing tool, we used improvement percentage metrics, as outlined in Eq. 22.4.

$$\text{IP}_1 = \frac{\text{Metric}_B - \text{Metric}_A}{\text{Metric}_B} \times 100\% \tag{22.4}$$

where Metric_A and Metric_B represent the accuracy metrics for the forecasting algorithm and the benchmark model, respectively.

The resulting data are visualized using box-and-whisker plots, chosen for their ability to highlight both the central tendency via the median and the variability through the box and whiskers across different methods and conditions. This approach provides a clear, comparative view of each method's performance and adaptability to changing data conditions, enabling us to observe accuracy improvements for each algorithm across varying scenarios and thereby assess them more effectively. Moreover, this visual approach allows us to detect subtler trends in algorithmic behavior, particularly in terms of consistency and sensitivity to shifts in data characteristics, such as noise levels or variability in wind speed. Mapping results across preprocessing methods provides insights into the most resilient and scalable techniques under varied conditions. This serves as a benchmark for selecting algorithms suited to dynamic wind energy forecasting.

The results shown in Figs. 22.2, 22.3, and 22.4 highlight the comparative performance across three types of methods: data decomposition, denoising, and dimensionality reduction.

Starting with data decomposition methods, hybrid approaches demonstrate the best overall accuracy, consistently outperforming single-method algorithms across all studied metrics. Specifically, VMD and WPT stand out among single methods, offering significant accuracy improvements across MAE, MAPE, and RMSE. EMD, however, exhibits a detrimental impact on accuracy for MAE and RMSE, likely due to model aliasing, which can impair performance.

For denoising methods, the effect on model accuracy is less pronounced than that observed with data decomposition, with most methods, including DAE, MODWT, SDAE, LPF, and WHTD, achieving less than a 20% improvement in accuracy.

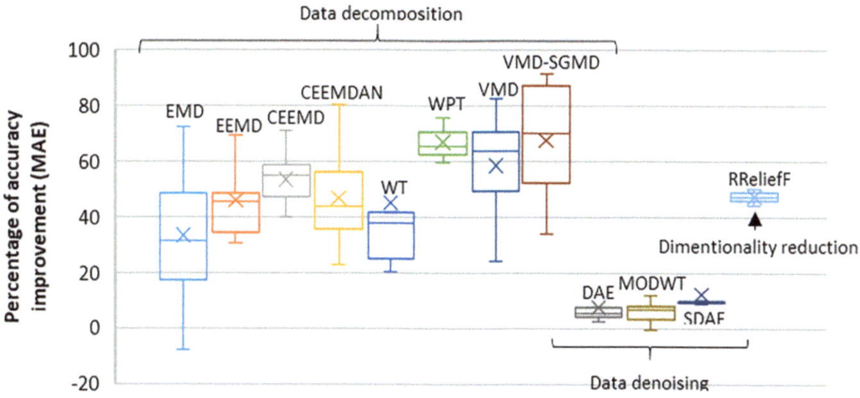

Fig. 22.2 Percentage of accuracy improvement for data preprocessing methods (MAE)

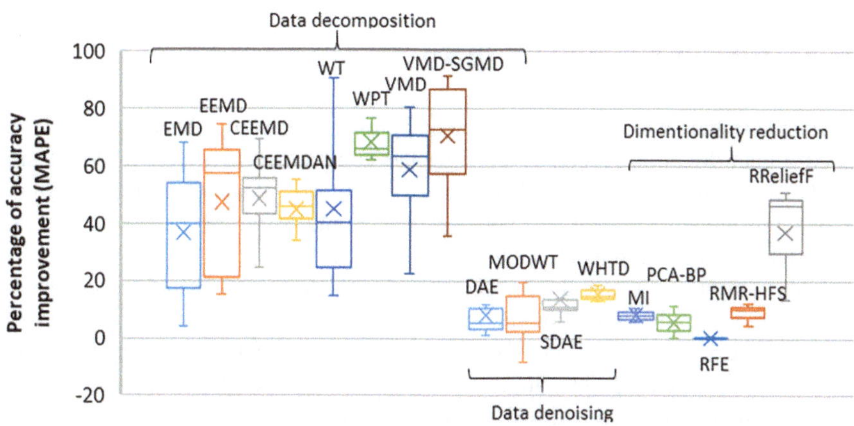

Fig. 22.3 Percentage of accuracy improvement for data preprocessing methods (MAPE)

Finally, in dimensionality reduction, most methods yield accuracy improvements below 13% for MAPE and RMSE. An exception is the RReliefF algorithm, which significantly exceeds other approaches, achieving an average improvement exceeding 45% for both MAPE and RMSE.

These results highlight the superiority of data decomposition algorithms compared to other data preprocessing techniques. This can be attributed to the fact that data decomposition breaks the signal into different modes. This separation enables prediction algorithms to focus on the specific characteristics of each mode, leading to more accurate forecasts. In contrast, dimensionality reduction and data denoising methods showed similar improvements in accuracy, suggesting that while these techniques help refine the data, their impact on model performance is more modest compared to data decomposition approaches.

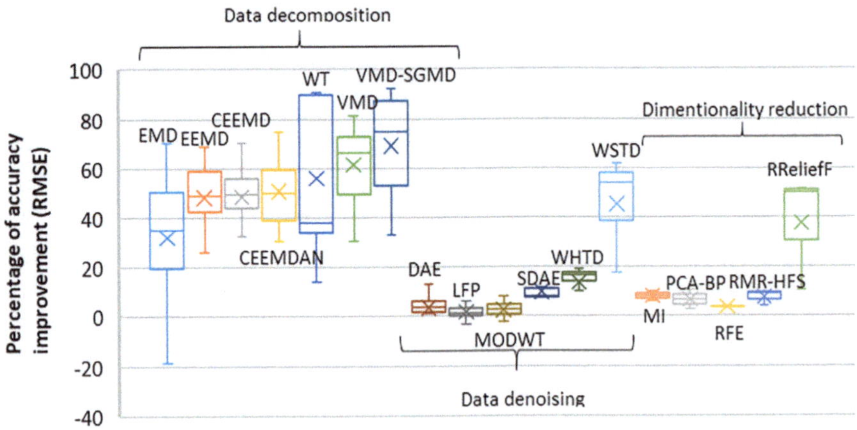

Fig. 22.4 Percentage of accuracy improvement for data preprocessing methods (RMSE)

22.4 Conclusions and Perspectives

With the growing interest in wind speed and power forecasting, this paper has developed a data-driven approach to analyze and compare various data preprocessing methods commonly used in the field. The goal was to evaluate their effectiveness in enhancing model accuracy and performance. Some of the results found are presented and discussed in the present investigation.

This study underscores the impact of various data preprocessing techniques on the accuracy of wind speed prediction, with data decomposition methods, in particular hybrid decomposition, showing the most substantial performance gains across all evaluated metrics. Other techniques also showed some improvements in accuracy, highlighting the critical role of advanced preprocessing tools and hybrid prediction algorithms in boosting predictive performance.

In future work, we plan to expand our dataset to validate these findings across a broader data spectrum. This will involve examining papers that explore the application of optimization techniques and data post-processing approaches, broadening the scope of our methodology. Furthermore, we plan to investigate probabilistic forecasting methods, which can provide a more comprehensive understanding of prediction uncertainties, as well as spatial correlation-based techniques. By expanding our focus to include these diverse approaches, we aim to develop a deeper understanding of the various facets of wind speed and wind power forecasting algorithms, comparing their strengths and weaknesses to enhance the reliability and efficiency of future forecasting models.

References

1. IRENA: Renewable Capacity Statistics 2024. With collaboration from J Prime, IA Ahmed, D Akande, N Elhassan, G Escamilla, Y Melnikov, A Whiteman (2024)
2. Annhesham A, Ahmed AM, Mahmoud AA (2017) Effect of stochastic nature of wind energy on power system. JEE 11(1):1. https://doi.org/10.26634/jee.11.1.13671
3. Yang B, Zhong L, Wang J, Shu H, Zhang X, Yu T, Sun L (2021) State-of-the-art one-stop handbook on wind forecasting technologies: an overview of classifications, methodologies, and analysis. J Clean Prod 283:124628. https://doi.org/10.1016/j.jclepro.2020.124628
4. Ahmadi M, Khashei M (2021) Current status of hybrid structures in wind forecasting. Eng Appl Artif Intell 99:104133. https://doi.org/10.1016/j.engappai.2020.104133
5. Wang Y, Yu Y, Cao S, Zhang X, Gao S (2020) A review of applications of artificial intelligent algorithms in wind farms. Artif Intell Rev 53(5):3447–3500. https://doi.org/10.1007/s10462-019-09768-7
6. Liu H, Chen C (2019) Data processing strategies in wind energy forecasting models and applications: a comprehensive review. Appl Energy 249:392–408. https://doi.org/10.1016/j.apenergy.2019.04.188
7. Liu H, Li Y, Duan Z, Chen C (2020) A review on multi-objective optimization framework in wind energy forecasting techniques and applications. Energy Convers Manag 224:113324. https://doi.org/10.1016/j.enconman.2020.113324
8. Xie Y, Li C, Li M, Liu F, Taukenova M (2023) An overview of deterministic and probabilistic forecasting methods of wind energy. iScience 26(1):105804. https://doi.org/10.1016/j.isci.2022.105804
9. Chen Y, Bai M, Zhang Y, Liu J, Yu D (2023) Multivariable space-time correction for wind speed in numerical weather prediction (NWP) based on ConvLSTM and the prediction of probability interval. Earth Sci Inform 16(3):1953–1974. https://doi.org/10.1007/s12145-023-01036-1
10. Giebel G, Brownsword R, Kariniotakis G (eds) (2004) State of the art in short-term prediction of wind power—from an offshore perspective. 2004 Sea TechWeek, Brest, France
11. Alhussan AA, El-kenawy E-SM, Abdelhamid AA, Ibrahim A, Eid MM, Khafaga DS (2023) Wind speed forecasting using optimized bidirectional LSTM based on dipper throated and genetic optimization algorithms. Front Energy Res 11. https://doi.org/10.3389/fenrg.2023.1172176
12. Khabbouchi I, Salem IB, Guellouz MS, Ritschel U (2023) Machine learning and deep learning for wind power forecasting. In: 2023 IEEE International Conference on Artificial Intelligence & Green Energy (ICAIGE). IEEE, Sousse, pp 1–6
13. Deng Y, Chong K, Wang B, Zhou Q, Lu Z (2022) Coupling framework for a wind speed forecasting model applied to wind energy. Sci China Technol Sci 65(10):2462–2473. https://doi.org/10.1007/s11431-022-2111-4
14. Bommidi BS, Kosana V, Teeparthi K, Madasthu S (2022) A hybrid approach to ultra short-term wind speed prediction using CEEMDAN and Informer. In: 22nd National Power Systems Conference (NPSC). IEEE, New Delhi, pp 207–212
15. Liu T, Qi S, Qiao X, Liu S (2024) A hybrid short-term wind power point-interval prediction model based on combination of improved preprocessing methods and entropy weighted GRU quantile regression network. Energy 288:129904. https://doi.org/10.1016/j.energy.2023.129904
16. Zhang F, Xi J, Sun X, Guo Z, Ren X (2020) Real-time wind power prediction based on WD-GA-SVM combined model. IOP Conf Ser: Earth Environ Sci 558:52019. https://doi.org/10.1088/1755-1315/558/5/052019
17. Borisagar KR, Thanki RM, Sedani BS (eds) (2019) Speech enhancement techniques for digital hearing aids. Springer, Cham

Chapter 23
Environmentally Friendly Rankine Cycle for Power Generation

Shaher Nasser Alotaibi, Sultan Binhathlain, Faleh Alazmi, Rayan Alshalawi, Mohammad Alnazha, and Muhammad Asad

Abstract This paper evaluates the design, analysis, and implementation of an environment-friendly Rankine cycle system powered by Concentrated Solar Power (CSP). This work aims to shift from conventional fossil fuel-based energy sources to renewable and clean solar energy, minimizing carbon footprint and promoting sustainable power. The design contains thermal storage, an advanced heat transfer mechanism, and better working fluids for improved output efficiency. A small-scale prototype was also modelled and tested, showing the feasibility of CSP integration in Rankine cycles for remote and off-grid applications. Theoretical calculations showed a thermal efficiency of 20.85% along with a net power output of 11.6 kW. Stress, thermal, and vibrational analyses were also conducted to evaluate the performance, structural integrity, and operational stability. The obtained results indicated that the designed system could work perfectly under operational loads, controlling the best temperature ranges, and work perfectly under varying conditions. The work shows the economic and environmental benefits of these CSP-integrated systems, including reduced operational costs and carbon footprints. Moreover, by properly addressing issues like scalability and intermittency, this work aims to contribute to the global transition toward clean, sustainable, and renewable energy systems. The main goals of this research are to make a CSP-integrated Rankine cycle system that incorporates thermal energy, reduces carbon emissions, and validates the design using thermal, stress, and vibrational analyses. The study aims to demonstrate the environmental and economic benefits of this system for off-grid and remote applications.

Keywords Concentrate solar power · Rankine cycle · Sustainable power generation · Thermal efficiency · Renewable energy

S. N. Alotaibi (✉) · S. Binhathlain · F. Alazmi · R. Alshalawi · M. Alnazha · M. Asad
Prince Mohammad Bin Fahd University, Dhahran, Saudi Arabia
e-mail: 202101331@pmu.edu.sa

L. Chen (ed.), *Clean Energy Research*, Green Energy and Technology,
https://doi.org/10.1007/978-3-031-99312-1_23

23.1 Introduction

The global energy sector has seen a major shift as the world is minimizing its reliance on conventional fossil fuels to tackle the increasing concerns of greenhouse gas emissions and climate change. Conventional power systems, including the Rankine cycle, mainly rely on natural gas, coal, or oil as the main input fuel for power generation. This adds to emissions and environmental degradation. To better handle this, renewable energy sources like solar energy have emerged as the main alternative for these fuels to provide sustainable, clean, and never-ending energy for power purposes. In solar panels, Concentrated Solar Power (CSP) has gained much attention in the new sector because of its capability to provide more energy in the form of heat input. [1]. Having the capability of integrating and concentrating more sunlight onto a receiver, such systems can handle steam at high temperatures of 400 ° C, making them ideal for Rankine cycles. However, the intermittent nature of solar energy input poses a major concern for using it as an alternative main energy source. To minimize this issue's impact, thermal energy storage (TES) systems have been linked with such systems to store excess heat during high irradiance hours of sunlight, utilize it under non-sunlight or cloudy conditions, and give an uninterrupted output supply. [2]. Despite all such work and advancements, integrating CSP with Rankine cycles still poses a challenge, especially in system scalability, thermal efficiency, and operational cost. Such issues need to be resolved to further widespread their adoption. [3]. It has been shown [4] that CSP included Rankine cycles, having the use of organic fluid, hybrid systems, or regenerative heating with some backup. However, there is still a need for more research and iteration to optimize the system performance, especially in off-grid or small-scale applications. This work focuses on the design, analysis, and working of the eco-friendly Rankine cycle coupled with CSP.

The inclusion of CSP in Rankine Cycle has gained much attention in the last few years because of its high potential for use as an alternative energy source compared to conventional fossil fuels for power generation. The Rankine cycle, run by natural gas, coal, or similar fuels, has been utilized in the past for power supply. However, environmental emissions and climate change have made researchers look for alternative environments and sustainable power generation technology. CSP, concentrating sunlight using lenses or mirrors, is a promising alternative because of its high potential for generating steam to utilize in power generation. Kalogirou [5] provided a detailed review of thermal energy applications, showing CSP's potential to replace traditional fossil fuel-based power generation systems. The study also explained the environmental pros of using CSP, like reduced emissions, but also explained challenges like intermittent energy supply. Similarly, Obernberger and his colleagues [6] compared a solar-powered Rankine cycle with a biomass-fired power system, showing the importance of thermal energy storage in them to maintain power supply in low irradiance hours. Recent works have focused on resolving these intermittency and thermal efficiency problems. Santosh et al. [7] modeled a low-temperature solar-based Rankine cycle containing Parabolic troughs, showing the importance of optimizing condenser and heat exchanger designs to enhance output performance. Further, Nitish et al. [8] analyzed the impact of combining regenerative heating with CSP, showing an increase in thermal efficiency by preheating input water using solar energy. Wang [9] Also conducted a comprehensive simulation-based analysis of working fluids for CSP-attached Rankine cycles, concluding that organic fluids give more efficient results compared to water as a working fluid, especially in low-temperature applications.

In these cycles, selecting the optimal working fluid is of high value. Rosen and Dincer [10] explained the benefits of using renewable energy sources like CSP in Rankine cycles. They found that renewable integration can reduce emissions by 50% compared to conventional fossil fuels. Mohammadi et al. [11] analyzed the hybrid systems containing backup supply systems like biomass or fossil fuels to minimize the intermittency concern. Their results suggested that these systems reduce the variability issue of solar systems, increasing reliability. Moreover, material selection also plays an important role in these system designs. Galindo et al. [12] analyzed solar selective coatings for minimal emissions and high solar absorption, which are needed for the high thermal efficiency of such systems. Their work showed the high potential of advanced materials in improving the performance of such systems. Emily et al. [13] also conducted a detailed study on solar Rankine cycle performance in an arid environment, finding that such systems can give reliable output even under harsh environments.

Thermal energy systems play a major role in resolving the intermittency issues in such systems. Several studies have also been based on using molten salt or phase-changing materials in CSP systems. Salem et al. [14] have conducted a comprehensive review of TES involvement in improving working hours and efficiency of solar organic Rankine cycles. In addition, economic viability and scalability are also major concerns. Various modular designs have been proposed for remote and off-grid use, where traditional energy infrastructure is not available or impractical. Such systems are a viable option, allowing incremental energy increase as per demand. Furthermore, the addition of Programmable Logic Controllers has also optimized the performance by adjusting loads as per changes in environmental conditions.

Despite such advancements, several challenges remain in optimizing CSP-integrated Rankine cycles for high use. Issues like intermittency, high capital, and material degradation are the main areas that need research and consideration [15, 16]. However, growing work in this field shows the potential of the CSP-attached Rankine cycle as an alternative, sustainable, efficient, and eco-friendly source for power generation. Based on all the above literature, gaps, and issues, the following are the main aims of our research:

- Developing a CSP integrated Rankine cycle to minimize carbon emissions, replacing fossil fuels
- Incorporating a thermal energy storage system to address the intermittency issue and confirm continuous energy generation
- Optimizing thermal efficiency through design sizing, calculations, and proper material selection

- Designing a modular and scalable system for remote and off-grid applications
- Validated system feasibility through thermal, stress, and vibrational analyses, confirming operational stability and structural integrity under different conditions

23.2 System Design and Methodology

Design Constraints and Objectives

This System is designed with the following objectives and constraints in consideration:

Sustainability Using eco-friendly materials along with efficient heat transfer systems, this system utilizes renewable (solar) energy to minimize the environmental impact. Later, the material utilized for each part, along with solar panel addition as the main power source, will result in minimal emissions, resulting in efficient and clean energy generation.

Geometrical Constraints This system is designed to be portable and compact and utilized in remote or off-grid applications. For this, the solar collector was limited to an area of 1–2 m^2 to confirm scalability.

Economic Viability The system aims to minimize cost by utilizing affordable parts and manufacturing operations. The total cost of this system was capped at SAR 4900. This not only makes the design modular but also economical, as it can be utilized in remote locations as well as small-grid applications.

Safety and Reliability The system contains safety systems to prevent operational risks like overheating and leakage. All system parts are designed to comply with ASME Pressure Vessel and Boiler standards.

System Components and Specifications

The system consists of various key components, contributing to optimizing system energy conversions and output efficiency. Table 23.1 *contains the specifications of all such components.*

Table 23.1 System component specifications

Components	Specifications	Purpose
Concentrated solar power (CSP) system	Main area: 1–2 m^2; output temperature: 400–500 °C	Captures solar energy to heat the working fluid (water)
Steam turbine	Operating pressure: 2–10 bar; output power: 500 W, radial or axial flow	Transforms high-pressure steam into mechanical output energy
Solar receiver (boiler)	Max pressure: 5–10 bar; capacity: 1–2 l; material: stainless steel	Converts solar energy into thermal output, helping to generate high-pressure steam
Condenser	Material: stainless steel or copper; cooling capacity: 500 W	Cooling exhaust steam converting into liquid
Pump	Power: 50 W, Flow rate: 1–2 l/min; pressure: 5–10 bar	Circulates the flow of working fluid in the system
Thermal storage	Material: Phase changing materials or molten salt	Stores extra heat as backup for low irradiance durations

Theoretical Calculations

The design of the Rankine Cycle for Power Generation involves various theoretical calculations that are crucial for ensuring the system's efficiency and performance. The main calculations focus on the thermodynamic properties of the working fluid, energy balance across the system, and performance metrics such as thermal efficiency and power output.

Assumptions and Given Data

Boiler Pressure: 4 *bar*; *Condenser Pressure*: 2 *bar*; *Working Fluid*: *Water*; *Flow Rate of Water* ($\dot{m}$): 0.025 *kg*; *Pump Efficiency* (η_p): 75%; *Turbine Efficiency* (η_t): 85%; *Generator Efficiency* ($\eta_{generator}$): 80%

Pump Work

$$W_{pump} = \dot{m}\left(h_2 - h_1\right) = 0.025\left(508 - 504\right) = \mathbf{0.1\,kW} \tag{1}$$

Boiler Heat Addition

$$Q_{in,total} = \dot{m}\left(h_3 - h_2\right) = 0.025\left(2730 - 508\right) = \mathbf{55.55\,kW} \tag{2}$$

Turbine Work

$$W_{turbine} = \dot{m}\left(h_3 - h_4\right) = 0.025\left(2730 - 2262.5\right) = \mathbf{11.69\,kW} \tag{3}$$

Heat Rejection in Condenser

$$Q_{out,total} = \dot{m}\left(h_4 - h_1\right) = 0.025\left(2262.5 - 504\right) = \mathbf{43.96\,kW} \tag{4}$$

Network and Thermal Efficiency

$$w_{net} = w_{turbine} - w_{pump} = 11.69 - 0.1 = \mathbf{11.59}\,kw \tag{5}$$

$$\eta thermal = \frac{W_{net}}{Q_{in,total}} = \frac{11.59}{55.55} = \mathbf{0.2085} \text{ or } \mathbf{20.85\%} \tag{6}$$

Boiler Design

Internal Pressure, P: 4 *bar* = 400 *kPa*; *Dimensions*: (*a* = 170 *mm* = 0.17 *m* (*shorter side*); *b* = 170 *mm* = 0.17 *m* (*longer side*)); *Material*: *Carbon steel*; *Allowable Stress, σ_allowable*: 250,000 *kPa*; *Joint Efficiency, E*: 0.85; *Minimum Safety Factor, SF_min*: 2.0

Wall thickness:

$$t = \frac{\left(P^{*}\left(a+b\right)\right)}{2^{*}\sigma_{\text{allowable}}{}^{*}\left(E-P\right)} = \frac{\left(400^{*}\left(0.17+0.17\right)\right)}{\left(2^{*}250{,}000^{*}0.85\right)-400} = \mathbf{0.32\ mm} \tag{7}$$

$$\sigma_{hoop} = \frac{\left(P*\left(a*b\right)\right)}{2*t*\left(b+a\right)} = \mathbf{106{,}300\,kPa} \tag{8}$$

$$\sigma_{longitudinal} = \frac{P*b}{2*t} = \mathbf{106{,}250\,kPa} \tag{9}$$

$$SF = \frac{\sigma_{allowable}}{\sigma_{max}} = \mathbf{2.35\left(Safe\ Design\right)} \tag{10}$$

Similarly, when keeping the internal pressure of the **condenser** at 400kPa, calculations are performed, and it shows an SF value of 2.21.

Turbine Design

Turbine Diameter: *D* = 100 *mm* = 0.1 *m*; *Steps per Revolution*: 200; *Rotational Speed*: *N* = 45 *RPM*; *Turbine Efficiency*, *ηt*: 85% = 0.85; *Power Output*: *Wturbine* = 11.69 *kW*

$$\omega = \frac{2\pi N}{60} = \frac{2\pi\left(45\right)}{60} = \mathbf{4.712\,rad/s} \tag{11}$$

$$\mathrm{T}(\text{Torque}) = \frac{\mathrm{W}}{\omega} = \frac{11.69(1000)}{4.712} = \mathbf{2479.6\,Nm}$$

$$\tau(\text{Shear Stress}) = \frac{\mathrm{T.r}}{\mathrm{J}} = \mathbf{3950\,kPa} \tag{12}$$

Pump Design

Flow Rate: ṁ = 0.025 kg/s; *Pump Efficiency*, ηp: 75% = 0.75; *Pressure Difference*: $\Delta P = 200\ kPa$

$$W_{\text{pump}} = \frac{\dot{m}^{*}\Delta P}{\eta p} = \frac{(0.025)(200{,}000)}{0.75} = \mathbf{6.67\,kW} \tag{13}$$

System Design and Efficiency

Overall, this cycle is designed to maximize thermal efficiency by optimizing the working fluid and maximizing heat transfer mechanisms. This system utilized a parabolic solar approach to achieve a high temperature of 400–500C, achieving thermal efficiency of 20.85%, with a high output of 11.6kW. Further, this system contained a regenerative heating system to preheat the working fluid, resulting in higher output and efficiency. Moreover, the utilization of PSMs in thermal storage systems confirms the energy supply even during low irradiance duration, solving the intermittency problem.

Subsystem Selection and Assembly

The system components were manufactured using economical and eco-friendly materials and practices. Their assembly involved the integration of a solar collector, condenser, turbine, pump, and condenser into a scalable and modular design. The final assembly is shown below (Fig. 23.1).

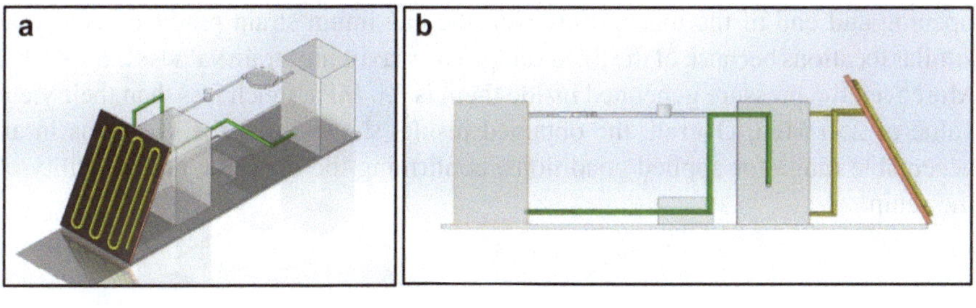

Fig. 23.1 Assembly isometric (**a: left**) and side view (**b: right**)

23.3 Modelling and Analysis

Next, the performance of the eco-friendly Rankine cycle system was evaluated through comprehensive modelling and analysis. This section shows the deformation and stress analysis, thermal performance analysis, and vibrational analysis. All such tests were conducted to confirm the structural integrity, optimal performance, and thermal efficiency of the system. Finite elemental analysis for all parts was performed using ANSYS software. Carbon steel is utilized for boilers, condensers, and heat exchangers. The main base is made using steel. While turbine and piping is made using aluminum. Further, the constraints applied included pressure loads for tank walls and fixed support for reflector tubes. Further, Fine Mesh was applied to all parts to get accurate results, with an average size of 1 mm for the piping system and 2 mm for the tank.

Stress Analysis

Tank First, a stress analysis of the tank was performed. For this, a load of 10 kN was applied on the walls, and a 25 kN load was applied on the bottom. In the results, it can be seen that a maximum strain of 2.149E-4 was produced. Moreover, maximum deformation is formed on the side faces of the tank with a maximum value of 2.187 mm, as illustrated in Fig. 23.2.

Piping System Another main subsystem responsible for transporting high-temperature working fluid is the piping system. Due to this, they were also analyzed for stress and deformation distribution. Results showed the maximum deformation inside the piping system is 0.0459 mm, produced at the joints of pipes, shown in Fig. 23.3. However, the generated deformation and stress are still well within the safe limit, confirming the safe and optimized operation under actual working conditions (Fig. 23.4).

Reflector Tubes For the reflector tubes, the sides were fixed to fully replicate the actual operational conditions. Based on the results shown in Fig. 23.4 below, it can be seen that the maximum deformation produced is 3.67×10^{-4} m, produced at the opening and end of the tubes. Moreover, the maximum strain produced is also at similar locations because of the fixed setup. The maximum strain value is 2.51×10^{-5}. Moreover, the pressure generated inside them is 3.6 MPa, much less than their yield value of 250 MPa. Overall, the obtained results showed that the system is in an acceptable range for applied conditions, confirming the integrity and reliability of the setup.

Fig. 23.2 Stress distribution (**2a: up**) & deformation distribution (**2b: Down**) for tank

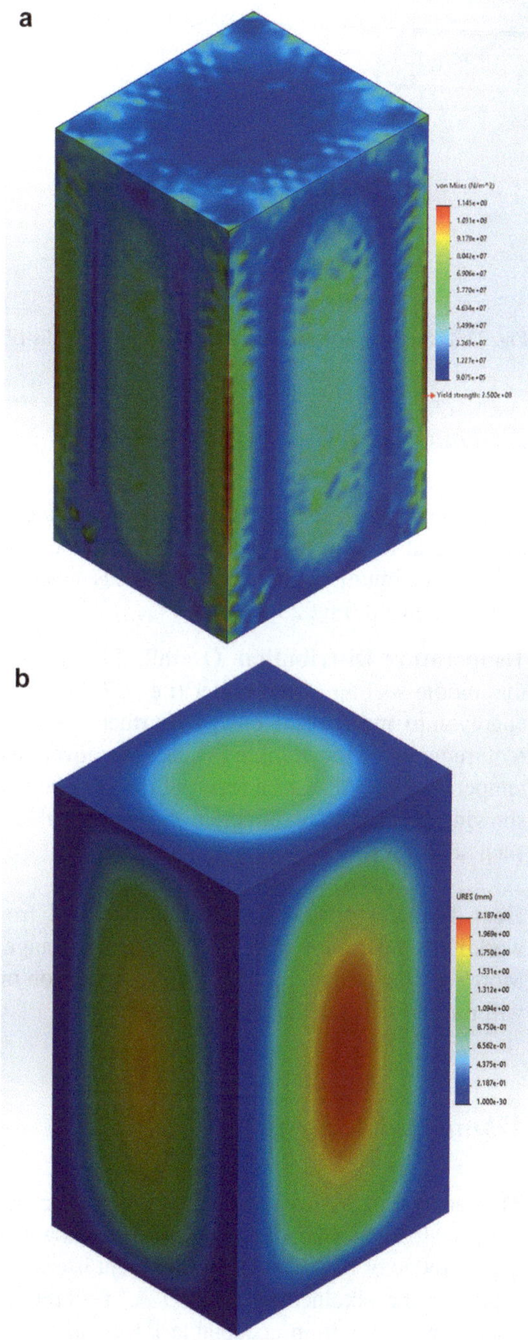

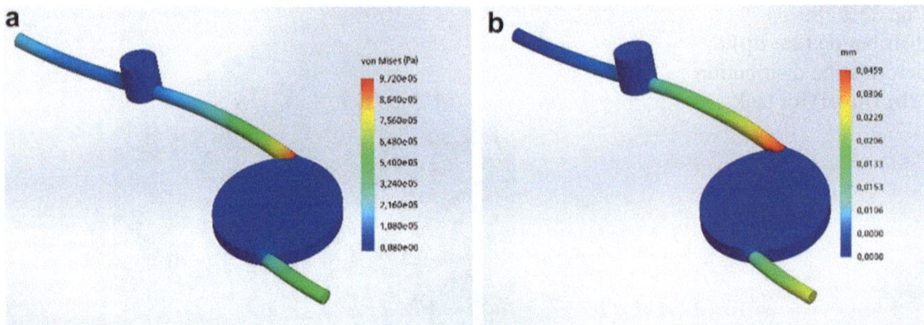

Fig. 23.3 Stress distribution (**a: left**) & deformation distribution (**b: right**) for piping system

Thermal Analysis

This thermal analysis was conducted to evaluate the heat transfer, distribution, and overall heat flow in the system. The obtained results highlighted the regions containing maximum temperature and areas where potential heat loss can be minimized, shown in Fig. 23.5.

Temperature Distribution Overall, the maximum temperature was observed in the middle section of the boiler (i.e., 378 K), which is needed to convert thermal energy into mechanical energy. Further, it also aligns with the thermodynamic requirements of the Rankine cycle. Moreover, **Reflector tubes** gave a maximum temperature of 308 K as a result of direct exposure to concentrated solar radiation, showing the need for effective thermal insulation to minimize the heat loss in such areas.

Heat Transfer Efficiency Results obtained from thermal analysis showed the working temperature within an acceptable range, along with minimal heat loss. Moreover, the utilization of advanced insulation materials in the concerned regions will further increase the output efficiency.

Vibrational Analysis

This analysis was performed to evaluate the effectiveness of the system under dynamic loading, like vibrations or wind, shown in Fig. 23.6. Such an analysis concerned reflector plates, as their alignment is crucial for solar collector performance. Based on the obtained results, 7.9 A of vibration amplitude was observed by the plate, which is within acceptable limits and confirms the stable operation of the

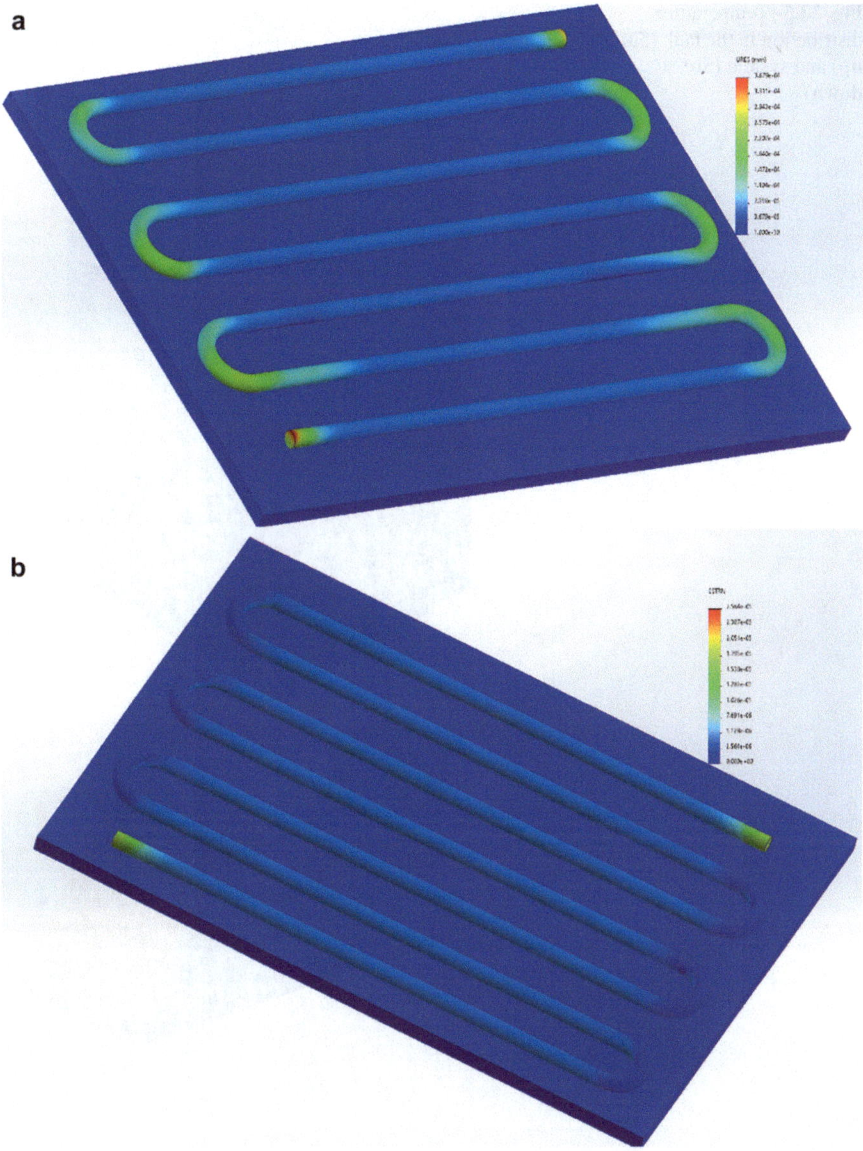

Fig. 23.4 Stress distribution (**a: up**) & deformation distribution (**b: right**) of reflector tubes

system under varied loads. Further, the fixed configuration of the system minimized instability and displacement, showing the system's ability to withstand operational vibrations. Moreover, this also eliminated the probability of resulting in resonance or failure due to fatigue.

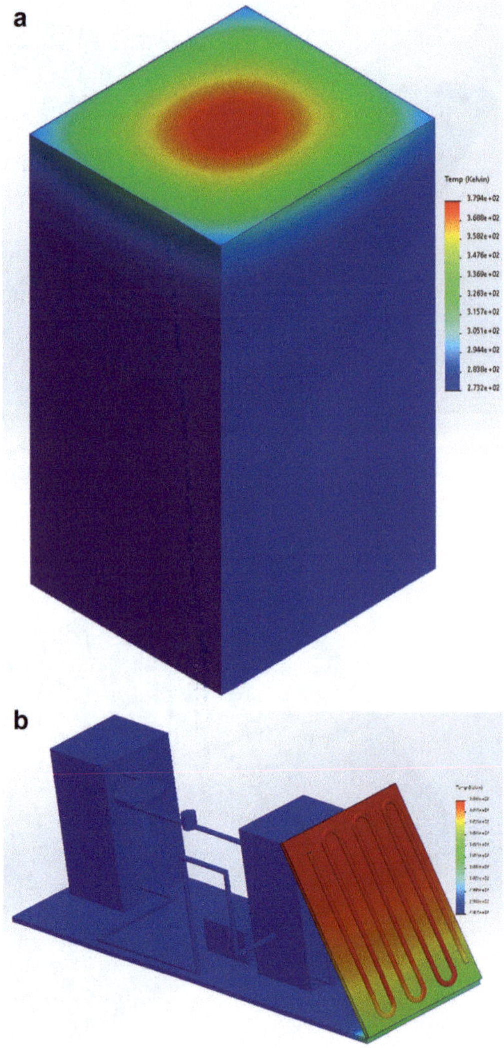

Fig. 23.5 Temperature distribution in the tank (**5a: up**) and system (**5b: down**)

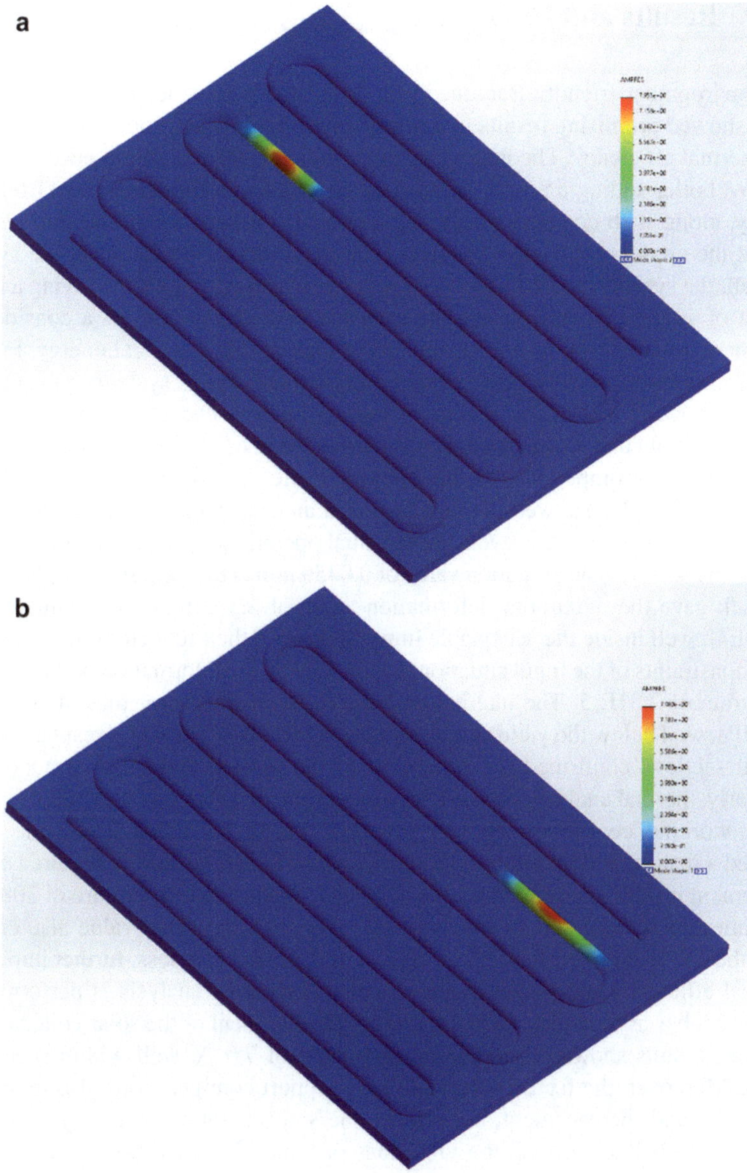

Fig. 23.6 Vibrational analysis 2nd amplitude (**6a: up**) and 3rd amplitude (**6b: down**)

23.4 Results and Discussion

The environment-friendly Rankine cycle integrated with concentrated solar power input showed promising results in terms of structural integrity, operational stability, and thermal efficiency. Theoretical calculations show that the pump needs work for 0.1 kW, boiler adding 55.5 kW of head to system, and turbine making 11.6 kW of energy, along with condenser rejecting 43.96 kW of heat. Such calculations confirmed the appropriate sizing determination and efficient operation of the system. Overall, the system showed a thermal energy conversion of 20.85%, giving a power output of 11.59 kW, showcasing the system's potential to convert a considerable portion of solar energy into mechanical and then usable electrical energy. Further, the attachment of modern heat transfer and thermal storage systems enhanced the working performance of the system by storing the extra energy in order to use it as a backup supply under low irradiance or cloudy weather conditions. The stress analysis of various components of the Rankine cycle, i.e., piping system, tank, and reflector tubes, also showed the results within an acceptable range, confirming the desired performance of this cycle under actual operational loads. The piping system showed a maximum deformation value of 0.0459 mm at the joint sections. Similarly, the tank gave the maximum deformation value of 2.187 mm and strain value of 2.149E-4, well inside the acceptable limit. Moreover, their reflector tubes, given the fixed constraints of the input situation, gave a maximum deformation of 3.67E-4 mm and strain of 2.51E-5. The internal pressure inside the reflector tubes falls around 3.6 MPa, well below the yield strength value of 250 MPa. The stress results obtained for all sections confirmed the reliability of the system under actual conditions. Similarly, thermal analysis showed the temperature distribution in the whole system under working conditions. Results revealed that under actual loads, the boiler showed a maximum temperature of 378 K, which is needed for the required energy conversion inside. The reflector tubes showed a maximum temperature of 308 K, as they are directly exposed to concentrated solar radiation. This value also emphasizes the need for proper thermal insulation to minimize heat loss, further improving thermal efficiency and power output. Lastly, vibrational analysis is performed on reflector tubes as they play a crucial role in the alignment of the solar collector. The obtained results showed a maximum amplitude of 7.9 A, well within acceptable limits. Moreover, the fixed configuration of support configurations also minimizes instability and displacement, confirming the safe operation of the system under dynamic loads like surrounding vibrations or high wind conditions. This analysis confirmed the cycle's ability to withstand all stresses and vibration without any probability of resonance or fatigue failure. Economic Viability & Sustainability: The viability of this CSP-integrated Rankine cycle is confirmed by its scalability and low system cost of just SAR 4900, making it viable for remote and small-scale applications. Further, the utilization of renewable solar energy minimized carbon emissions, making it more environmentally sustainable. Overall, the system's modular design coupled with incremental expansion in capacity made it more suitable for a variety of applications.

23.5 Conclusion

The eco-friendly Rankine cycle system coupled with CSP showed a major step towards renewable and sustainable energy solutions. This work successfully showed the feasibility of changing conventional fossil fuel-based energy systems with a sustainable CSP input-based system, achieving a high thermal conversion efficiency of 20.85% and a net power output of 11.59 kW. The model's modular and scalable design makes it more preferred for remote or off-grid energy applications, addressing major issues like energy intermittency or scalability. Moreover, by adding advanced heat transfer or thermal storage systems, this model can provide continuous energy even during hours of low irradiance, further increasing its demand, reliability, and efficiency. Theoretical calculations, including boiler heat addition, pump & boiler work, and condenser heat rejection, made a strong foundation for component sizing and system optimization. Moreover, the comprehensive thermal, stress, and vibrational analysis also confirmed the operational stability and structural integrity of the system under varying loading conditions. The stress results showed minimal deformation, stress, and strain values, while the thermal analysis also showed optimal temperature distribution and heat transfer. Further, vibrational analysis confirmed the system's ability to withstand varying loads, confirming its reliability. Overall, the findings of this study underscore the economic and environmental benefits of CSP-linked Rankine cycle, along with reduced carbon footprints and operational costs. Further, the developed CSP-integrated Rankine cycle has a variety of applications in off-grid and remote applications, mainly in rural communities for agricultural use or disaster relief scenarios. Moreover, the system's modular design makes it more viable for linking with current power grids to supplement energy, mainly in peak hours. Lastly, future works should focus on improving thermal efficiency by integrating AI or IC systems, advanced materials, or innovative heat exchanger design. By focusing on such areas, this system can give better results, paving the way for more adoption. This work adds to the global transition toward sustainable, clean, and renewable energy systems, providing a better solution to the current challenges of energy security and climate change.

References

1. Alami AH, Olabi AG, Mdallal A, Rezk A, Radwan A, Rahman SMA, Shah SK, Abdelkareem MA (2023) Concentrating solar power (CSP) technologies: status and analysis. Int J Thermofluids 18:Article 100340
2. Ugo Pelay LLYFDS (2018) Thermal energy storage systems for concentrated solar power plants. Renew Sustain Energy Rev 79:82–100
3. Xiao Ren JLGG (2022) An innovative concentrated solar power system driven by high-temperature cascade organic Rankine cycle. J Energy Storage 52
4. A. M. A. E. Elysia J. Sheu (2012) A review of hybrid solar-fossil fuel power generation systems and performance metrics. J Solar Energy Eng 134(4):1–17

5. Kalogirou SA (2004) Solar thermal collectors and applications. Prog Energy Combust Sci 30(3):231–295
6. Thek IO a G (2013) Electricity from biomass—combustion plants. In: Biomass combustion science, technology and engineering. Woodhead Publishing, pp 243–266
7. KA Santhosh Kumar B (2019) Thermodynamic analysis of solar organic rankine cycle by using working fluid for low temperature application. Int J Recent Technol Eng (IJRTE) 8(1)
8. Desai NB (2015) Optimization of concentrating solar thermal power plant based on parabolic trough collector. J Clean Prod 89:262–271
9. Wang H (2009) Selection of working fluids for low-temperature solar thermal power generation organic rankine cycles. Dongli Gongcheng/Power Engineering 29(3):287–291
10. Ibrahim Dincer MAR (2010) Thermal energy storage: systems and applications, 2nd edn. Wiley
11. Kasra Mohammadi SKHKK (2020) A comprehensive review of solar only and hybrid solar driven multigeneration systems: classifications, benefits, design and prospective. ETASR 49
12. Ramón Escobar Galindo MKKN (2021) Solar selective coatings and materials for high-temperature solar thermal applications. Sustain Mater Solutions Solar Energy Technol:383–427
13. Heejin Cho PJMES (2017) Performance evaluation of a solar-powered regenerative organic rankine cycle in different climate conditions. Energies 10(1)
14. Mohamed Salem MFAIM (2021) Applications of thermal energy storage in solar organic rankine cycles: a comprehensive review. Process Energy Syst Eng 9
15. Omar Aboelwafa S-EKFAS (2018) A review on solar rankine cycles: working fluids, applications, and cycle modifications. Renew Sustain Energy Rev 82(1):868–885
16. Shaaban S (2016) Analysis of an integrated solar combined cycle with steam and organic Rankine cycles as bottoming cycles. Energy Conv Manag:1003–1112

Chapter 24
Validating Ventilation-Thermal Systems for School Classrooms in Mediterranean Climate

Jesús Llanos-Jiménez, Alicia Alonso, and Juan José Sendra

Abstract In recent years, there has been debate about the reliability of the results obtained from thermal-energy simulations of buildings, specifically about the so-called performance gap, or difference in behavior between the real and the simulated. The validation of simulation models of ventilation and thermal systems of an archetypical school classroom in a Mediterranean climate in southern Spain is the main objective of this work. For this purpose, first, hygrothermal and air quality variables have been monitored in situ in this classroom, with different ventilation and thermal treatment systems; virtual models have been generated with TRNSYS and, finally, the models have been adjusted and validated. The ventilation and thermal systems analyzed were: (1) Natural ventilation combined with heating by radiators, the most common solution in southern Spain; (2) mechanical ventilation combined with heating by radiators in winter or with adiabatic cooling in summer, a novel efficient solution that educational authorities are promoting to mitigate the increasingly frequent overheating of classrooms. The results for validated models show low discrepancies between measured and simulated, with average tolerances between 0.3 and 1 °C for temperature and between 32.8 and 106.4 ppm for CO_2. These error margins comply with the standards and are acceptable for a study of indoor environmental conditions. These models enable the suitability assessment of various energy retrofitting options to address future climate change challenges, not only regarding the building's thermal envelope but also its HVAC systems.

Keywords Efficient ventilation strategies · Building modelling and simulation · Model validation

J. Llanos-Jiménez (✉) · A. Alonso · J. J. Sendra
Instituto Universitario de Arquitectura y Ciencias de la Construcción, Escuela Técnica Superior de Arquitectura, Universidad de Sevilla, Seville, Spain
e-mail: jllanos@us.es

L. Chen (ed.), *Clean Energy Research*, Green Energy and Technology,
https://doi.org/10.1007/978-3-031-99312-1_24

24.1 Introduction

The school building stock in southern Spain is largely outdated, especially in terms of energy efficiency. It also lacks the necessary systems to ensure minimum conditions of thermal comfort (TC) and indoor air quality (IAQ) required by the current Spanish building [1]. In fact, several studies have shown that poor IAQ, associated with high levels of CO_2 and other pollutants, can even cause health problems [2, 3]. There is a need to establish energy and environmental retrofitting strategies for schools that are compliant with regulations and anticipate the impacts of climate change, such as increased heat waves and air pollution.

The objective of this work is to generate properly validated virtual diagnostic models that allow a holistic, objective and plausible assessment of both TC and IAQ. The most common validation process is typically based on temperature (T) adjustment in the case of building partial models. This work also includes the validation of CO_2 as a novelty, since it is the main indicator of ventilation, which also has an impact on energy consumption. Recent studies focusing on school simulations [4, 5] have shown that retrofit strategies combining controlled natural ventilation and mechanical ventilation with heat recovery or T / CO_2 control demand significantly improve IAQ as well as energy consumption. Furthermore, the use of cooling systems in schools in Mediterranean climates is not common but may be necessary given the increase in heat waves because of climate change. This work presents the validation, in addition to natural and mechanical ventilation protocols and systems, of an adiabatic cooling system that has been little implemented so far in non-industrial buildings. These models enable the suitability assessment of various energy retrofitting options to address future climate change challenges, not only regarding the building's thermal envelope but also its HVAC systems.

24.2 Methods

An archetypical school classroom in a Mediterranean climate (Csa, according to the Köppen-Geiger climate classification) in southern Spain has been chosen as the case study. This classroom was modelled and simulated in TRNSYS, using the same ventilation systems and protocols as those monitored in situ. Finally, it was validated based on the results obtained from both the simulation and monitoring.

Case Study

The selection of this case study resulted from previous research: first, a multiparametric statistical analysis to define archetypal schools in the region according to construction date, morphology, orientation, façade type, window-to-wall ratio, solar

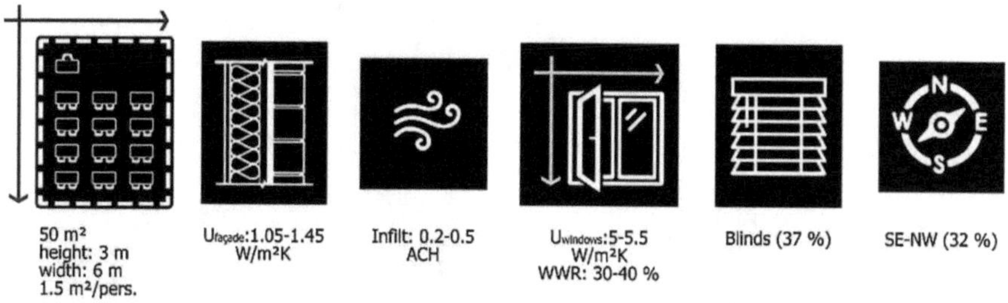

Fig. 24.1 Graphical summary of the case study characteristics

shading and HVAC systems [6]; and second, an environmental characterization study of representative classrooms from several schools matching the previously defined archetype criteria [7]. The geometrical and constructive characteristics are summarized in Fig. 24.1 and detailed below:

- Constructive solutions. These are simple and traditional, without thermal insulation, as 94% of the schools in the region were built before the Spanish Building Code [8]. The thermal transmittance of façades ranges from 1.05 to 1.45 W/m^{2}K, while that of windows ranges from 5 to 5.5 W/m^{2}K.
- Infiltrations. Blower door tests were conducted, obtaining air change per hour (ACH) values between 0.2 and 0.5.
- Window-to-Wall Ratio (WWR). As well as the predominant design in 45% of schools in the region, values range between 30% and 40%.
- Solar shading. Blinds are used predominantly in 37% of schools in the region.
- Classroom Orientation (according to windows). The predominant orientation is Northwest–Southeast, aligning with 32% of schools in the region.
- Geometric Definition. The morphology, dimensions, and occupancy density are representative, as classroom designs commonly adhere to the Design and Construction Standards for Educational Buildings in the region [9]. These standards specify minimum requirements for secondary school classrooms: 50 m^2 (1.5 m^2 per person) and a width of 6 m.

Monitoring

Several tests were conducted to monitor hygrothermal and air quality parameters in the case study under different ventilation protocols and systems. Those with natural ventilation include single-sided and cross natural ventilation with varying window (100%, 50%, 25%) and door (one or two doors) openings. Those with mechanical ventilation include single-drive at different airflow rates, with and without heating via radiators and an adiabatic cooling system. Each test involved monitoring for the full school day (morning schedule), with up to four tests conducted: two with natural ventilation during winter and summer, and two with mechanical ventilation (also

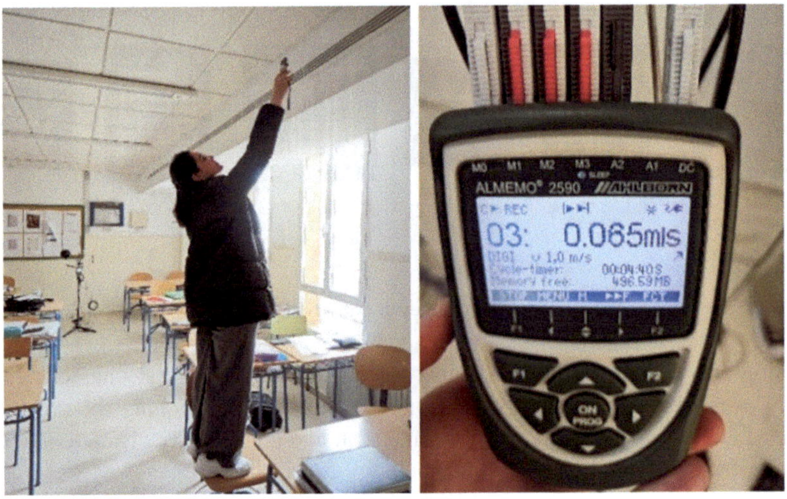

Fig. 24.2 Photos of a monitoring day, sensor location and display

Table 24.1 AHLBORN ALMEMO 2590-4AS Datalogger and accessories features

Variables		Units	Range	Accuracy (and conditions)
T	Air temperature	°C	−20 to +80	5–60 °C T: ±0.4 °C 5 > T > 60 °C: ±0.7 °C
RH	Relative humidity	%	0 to 98	10–90% RH: ± 2% 10% > RH > 90%: ±4%
CO_2	Carbon dioxide	ppm	0 to 5000	± 50 ppm +3%

in both seasons). A portable AHLBORN ALMEMO Datalogger 2590-4AS, mounted on a tripod at breathing zone height in the central rear part of the classroom (Fig. 24.2) was primarily used for this purpose. The parameters measured include temperature (T), relative humidity (RH), and carbon dioxide (CO_2), with their measurement ranges and accuracies detailed in Table 24.1.

Modelling and Validation

The model was developed using the dynamic-transient system simulation software TRNSYS 18 v.6.0.0.7. The 3D modeling tool was Sketchup 8. Figure 24.3a) shows the geometry modelling, including a classroom, corridor, and another classroom, with an initial presetting of indoor and outdoor surfaces, such as walls, floors, ceilings, and the definition of airnodes. Once configured, this geometry was imported into TRNSYS, and then the constructive details of each element were defined. Figure 24.3b) shows the TRNBuild settings of the model, where the schedules (occupancy, lighting, heating and ventilation), the constructive solutions of the

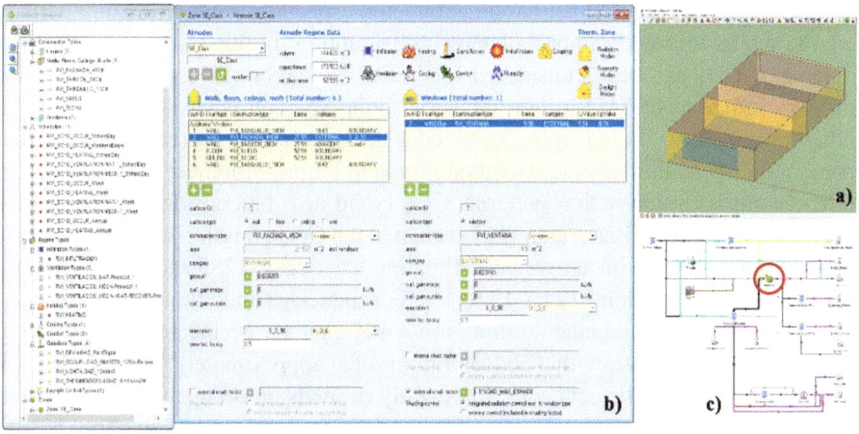

Fig. 24.3 TRNSYS model

façade and other facings, windows and the thermal loads are configured. Specifically, the configuration sheet of one airnode with 144.6 m^3 and 52.6 m^2 corresponding to one of the classrooms is shown. In the yellow box, the facade (21.6 m^2) is selected, which has been assigned a thermal transmittance of 1.435 W/m^2K, and within this, the window configuration, with a surface area of 9.5 m^2. Figure 24.3c) shows the general interface with the connections between the different system components: the component containing the geometric, constructive, and operational information of the model (TRNBuild) is outlined in red; climate file and a series of unit conversion equations and model inputs definitions (mainly for defining complex schedules) in left side; the result output printouts in right side; the component responsible for calculating CO_2 levels is located at the bottom and is linked to the model through parameters such as classroom volume, occupancy, ventilation rate, and their corresponding schedules.

In order to achieve the validation of the model, discrepancies between simulated and monitored variables were evaluated, specifically T and indoor CO_2. According to the main international building model validation standard, ASHRAE Guideline 14 [10], the indicators used for evaluating these discrepancies are the Coefficient of Variation of the Root Mean Square Error (CV(RMSE)) and the Normalized Mean Bias Error (NMBE). For CV(RMSE), the threshold for hourly evaluation is <30%, while for NMBE, the threshold is <10%. Based on these requirements, a manual adjustment process was followed, adjusting operation schedules, thermal loads, and ventilation. The basic school week schedule and annual calendar were established. Additionally, the heating schedule was set to activate during the first three morning hours when outdoor temperatures remain below 10–15 °C for several days, while the ventilation schedule was adjusted based on the opening protocols tested during the monitoring campaigns. Climate files, created using the Meteonorm v.8 tool, were adjusted to match the monitoring outdoor environmental conditions using hourly data from the Spanish Meteorological Agency (AEMET) [11].

Temperature is a common validation parameter, especially when analyzing parts of a building, as it directly impacts energy consumption. In this study, the additional parameter CO_2 has been considered for model validation. In general, CO_2 is the primary indicator of IAQ in non-industrial buildings and is easier to model than other pollutants, as there is more knowledge about emission sources (mainly occupancy) and their dilution rate depending on ventilation. Moreover, due to its close relationship with ventilation patterns, CO_2 could also be considered to influence energy consumption. Batterman [12] evaluated the CO_2 emission rate per person in various schools based on age range, estimating it between 0.285 and 0.319 l/min for secondary school students (ages 12 to 16). For the CO_2 simulation, TRNSYS Type 388 was used. This component allows modeling the CO_2 exchange between a supply airflow (outdoor air, with a concentration set at approximately 430 ppm) and an indoor space, where the concentration mainly depends on occupancy and its previously mentioned CO_2 emission rate. The calculation developed by this tool is validated [13] using the static state equation Eq. (24.1) from the international standard DIN EN 16798–1 [14], which is a reference for building ventilation.

$$C_{\text{Node}} = \left(C_{\text{Node},t-1} - C_{\text{sup}} - \frac{\left(G_{\text{pers}} + G_{\text{misc}}\right)}{n\varepsilon} \right) \bullet e^{-n\varepsilon\Delta t} + C_{\text{sup}} + \frac{\left(G_{\text{pers}} + G_{\text{misc}}\right)}{n\varepsilon} \quad (24.1)$$

CO_2 emission rate per occupant, G_{pers} (m^3/h): 0.01764 m^3/h per person according to Batterman [12].

- Total occupancy of the space, G_{pers} (people): an average of 28 people is used.
- Emissions from other sources, G_{misc} (ppm): not identified.
- Volume (m^3): airnodes volume has been used, 144.6 m^3.
- Supply air volume, C_{sup} (m^3/h): the airflow rates are introduced based on ventilation schedules and adjustments of ACH for tested ventilation systems.

CO_2 concentration in supply air, C_{sup} (ppm): 430 ppm.

- Ventilation effectiveness ($n\varepsilon$): according to Table 24.1 of the tool's user manual [13], a value of 2 is set for natural ventilation strategies and 1 for mechanical ventilation systems.

24.3 Results

Validation of Systems in Relation to Indoor Environmental Conditions

The ventilation schedules were primarily adjusted. Additionally, inputs linked to both the CO_2 calculation component and the general model component were established to define occupancy and ACH. Occupancy simultaneously affects thermal loads and CO_2 emissions. ACH influences CO_2 dilution and also the thermal mass

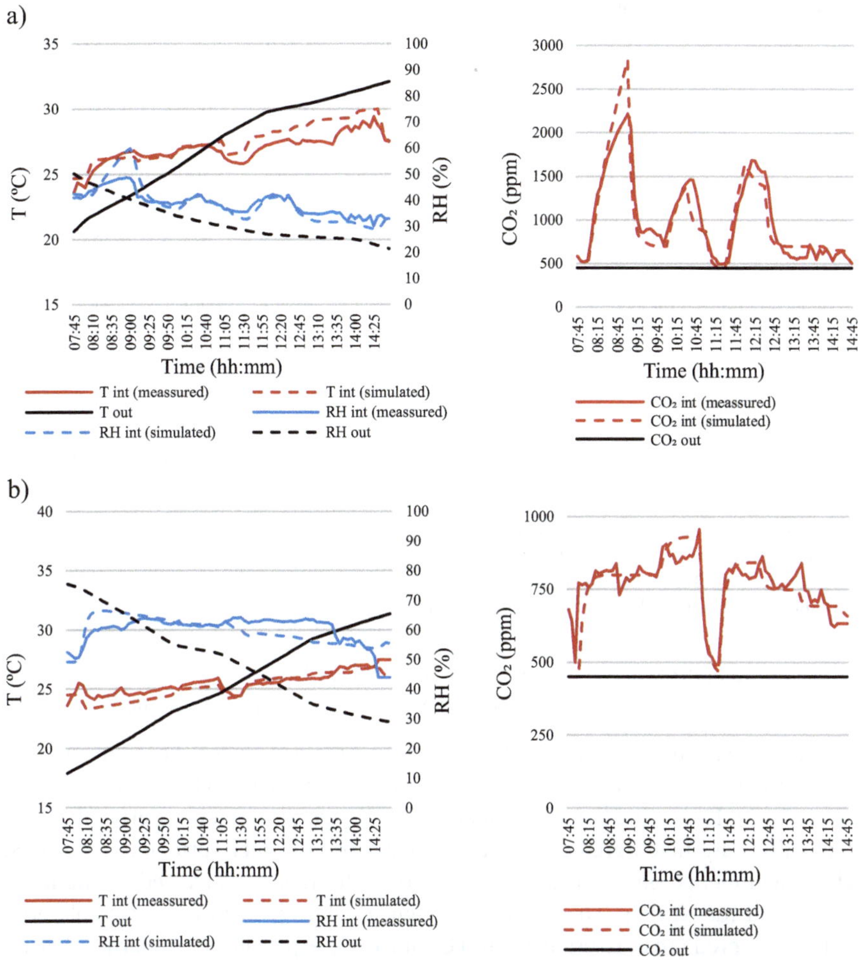

Fig. 24.4 Graphical analysis of discrepancies between monitored and simulated variables for (**a**) natural ventilation and (**b**) mechanical ventilation, single drive + adiabatic cooling

exchange between the indoor and outdoor. These factors were manually adjusted based on the monitoring results until model validation was achieved (Fig. 24.4 and Table 24.2). Table 24.2 also shows the median values and the 95th percentile of the differences (Δ) between the simulated and monitored variables.

Natural Ventilation System As a result of this iterative process of discrepancy analysis to achieve model validation, it is concluded that under natural ventilation protocols in summer, ACH values range between 7.5 (with 100% of the operable glazed area and 2 doors opened) and 0.4 (25% windows opened). In winter, probably due to the greater temperature difference between the indoor and outdoor, ACH

Table 24.2 Numerical analysis of discrepancies between monitored and simulated variables

Systems	Season	Variables	CV(RMSE) <30%	NMBE ±10%	Median Δ	Max Δ (p 95)
Natural ventilation (NV)	H	T	1.6	4.7	0.9 °C	1.4 °C
		RH	13	11.3	7.8%	14.9%
		CO_2	14.3	11.4	106.4 ppm	336.7 ppm
	C	T	1.9	2.6	1 °C	1.4 °C
		RH	8.4	6.2	4%	8.3%
		CO_2	19	14.2	59 ppm	206.4 ppm
Mechanical ventilation single drive (MV)	H[a]	T	4	3	0.5 °C	1.2 °C
		RH	9.1	8.5	4.3%	8.6%
		CO_2	18	12.7	88 ppm	444.4 ppm
	C[b]	T	3.1	2	0.5 °C	1.6 °C
		RH	7.3	5.9	4%	11.6%
		CO_2	8.7	5.8	33.4 ppm	141 ppm

H heating season, *C* cooling season

[a]The heating system was turned on during the testing day of this system and, therefore, was used to determine its thermal load contribution

[b]The adiabatic cooling system was in operation throughout the entire testing day and its validation is explained in detail in the corresponding section

values could approximately double, reaching values between 7.5 (with 50% of the operable glazed area and 1 door opened) and 1.2 (25% windows opened).

Discrepancies ranging from 1.6 to 19% were obtained under the application of CV(RMSE), thus meeting the 30% requirement set by the ASHRAE Guideline 14 (Table 24.2). Regarding the adjustment of the CO_2 parameter, considering that CO_2 fluctuations are much greater and faster than those of other variables, a plausible adjustment can be observed despite occasional discrepancies exceeding the limits established for NMBE. The median difference values present an acceptable error margin for predictive assessment of TC and IAQ, especially considering that the typical accuracy of CO_2 sensors is 50 ppm. The median discrepancies range from 0.9 to 1 °C and from 59 to 106.4 ppm.

Mechanical Ventilation System In the case of mechanical ventilation at different airflow rates, the results show an ACH range between 10 (100% airflow) and 6 (60% airflow). Additionally, this system is supplemented in summer with a thermal conditioning system using efficient adiabatic or evaporative cooling. To model its operation, a simplified equation, Eq. (24.2), was used to calculate the supply air temperature ($T_{ADIAB,imp}$) by applying a machine performance coefficient (η) to the optimal thermal difference between the dry bulb temperature (T_{DB}) and the wet bulb temperature (T_{WB}). According to distributors, direct evaporation (the chosen option) achieves performance closer to 90%; however, in order to validate the system, it has been set to 45%.

$$T_{ADIAB,imp} = T_{DB} - \eta \cdot (T_{DB} - T_{WB}) \tag{24.2}$$

- $T_{ADIAB,imp}$ = Supply air temperature.
- T_{DB} = Dry bulb temperature.
- T_{WB} = Wet bulb temperature.
- η = Performance of the cooling process,

The moisture addition in the supply air due to the evaporative cooling process performed by this system has also been adjusted, resulting in a value of 0.0115 kg of water vapor per kg of dry air.
Table 24.2 discrepancies, ranging from 3.1% to 18%, were obtained under the application of CV(RMSE), meeting the limits, and from 2% to 12.7% for NMBE, slightly exceeding the limits in the case of CO_2. Additionally, the median values and 95th percentile of the differences (Δ) between the simulated and monitored variables are also acceptable, with median discrepancies ranging up to 0.5 °C for temperature and from 33.4 to 88 ppm for CO_2.

24.4 Conclusions

An archetypal school classroom model has been defined and various natural and mechanical ventilation protocols have been monitored and later validated through a discrepancy analysis. After a proper definition of construction and operation characteristics, the following main conclusions were found:

- TRNSYS models for natural ventilation (both simple and cross) and single-drive mechanical ventilation systems have been validated. These models include configurations with and without radiator heating in winter and an efficient adiabatic cooling system in summer.
- The performance gap discrepancy between simulation and monitoring is low, with better alignment under controlled mechanical ventilation protocols. Additionally, the air change rates in natural ventilation strategies could double in winter compared to summer due to a higher thermal difference between the indoor and outdoor.
- Temperature discrepancies considering natural ventilation ranging from 0.9 to 1 °C, and CO_2 discrepancies between 59 and 106.4 ppm.
- Temperature discrepancies considering mechanical ventilation ranged up to 0.5 °C, and CO_2 discrepancies between 33.4 and 88 ppm.

These results form the basis for using the validated model in the search for energy retrofitting strategies for schools, balancing energy efficiency, thermal comfort and indoor air quality in both present and future climate scenarios. The main limitation is the adaptation of this model to other typologies, as it has been defined both geometrically and operationally based on the conditions of an archetypal school, a highly specific use that differs significantly from others.

Acknowledgments and Funding The authors wish to acknowledge the financial support provided by Grant (PID2020-117722RB-I00) "Retrofit ventilation strategies for healthy and comfortable schools within a nearly zero-energy building horizon" funded by MICIU/AEI/10.13039/501100011033/). In addition, the author J.Ll. gratefully acknowledges the funding received by the Ministry of Universities through the University Teacher Training Contract Grant Programme (FPU20/04393).

References

1. Reglamento de Instalaciones Térmicas en los Edificios (RITE) (2021) Ministerio de Industria—Gobierno de España, Madrid
2. Valls Roche I, Ubalde-Lopez M, Daher C, Nieuwenhuijsen M, Gascon M (2024) The health-related and learning performance effects of air pollution and other urban-related environmental factors on school-age children and adolescents-a scoping review of systematic reviews. Curr Environ Health Reports 11:300–316
3. De Bont J, Jaganathan S, Dahlquist M, Persson A, Stafoggia M, Ljungman P (2021) Ambient air pollution and cardiovascular diseases: an umbrella review of systematic reviews and meta-analyses. J Intern Med 291(6):779–800
4. Calama-González CM, León-Rodríguez AL, Suárez R (2019) Indoor air quality assessment: comparison of ventilation scenarios for retrofitting classrooms in a hot climate. Energies 12(4607)
5. Grassie D, Dong J, Schwartz Y, Karakas F, Milner J, Bagkeris E, Chalabi Z, Mavrogianni A, Mumovic D (2023) Dynamic modelling of indoor environmental conditions for future energy retrofit scenarios across the UK School Building Stock. J Build Eng 63(105536):105536
6. Llanos-Jiménez J, Suárez R, Alonso A, Sendra JJ (2022) Archetypes of public secondary schools in Mediterranean climate. Indoor air quality and comfort field studies. In: Proceedings of the 42nd AIVC, 10th TightVent and 8th venticool conference: ventilation challenges in a changing world. AIVC, Rotterdam, pp 808–8017
7. Llanos-Jiménez J, Suárez R, Alonso A, Sendra JJ (2024) Objective and subjective indoor air quality and thermal comfort indices: characterization of Mediterranean climate archetypal schools after the COVID-19 pandemic. Indoor Air 2024(2456666)
8. Código Técnico de La Edificación, Documento Básico de Ahorro de Energía (CTE DB HE) (2006) Ministerio de Vivienda—Gobierno de España, Madrid
9. Normas de Diseño y Constructivas para los Edificios de Uso Docente (2003) Agencia Pública Andaluza de Educación, Junta de Andalucía
10. ASHRAE guideline 14—measurement of energy, demand, and water savings (2014) ASHRAE, Atlanta
11. Datos Climatológicos—Agencia Estatal de Meteorología (AEMET). https://www.aemet.es/es/serviciosclimaticos/datosclimatologicos. Last accessed 05 Dec 2024
12. Batterman S (2017) Review and extension of CO_2-based methods to determine ventilation rates with application to school classrooms. Int J Environ Res Public Health 14
13. Lam J (2022) TRNSYS user manual type 388 CO_2 concentration model. Transsolar Energietechnik GmbH, Stuttgart
14. EN 16798-1 Energy performance of buildings—Ventilation for buildings—Part 1: Indoor environmental input parameters for design and assessment of energy performance of buildings addressing indoor air quality, thermal environment, lighting and acoustics (2019)

Chapter 25
The N-Shaped Effect of Urbanization and Industrial Structure on Water Pollutant Emissions: New Evidence Using the Panel Kink Model

Paravee Maneejuk

Abstract This study examines the connections between urbanization, industrial structure, and water pollutant emissions in China using a panel kink regression model. The study finds that urbanization generally contributes to increased water pollutant emissions in China, except within a specific range of urbanization levels. The industrial structure also positively affects water pollutant emissions when industrial activity represents a certain proportion of GDP. Notably, the study does not find evidence of an inverted U-shaped or N-shaped Environmental Kuznets Curve (EKC) for water pollutants in China. However, it identifies an inverted N-shaped relationship between industrial structure and water pollutant emissions. This pattern reflects distinct phases of China's economic development: an initial phase with primary industry dominance and minimal emissions, a second phase with increased emissions from secondary industries, particularly heavy industries, and a third phase with a shift towards tertiary industries, which reduces water pollutant emissions.

Keywords China · EKC · Inverted N-shaped · Industrial Structure · Urbanization · Water pollution

25.1 Introduction

Water resources are fundamentally crucial for both social and economic development, serving as a vital component of society and nature alike [1]. However, the world currently faces a critical dilemma of water scarcity. Despite the Earth being known as the "Blue Planet," only 3% of the global water supply is freshwater, and a mere 0.5% of this is currently accessible and usable for human needs [2]. This

P. Maneejuk (✉)
Center of Excellence in Econometrics, Faculty of Economics, Chiang Mai University, Chiang Mai, Thailand

L. Chen (ed.), *Clean Energy Research*, Green Energy and Technology,
https://doi.org/10.1007/978-3-031-99312-1_25

scarcity is further exacerbated by increasing water pollution, primarily driven by rapid industrialization and urbanization. These developmental processes are significantly worsening water-related problems, leading to severe pollution that poses substantial threats to public health and economic stability [3].

In light of these challenges, our study aims to explore strategies to manage the balance between economic development and environmental preservation, with a particular focus on reducing water pollutant discharge. By employing a panel kink model, we seek to provide new evidence on the nonlinear effects of urbanization and industrial development on water pollutant emissions. This approach allows us to identify potential thresholds or turning points in the relationship between development and water pollution, offering valuable insights for policymakers and urban planners.

A significant approach to assessing sustainable development involves scrutinizing the connection between economic growth and environmental pollution, a relationship elegantly described by the environmental Kuznets curve (EKC) concept. According to the EKC, environmental pollution initially escalates as per capita income rises. However, as economies continue to grow, a heightened demand for environmental quality emerges, leading to a decline in environmental degradation. Initially proposed by Grossman and Krueger in 1991, this concept depicts the relationship between economic growth and environmental deterioration as an inverted U-shaped curve. Many researchers have provided empirical support for this inverted U-shaped relationship between income and environmental degradation. For instance, Leitão [4] verified this pattern across 94 countries with varying development levels using a Quadratic Panel regression model. Similarly, Culas [5] identified the inverted U-shaped EKC in nine Latin American countries. In essence, if this inverted U-shaped pattern is substantiated, sustainable development can be attained by surpassing the threshold or turning point of the curve through further economic growth.

Subsequently, a novel perspective on the nexus between economic growth and environmental degradation emerges with the N-shaped EKC theory. This theory suggests that the previously assumed inverted U-shaped EKC hypothesis may not hold long term. Instead, it posits that beyond a certain income threshold, increased income could once again result in a positive relationship between economic growth and environmental degradation. While the inverted U-shaped EKC suggests that environmental improvements occur as economies grow, studies have indicated that the relationship might follow an N-shaped pattern, as noted in research by Torras and Boyce [6], Álvarez-Herranz and Balsalobre Lorente [7], and Shehzad et al. [8].

Despite being considered a relatively recent phenomenon, the N-shaped EKC has roots dating back to the 1990s. Researchers such as Grossman and Krueger [9] and Panayotou [10] were among the early pioneers in identifying an N-shaped relationship between economic development and sulfur dioxide (SO2) emissions. This pattern has also been observed in other regions and countries. For instance, Austria exhibited an N-shaped EKC, as demonstrated by Friedl and Getzner in 2003 [11]. Moreover, a study encompassing 28 OECD countries conducted by Álvarez et al. in 2015 further substantiated the existence of the N-shaped EKC. Recent research in

Algeria by Shehzad, Zeraibi, and Zaman [8] also supports this pattern. In the case of China, evidence from Huan et al. [12] highlights an N-shaped relationship between industrial production and environmental quality. Their investigation, covering the period from 1972 to 2020, employed the ARDL bounds test and confirmed the presence of this N-shaped impact of industrial production on environmental quality in both the long and short run in the Chinese economy. This growing body of research underscores the complexity of the relationship between economic growth and environmental degradation, challenging the conventional wisdom of a simple inverted U-shaped EKC.

As previously mentioned, while substantial evidence supports the existence of the N-shaped Environmental Kuznets Curve (EKC), there appears to be a gap in the literature regarding examining the N-shaped relationship between urbanization, industrial structure, and environmental pollution. In this study, we propose the utilization of a panel kink regression model, as outlined in works by Zhang et al. [13] and Tibprasorn et al. [14], to investigate this N-shaped phenomenon.

This study aims to address pivotal questions of significance to academics, policymakers, and environmental authorities in China. It seeks to determine whether urbanization and industrial structure contribute to water pollution in the country or if they can be harnessed to reduce water pollution. Additionally, the study investigates whether the N-shaped EKC phenomenon, observed in other regions, is evident in China. By exploring these interconnected inquiries, this study offers valuable insights that can inform policy decisions, guide sustainable development initiatives, and enhance our comprehension of the relationship between economic growth, urbanization, industrial structure, and environmental sustainability in China.

25.2 Environmental Kuznets Curve

The concept of the Environmental Kuznets Curve (EKC) was introduced by Grossman and Krueger [15] during discussions related to the North American Free Trade Area. Concerns were raised about the potential environmental impact of free trade on Mexico and the United States, which led to exploring the relationship between environmental quality and income levels. This relationship between pollution and income can be summarized as follows: In lower-income settings, pollution increases as income levels rise, while in higher-income contexts, pollution decreases as income grows. This phenomenon is represented by an inverted "U-shaped" curve known as the Environmental Kuznets Curve (EKC) [16].

Moreover, the extension of the "inverted U-shaped" to the "N-shaped" curve presents a more complex dynamic. It implies that one variable, such as urbanization, may initially correspond with a reduction in another variable, like pollution, as seen on the left segment of the "N." However, as the influential variable undergoes further change, it can subsequently incite an increase in pollution during the middle portion of the curve before culminating in a subsequent downturn on the right-hand side. These curve shapes encapsulate the intricate interplay between economic advancement, environmental impact, and the thresholds that govern the trajectory of environmental conditions.

The N-shaped EKC challenges the conventional inverted U-shaped EKC hypothesis by proposing that, in the long run, beyond a certain income level, increased income could once again lead to a positive relationship between economic growth and environmental degradation, as noted by De Bruyn et al. in 1998 [17]. Torras and Boyce [6] argue that this N-shaped relationship occurs when the scale effect outweighs the composition and technical effects. Such an outcome might be attributed to constraints in further improving the distribution of industries or diminishing returns on technological changes, as discussed by both Torras and Boyce [6] and Álvarez-Herranz and Balsalobre Lorente [7]. Consequently, the study of the EKC has spurred numerous scholars to explore the relationship between environmental pollution and income, with a particular focus on factors like wastewater discharge.

25.3 Data and Methodology

Data

We utilize secondary data sourced from the China Statistics Yearbook. We consider pollution levels in 30 Chinese provinces within the timeframe of 2005 to 2021, using wastewater as a key pollution indicator. The 30 provinces under examination encompass Beijing, Tianjin, Hebei, Shanxi, Inner Mongolia, Liaoning, Jilin, Heilongjiang, Shanghai, Jiangsu, Zhejiang, Anhui, Fujian, Jiangxi, Shandong, Henan, Hubei, Hunan, Guangdong, Guangxi, Hainan, Chongqing, Sichuan, Guizhou, Yunnan, Shaanxi, Gansu, Qinghai, Ningxia, and Xinjiang. All data used in this study is sourced from the China Environmental Statistics Yearbook. The independent and control variables used in this study are explained in Table 25.1.

Panel Kink Regression

In this study, we aim to examine the N-shaped EKC on the relationship between Urbanization, Industry structure and water pollution represented by wastewater; thus, the model takes the form as

$$\begin{aligned}\ln W_{\text{it}} = {} & \beta_1^{(1)}\left(\ln IS_{it} \le \gamma_{1,1}\right) + \beta_1^{(2)}\left(\gamma_{1,1} < \ln IS_{it} \le \gamma_{1,2}\right) \\ & + \beta_1^{(3)}\left(\ln IS_{it} \ge \gamma_{1,3}\right) + \beta_2^{(1)}\left(\ln U_{\text{it}} \le \gamma_{2,1}\right) \\ & + \beta_2^{(2)}\left(\gamma_{2,1} < \ln U_{it} \le \gamma_{2,2}\right) + \beta_2^{(3)}\left(\ln U_{it} \ge \gamma_{2,3}\right) \\ & + \beta_3 \ln T_{it} + \beta_4 \ln EI_{it} + \beta_5 \ln TO_{it} + \alpha_i + \varepsilon_{it},\end{aligned} \tag{25.1}$$

Table 25.1 Data description

Variable	Label	Description
Wastewater	*W*	Wastewater discharge (unit: Ten thousand tons)
Industry structure	*IS*	The sum of the secondary and tertiary industries as a percentage of GDP (unit: %)
Urbanization	*U*	Population ratio of urban to the total population (unit: %)
Technology	*T*	The sum of the number of domestic patent applications (unit: Pieces)
Environmental investment	*EI*	Environmental governance investment accounted for GDP (unit:%)
Trade openness	*TO*	Total import and export volume as a share of GDP(unit: %)

where$\beta_1^{(j)}$ and $\beta_1^{(j)}$ with $j = 1, ..., 3$ are the nonlinear slope parameters. Thus, if $\beta_1^{(1)} > 0, \beta_1^{(2)} \langle 0, \beta_1^{(3)} \rangle 0$ Industry structure and wastewater are confirmed to have N-shaped EKC. Likewise, if $\beta_2^{(1)} > 0, \beta_2^{(2)} \langle 0, \beta_2^{(3)} \rangle 0$ Urbanization and wastewater are confirmed to have N-shaped EKC. $\beta_3 - \beta_5$ represent the linear slopes of control variables. To estimate this model, Pooled OLS, fixed effects, and random effects estimators are considered.

25.4 Empirical Results

Descriptive Statistics

Table 25.2 provides a concise overview of the statistical characteristics of six distinct variables. Due to space limitations, the results of stationarity tests are not included. However, a cross-sectional dependence unit root test has been conducted, confirming that all variables reject the null hypothesis of non-stationarity.

Model Comparison

Table 25.3 presents the Akaike Information Criterion (AIC) values for different panel kink models, each characterized by a varying number of regimes (1 Regime, 2 Regime, and 3 Regime), and employs three distinct estimation methods (Pooled OLS, Fixed Effect, and Random Effect). Notably, the 3 Regime model consistently emerges with the lowest AIC value among the considered models, signifying its superior goodness of fit based on this criterion. Consequently, our analysis indicates that the three-regime panel kink models estimated using the Fixed Effect approach are the most suitable.

Table 25.2 Statistical data analysis

Variable	Mean	Median	Max	Min	Std.Dev
lnW	11.952	12.023	13.752	9.625	0.839
lnIS	−0.119	−0.117	−0.004	−0.410	0.066
lnU	−0.668	−0.677	−0.110	−1.314	0.253
lnT	−1.654	−1.931	0.568	−4.086	0.975
lnEI	−4.410	−4.419	−3.161	−5.809	0.460
lnTO	9.185	9.217	12.715	4.369	1.613

Table 25.3 Model selection

AIC	1 Regime	2 Regime	3 Regime
Pooled OLS	−289.204	−401.245	−426.084
Fixed effect	−809.109	−891.098	**−923.029**
Random effect	−712.749	−723.093	−787.313

Note: The bold number indicates the best-fit model

Examination of the N-Shaped EKC for Urbanization, Industrial Structure, and Water Pollution

The impacts of industrial structure and urbanization on wastewater are categorized into three regimes: Regime 1 corresponds to the region characterized by $\ln IS < \gamma^1_{\ln IS}$ and $\ln Ur < \gamma^1_{\ln Ur}$. Regime 2 encompasses the region characterized by $[\gamma^1_{\ln IS} \le \ln IS \le \gamma^2_{\ln IS}$ and $\gamma^1_{\ln Ur} \le \ln Ur \le \gamma^2_{\ln Ur}$. Regime 3 pertains to the region characterized by $\ln IS > \gamma^2_{\ln IS}$ and $\ln Ur > \gamma^2_{\ln Ur}$.

Based on the findings presented in Table 25.4, the fixed effect model demonstrates the best fit for the data. The analysis reveals that the variable *lnIS* exhibits negative coefficients in both regime 1 and regime 3, while the coefficients for regime 2 are positive and statistically significant. This investigation also explores the impact of the industrial structure under three distinct regimes, which are delineated by two threshold points, specifically, $\gamma^2_{\ln IS} = -0.1977$ and $\gamma^1_{\ln IS} = -0.0173$. The coefficients within these three regimes exhibit substantial variations: negative coefficients are evident in regimes 1 and 3, while a positive coefficient is notable in regime 2. This suggests that the N-shaped Environmental Kuznets Curve (EKC) does not hold true for the relationship between industrial structure and water pollution. Instead, our findings support the presence of an inverted N-shaped EKC, which aligns with the conclusions drawn by Fakher et al. [18].

The estimated results reveal distinct impacts of *lnIS* (natural logarithm of an unspecified variable) on wastewater discharge, depending on its value. When *lnIS* falls below −0.1977, it significantly reduces wastewater discharge by 1.5701% for every 1% increase. However, within the range of −0.1997 to −0.0173, *lnIS* has a significant positive impact, increasing wastewater discharge by 0.2959% with a 1% increase. This shift suggests that as the secondary industry and heavy industries grow, wastewater discharge rises, intensifying water pollution. Conversely, if *lnIS*

Table 25.4 Estimated results from three-regime panel kink regression

Variable	Fixed effect	Pooling	Random effect
$\ln IS < \gamma^1_{\ln IS}$	−1.5701*** (0.5634)	−0.7728 (0.4660)	−1.5907* (0.7102)
$\gamma^1_{\ln IS} \leq \ln IS \leq \gamma^2_{\ln IS}$	0.2959*** (0.0900)	0.3489*** (0.1500)	0.8939*** (0.1795)
$\ln IS > \gamma^2_{\ln IS}$	−6.8786 (9.7283)	−5.4237*** (1.6370)	−0.5312 (0.4471)
$\gamma^1_{\ln IS}$	−0.1977*** (0.0085)	−0.0669 (0.0548)	−0.2547** (0.0090)
$\gamma^2_{\ln IS}$	−0.0173*** (0.0017)	−0.0954*** (0.0068)	−0.2150*** (0.0047)
$\ln U < \gamma^1_{\ln U}$	1.2477*** (0.1765)	−0.2724 (0.1988)	0.9334*** (0.2290)
$\gamma^1_{\ln U} \leq \ln U \leq \gamma^2_{\ln U}$	0.0045 (0.0180)	0.0018 (0.0348)	−0.0241 (0.0193)
$\ln U > \gamma^2_{\ln U}$	2.2064** (1.0564)	1.5138*** (0.3174)	0.1525 (0.2064)
$\gamma^1_{\ln U}$	−0.9464*** (0.2836)	−0.3765*** (0.0080)	−0.9765*** (0.0749)
$\gamma^2_{\ln U}$	−0.3209 (44.2505)	−0.7765** (0.3179)	−0.6430*** (0.0774)
lnT	0.1232*** (0.0083)	0.4696*** (0.0160)	0.1741*** (0.013))
lnEI	0.0807*** (0.0189)	0.1385*** (0.0445)	0.0471** (0.0220)
lnTO	0.0267 (0.0203)	0.2951*** (0.0304)	0.0220 (0.0216)

Note: "*," "**," "***" indicate significance at 10%, 5% and 1% levels

exceeds −0.0173, it negatively affects wastewater discharge, reducing it by 6.8786% for every 1% increase in *lnIS*.

During the early stages of industrialization and rapid industrial growth, pollution tends to escalate, reflecting a negative phase. This is often marked by unregulated industrial activities prioritizing production over environmental concerns. However, as industrialization progresses and income or development reaches a certain threshold, environmental awareness and regulatory pressures typically increase. Industries respond by adopting cleaner technologies and practices, reducing pollution levels, constituting a positive phase. However, in some cases, as industrialization advances further, pollution might rise again, marking a return to a negative phase. The specific shape and timing of this curve can vary depending on factors like pollution types, government policies, technological advancements, and societal attitudes, but it provides a useful framework for understanding how pollution trends may change throughout different stages of industrial growth.

Regarding urbanization (*lnU*), it has a positive impact on wastewater in all three regimes. However, this impact is statistically significant only in regimes 1 and 3.

Comparing the coefficients across these three regimes, the smallest coefficient is in regime 2, the second smallest is in regime 1, and the largest is in regime 3, with a value of 2.2064.

The impact of lnU on wastewater is noteworthy. When lnU is less than −0.9496, there is a significant and positive effect, with a one-percentage-point increase in urbanization leading to a 1.2477 percentage-point increase in wastewater discharge. During regime 2, where lnU falls between −0.9464 and − 0.3209, there is no significant impact on wastewater, with a 1% increase in lnU contributing only a 0.0045% increase to wastewater discharge. However, the results indicate urbanization's significant and positive influence on wastewater after the kink point at −0.3209. This suggests that increasing urbanization in these three stages exacerbates water pollution.

25.5 Conclusion

The study's findings reveal that urbanization positively impacts water pollutant emission in all regimes except regime 2, which encompasses a specific range of urbanization levels. In contrast, industrial structure contributes positively to environmental degradation when industrial activity falls within a certain range of GDP proportions. Interestingly, the study concludes that there is no evidence of an inverted U-shaped or N-shaped EKC in China. However, it does identify an inverted N-shaped relationship between industrial structure and environmental degradation.

This inverted N shape is attributed to distinct phases in China's economic development: In the first stage, China is in the early phases of upgrading its industrial structure, with primary industry dominating. This phase does not significantly increase water pollutant emissions. The second stage represents the development phase of industrial structure upgrading, characterized by increased activity in secondary industries, particularly heavy industries, leading to heightened environmental pollution, notably through industrial wastewater discharge. In the third stage, China prioritizes the development of tertiary industries, reducing the GDP share of the secondary industry and consequently decreasing wastewater discharge to some extent. Moreover, the study finds that environmental investment and technology have a positive relationship with water pollutant emission, while trade openness does not exhibit a significant relationship with pollution levels.

Based on the study's findings, several policy recommendations can be formulated. Firstly, it is essential to tailor environmental regulations to account for the varying impacts of industrial structure and urbanization at different stages of development. Secondly, a focus on sustainable urban planning should be prioritized, incorporating green infrastructure, efficient public transportation, and sustainable building practices to mitigate the environmental consequences of urbanization. Thirdly, there should be investments in research and development of clean technologies to help industries reduce pollution, alongside incentives for their adoption.

Acknowledgments We are grateful for financial support from Chiang Mai University.

References

1. Jahan S, Singh A (2023) Causes and impact of industrial effluents on receiving water bodies: a review. Malays J Sci Adv Technol:111–121
2. Noor R, Maqsood A, Baig A, Pande CB, Zahra SM, Saad A et al (2023) A comprehensive review on water pollution, South Asia Region: Pakistan. Urban Clim 48:101413
3. Wang Y, Chen X, Borthwick AG, Li T, Liu H, Yang S et al (2020) Sustainability of global golden inland waterways. Nat Commun 11(1):1553
4. Leitão A (2010) Corruption and the environmental Kuznets curve: empirical evidence for sulfur. Ecol Econ 69(11):2191–2201
5. Culas RJ (2012) REDD and forest transition: tunneling through the environmental Kuznets curve. Ecol Econ 79:44–51
6. Torras M, Boyce JK (1998) Income, inequality, and pollution: a reassessment of the environmental Kuznets curve. Ecol Econ 25(2):147–160
7. Alvarez-Herranz A, Balsalobre-Lorente D (2015) Energy regulation in the EKC model with a dampening effect. J Environ Anall Chem 2(3):1–10
8. Shehzad K, Zeraibi A, Zaman U (2022) Testing the N-shaped environmental Kuznets curve in Algeria: an imperious role of natural resources and economic globalization. Resour Policy 77:102700
9. Grossman GM, Krueger AB (1995) Economic growth and the environment. Q J Econ 110(2):353–377
10. Panayotou T (1997) Demystifying the environmental Kuznets curve: turning a black box into a policy tool. Environ Dev Econ 2(4):465–484
11. Friedl B, Getzner M (2003) Determinants of CO2 emissions in a small open economy. Ecol Econ 45(1):133–148
12. Huan Y, Hassan MS, Tahir MN, Mahmood H, Al-Darwesh HRI (2022) The role of energy use in testing N–shaped relation between industrial development and environmental quality for Chinese economy. Energ Strat Rev 43:100905
13. Zhang Y, Zhou Q, Jiang L (2017) Panel kink regression with an unknown threshold. Econ Lett 157:116–121
14. Tibprasorn P, Maneejuk P, Sriboochitta S (2017) Generalized information theoretical approach to panel regression kink model. Thai J Math:133–145
15. Grossman GM, Krueger AB (1991) Environmental impacts of a North American Free Trade Agreement. Working Paper No. 3914
16. Panayotou T (1993) Empirical tests and policy analysis of environmental degradation at different stages of economic development. ILO; Technology and Employment Programme, Geneva
17. De Bruyn SM, van den Bergh JC, Opschoor JB (1998) Economic growth and emissions: reconsidering the empirical basis of environmental Kuznets curves. Ecol Econ 25(2):161–175
18. Fakher HA, Ahmed Z, Acheampong AO, Nathaniel SP (2023) Renewable energy, nonrenewable energy, and environmental quality nexus: an investigation of the N-shaped environmental Kuznets curve based on six environmental indicators. Energy 263:125660

Chapter 26
Does the Level of Inclusive and Sustainable Industrialization Matter in the Effect of Electric Vehicles on Air Pollution?

Woraphon Yamaka and Chanisara Kaewpimol

Abstract This study investigates the impact of electric vehicle (EV) adoption rates on air pollution, considering their varying levels of inclusive and sustainable industrialization. The study utilizes data from 45 countries worldwide over the period 2010 to 2020, employing a panel threshold model to analyze the relationships. The Sustainable Development Goal (SDG) 9 Index, a measure of sustainable industrialization, is used as the threshold variable to categorize countries into three groups: low, medium, and high levels of green industrial development. The findings reveal a nonlinear relationship between EV adoption and air pollution and the significant role of inclusive and sustainable industrialization. In countries with low and medium levels of green industrial development, increased EV usage relative to population surprisingly leads to a statistically significant increase in air pollution. Conversely, in countries with high levels of green industrial development, a rise in the proportion of EVs to the population significantly contributes to reducing air pollution. These results highlight the importance of considering a country's level of sustainable industrialization when evaluating the environmental impact of EV adoption. The study suggests that the effectiveness of EV adoption in reducing air pollution is contingent on a country's overall industrial sustainability.

Keywords Air pollution · Electric vehicle adoption · Green industrial development · Panel threshold approach

26.1 Introduction

The environmental policies of various countries aim to reduce air pollution by promoting the use of electric vehicles (EVs). This study seeks to examine the impact of EV usage on air pollution reduction and how this effect varies across countries with

W. Yamaka (✉) · C. Kaewpimol
Center of Excellence in Econometrics, Faculty of Economics, Chiang Mai University, Chiang Mai, Thailand

L. Chen (ed.), *Clean Energy Research*, Green Energy and Technology,
https://doi.org/10.1007/978-3-031-99312-1_26

different levels of environmentally friendly industrialization. Since the Industrial Revolution began in 1760, carbon dioxide emissions from industrial factories have increased significantly across many countries, contributing to abnormal natural phenomena, commonly referred to as anthropogenic global warming. Numerous empirical studies have extensively explored the causes of anthropogenic global warming, with most concluding that the rise in global temperatures is due to the combustion of fossil fuels, which elevates atmospheric carbon dioxide levels. As the concentration of CO_2 increases, global temperatures rise correspondingly [2]. Despite the well-established understanding of the link between rising CO_2 levels and global warming, efforts to control these emissions face several challenges, including advances in power generation, industrial, and transportation sectors. In recent decades, technological developments across these sectors have progressed significantly. Over the past 10 years, many governments worldwide have implemented policies aimed at reducing carbon dioxide and greenhouse gas emissions, such as developing low-carbon technologies to mitigate emissions from the transportation sector [3]. These policies present significant challenges for transportation, particularly as the number of vehicles on the road is projected to double by 2050 [20]. The relationship between the number of vehicles and CO_2 emissions is clear: the more vehicles on the road, the higher the CO_2 and greenhouse gas emissions in the atmosphere.

In response to this situation, the transportation sector has significantly advanced the development of vehicle technologies. A key example is the rapid progress in EVs, which are designed to reduce emissions from internal combustion engines currently in use [8, 21]. Li et al. [22] confirmed that increased adoption of EVs can help reduce greenhouse gas and CO_2 emissions from the transportation sector, particularly in countries with high emission levels. The growing popularity of EVs reflects their appeal as zero-emission vehicles, which offer substantial environmental benefits over traditional ICE vehicles [1]. As such, understanding the environmental impact of these different vehicle types is essential for developing effective strategies to reduce pollution and transition toward sustainable transportation.

Several studies have highlighted that a higher penetration of EVs can contribute significantly to reducing air pollutants. Sun et al. [15] noted that increasing EV usage has the potential to lower atmospheric carbon dioxide (CO_2) levels, with the reduction in pollution becoming more significant in countries with substantial renewable energy use and government policies that actively support EV adoption [11]. Beyond reducing CO_2, EVs also contribute to substantial decreases in the concentration of pollutants like PM2.5 and nitrogen oxides, by as much as 45.8% and 35%, respectively, as evidenced by studies conducted in Beijing, China [17]. Compared to conventional gasoline-powered vehicles, electric vehicles produce lower greenhouse gas emissions, particularly when the electricity used for charging is sourced from renewable energy [9]. Increasing the share of renewable energy in the electricity grid further enhances the efficiency of greenhouse gas reduction associated with EV usage. Moreover, Yang et al. [18] have examined various approaches to regulating vehicle emissions, such as implementing direct and indirect taxes on vehicles according to their CO_2 emissions levels and fuel taxes. Studies have found that tightening CO_2 emission standards for new vehicles is the most effective strategy to reduce emissions. Consequently, the broader adoption of EVs can help countries achieve stricter emission standards for new vehicles, leading to a more effective reduction of CO_2 emissions nationwide.

Although EVs have the potential to positively impact air pollution reduction, this outcome is not guaranteed. Different studies conducted by various authors present varying results, and negative impacts of EVs on pollution have also been identified. Most of these negative impacts are associated with the automobile manufacturing industry and the disposal of EV batteries [16]. For instance, the production of electric vehicles generally results in higher carbon dioxide emissions compared to the production of conventional gasoline-powered vehicles. This is primarily due to the greater use of nickel and copper in the manufacturing process, which are essential components of EV batteries. As a result, the EV manufacturing industry tends to generate higher pollution levels than the traditional automotive industry [8]. Moreover, beyond the emissions linked to battery production, the improper disposal of used EV batteries can further contribute to air pollution. Additionally, the source of electricity used to charge EVs can also impact pollution levels. As Jochem et al. [10] pointed out, in many countries, electricity production still relies heavily on coal, a major pollutant. Consequently, the electricity generated to charge EVs may still have a negative impact on air pollution levels [13].

The impact of EVs on air pollution is influenced not only by the carbon intensity of a country's electricity grid but also by its progress in achieving inclusive and sustainable industrialization. As outlined by Moro and Lonza [12] and Albrechtowicz [1], this relationship is further shaped by Sustainable Development Goal (SDG) 9—Industry, Innovation, and Infrastructure, which emphasizes sustainable industrialization as a key driver of economic growth, job creation, and environmental progress [4]. By analyzing the interplay between SDG 9 performance and per capita carbon emissions, we gain insights into how EV adoption affects air pollution across diverse national contexts. For instance, countries like Armenia (SDG 9 score: 82.5, CO_2 per capita: 2.40 metric tons), Albania (SDG 9 score: 75.3, CO_2 per capita: 1.54 metric tons), and Sri Lanka (SDG 9 score: 73.8, CO_2 per capita: 0.99 metric tons) demonstrate that strong SDG 9 performance, indicative of sustainable industrial practices, can enhance the positive impact of EV adoption. These nations likely leverage inclusive industrialization and clean energy technologies, making EVs more effective in reducing emissions and improving air quality. Conversely, countries such as Bahrain (SDG 9 score: 78.9, CO_2 per capita: 21.98 metric tons), Qatar (SDG 9 score: 75.2, CO_2 per capita: 31.73 metric tons), and Saudi Arabia (SDG 9 score: 71.3, CO2 per capita: 14.27 metric tons) also show high SDG 9 scores but with significantly higher CO_2 emissions per capita. This suggests that despite advancements in industrialization, the sustainability of their practices, including reliance on cleaner energy sources and inclusive infrastructure, is critical to maximizing the environmental benefits of EVs. Thus, adopting EVs alone is insufficient; broader progress toward inclusive and sustainable industrialization is necessary to achieve meaningful reductions in air pollution.

While previous studies have examined the environmental benefits of EVs, there is limited exploration of how EV adoption impacts air pollution across countries with varying levels of industrial development. Most prior research has focused on the carbon intensity of electricity grids, overlooking how differences in sustainable industrialization influence the effectiveness of EV adoption in reducing pollution. This gap is critical because sustainable industrialization plays a key role in shaping the long-term environmental impact of EV adoption.

This study addresses the research gap by using SDG 9 as a threshold variable to analyze the impact of EV adoption on air pollution across 45 countries from 2010 to 2020. By incorporating SDG 9, the research uniquely investigates how varying levels of industrial sustainability affect the extent to which EVs contribute to reducing emissions. This approach allows for a more nuanced understanding of how both clean energy and inclusive industrial practices combine to influence environmental outcomes. To achieve this, we employ a panel threshold regression model, which is well-suited for this analysis because it captures the nonlinear relationships between variables across different regimes of industrial development [7]. The threshold regression model allows us to examine whether the impact of EV adoption on air pollution differs based on a country's level of progress in sustainable industrialization, as represented by SDG 9. This model can reveal whether there are distinct thresholds where the effectiveness of EV adoption in reducing pollution changes, providing a clearer picture of how inclusive industrialization amplifies or limits the environmental benefits of EVs. The results of this study are expected to offer valuable policy insights, enabling governments to better target their efforts in promoting EV adoption and sustainable industrial practices.

26.2 Data and Methodology

Data

To investigate the impact of electric vehicle (EV) usage on air pollution levels across countries with varying levels of industrial development, this study examines 45 countries using secondary panel data. The countries included in the dataset are Australia, Austria, Belgium, Brazil, Bulgaria, Canada, Chile, China, Croatia, Cyprus, Czechia, Denmark, Estonia, Finland, France, Germany, Greece, Hungary, Iceland, India, Ireland, Israel, Italy, Japan, the Republic of Korea, Latvia, Lithuania, Luxembourg, Malta, Mexico, the Netherlands, New Zealand, Norway, Poland, Portugal, Romania, Slovakia, Slovenia, South Africa, Spain, Sweden, Switzerland, Türkiye, the United Kingdom, and the United States of America. The dataset spans from 2010 to 2020, utilizing annual data. The analysis focuses on EV usage per capita in each country as the independent variable, while the dependent variable is the level of carbon dioxide (CO_2) emissions. Three control variables are also incorporated: Gross Domestic Product (GDP) per capita, the urban population percentage, and the use of renewable energy in each country [15].

Given the study's use of the panel threshold model, an industrial development index was selected to identify the threshold points within the model. The SDG9 index was chosen for this purpose, focusing on three primary objectives: Industry, Innovation, and Infrastructure. SDG9 aims to measure a country's infrastructural development, industrial advancement, and innovation capacity. This index includes various components such as CO_2 emissions per unit of value added, manufacturing value added as a percentage of GDP, and the share of manufacturing employment in total employment. The sources of this data are provided in Table 26.1. Table 26.1 provides key statistics on our variables.

To enhance robustness, all data were log-transformed into their natural logarithm form (ln) and lagged by one period to address potential endogeneity issues. The study also checked for multicollinearity using the Variance Inflation Factor (VIF) test and analyzed the correlations between the variables, as shown in Table 26.2. Additionally, the data's stationarity was tested using the Levin-Lin-Chu (LLC), Hadri, and Im-Pesaran-Shin (IPS) tests. The results from these stationarity tests indicated significance at the 0.01 level (Table 26.3), confirming that the data is stationary and suitable for further analysis using the panel threshold model.

Table 26.1 Basic information about the variables

Variable	Explanation	Mean	SD	Min	Max	Source
CO_2	Carbon dioxide emissions per capita (tons)	7.053	7.765	1.004	29.320	OECD data
EV	Ratio of electric cars per 1000 inhabitants	0.330	1.140	0.000	14.294	IEA/European Commission/ SGD report
RENEW	Renewable energy consumption (% of total final energy consumption)	21.584	15.490	1.220	82.790	The World Bank
GDP	GDP per capita (current US)	34421.262	24136.366	1350.634	123678.702	The World Bank
URB	Urbanization (percentage of total population)	75.046	13.778	30.930	98.079	The World Bank
SDG9	SDG-9 industry index score	42.942	13.376	17.463	77.196	UNIDO

Table 26.2 Correlation matrix between variables and variance inflation factor (VIF)

	VIF test	lnEV	lnRENEW	lnGDP	lnURB
lnEV	9.120	1.000			
lnRENEW	1.043	0.030	1.000		
lnGDP	2.108	0.259	−0.065	1.000	
lnURB	1.759	0.169	−0.185	0.636	1.000

Table 26.3 Panel unit root test

Variable	LLC	IPS	HADRI
lnCO2	−3.229***	0.687***	21.288***
lnEV	−8.722***	−11.601***	22.565***
lnRENEW	−3.334***	0.733	25.870***
lnGDP	−7.355***	−3.574***	17.022***
lnURB	−33.152***	−37.380***	40.712***

Note: ***, **, * indicate significance at the 0.01, 0.05, and 0.10 levels, respectively

26.3 Methodology

Panel Threshold Regression

Hansen [7] proposed a model for estimation and inference for panel data. The model can be expressed as follows:

Single-threshold regression

$$\begin{aligned} \ln CO2_{it} &= \alpha + \beta_1 \ln EV_{it} I(SDG9_{it} \le \gamma_1) + \beta_2 \ln EV_{it} I(SDG9_{it} > \gamma_1) \\ &+ \theta_1 lnRENEW_{it} + \theta_2 \ln GDP_{it} + \theta_3 \ln URB_{it} + u_i + \varepsilon_{it} \end{aligned} \quad (26.1)$$

Double-threshold regression

$$\begin{aligned} \ln CO2_{it} &= \alpha + \beta_1 \ln EV_{it} I(SDG9_{it} \le \gamma_1) \\ &+\beta_2 \ln EV_{it} I(\gamma_1 < SDG9_{it} \le \gamma_2) \\ &+\beta_2 \ln EV_{it} I(SDG9_{it} > \gamma_2) + \theta_1 lnRENEW_{it} \\ &+\theta_2 \ln GDP_{it} + \theta_3 \ln URB_{it} + u_i + \varepsilon_{it} \end{aligned} \quad (26.2)$$

where $i = 1, \ldots, N$ represents the cross-sectional data, and $t = 1, \ldots, T$ indicates the time-series data. The regime-independent variables in this model are represented by lnEV_{it}, while the coefficients for the three subgroups or regimes are represented by β_1, β_2 and β_3. The model uses threshold values γor a threshold variable, SDG9, to segment the data into different regimes. The indicator function $I(\cdot)$represents sample splitting based on the value of the threshold variable SDG9. In this model, $u_i$is defined as the country-specific fixed effect, while ε_{it} represents the error term of the model. Additionally, the control variables lnRENEW, lnGDP, and lnURB are included with their respective coefficients θ_1, θ_2 and θ_3, accounting for renewable energy use, GDP, and urbanization effects, respectively.

26.4 Empirical Results

To determine the optimal number of regimes in our panel threshold model, we employed the Likelihood Ratio (LR) test, with results presented in Table 26.4. Using the SDG9 index as our threshold variable, which represents Inclusive and Sustainable Industrialization, we identified two significant thresholds at the 5% and 1% levels. The estimated threshold values are 26.643 and 51.660, with 95% confidence intervals of [22.712, 67.942] and [32.447, 64.796], respectively. These findings suggest that the relationship between electric vehicle (EV) adoption and CO_2 emissions varies across three distinct regimes of industrial development and sustainability. Countries below the lower threshold likely represent less industrialized or less sustainable economies, those between thresholds indicate moderate development, while those above the upper threshold suggest highly industrialized and sustainable nations. This segmentation provides crucial insights into how the impact of EV adoption on air pollution may differ based on a country's level of industrial development and sustainability practices.

The empirical results, as presented in Table 26.5, highlight the critical role of inclusive and sustainable industrialization (measured by the SDG9 Index) in moderating the effect of EV usage on air pollution levels.

The results presented in Table 26.5 reveal a complex relationship between electric vehicle (EV) adoption and CO2 emissions, varying significantly across different levels of industrial sustainability as measured by SDG9. Using a double panel threshold regression, the analysis divides the sample into three regimes: low SDG9 ($\leq$26.643), medium SDG9 (26.643 < SDG9 $\leq$ 51.660), and high SDG9 (>51.660). In the low SDG9 regime, an increase in EV use is associated with a slight increase in CO2 emissions (0.004 or 0.4%), significant at the 0.10 level. This counterintuitive finding aligns with research by Qiao et al. [14], who noted that EV benefits heavily depend on the carbon intensity of the electricity grid. In less developed countries, carbon-intensive electricity production, lifecycle emissions from EV manufacturing, and inefficient early-stage integration may contribute to this effect. These

Table 26.4 The result of the single and double threshold effect test

	Single	Double
LR-test	11.634***	10.924**
P-value	0.009	0.012
	Threshold value 1	Threshold value 2
	26.643 [22.712, 67.942]	51.660 [32.447, 64.796]

Note: ***, **, * indicate significance at the 0.01, 0.05, and 0.10 levels, respectively

Table 26.5 Double panel threshold regression results

	SDG 9		
	SDG9 ≤ 26.643	26.643 < SDG9 ≤ 51.660	SDG9 > 51.660
	Low SDG 9	Medium SDG 9	High SDG 9
lnEV	0.004*	0.001	−0.004***
	(0.002)	(0.001)	(0.002)
	−0.065***	−0.091***	0.240***
	(0.013)	(0.018)	(0.042)
lnRENEW	−0.065*** (0.013)		
lnGDP	−0.091*** (0.018)		
lnURB	0.240*** (0.042)		
R-squared			
Within	0.7934		
Between	0.0323		
Overall	0.0045		

Note: ***, **, * indicate significance at the 0.01, 0.05, and 0.10 levels, respectively.

factors can potentially offset the intended environmental benefits of EVs in the short term. The medium SDG9 regime shows a negligible relationship between EV use and CO_2 emissions (0.001 or 0.1%) with no statistical significance. This suggests a transitional phase where the benefits of EV adoption have yet to materialize fully. Holland et al. [9] supported this finding, observing that the environmental benefits of EVs can vary substantially across different contexts. Countries in this regime may be experiencing a balance between the emerging benefits of EVs and remaining systemic inefficiencies, or they may be in the process of upgrading their energy grids and EV infrastructure. In contrast, the high SDG9 regime demonstrates a significant reduction in CO_2 emissions with increased EV use (−0.004 or −0.4%), which is significant at the 0.01 level. This finding is consistent with the intended benefits of EV technology and is supported by Zhao et al. [19], who confirmed that EVs can significantly reduce greenhouse gas emissions in regions with low-carbon electricity. Countries in this regime likely benefit from cleaner electricity grids, more efficient charging infrastructure, and supportive policies that enhance the environmental impact of EVs.

In terms of control variables, renewable energy and GDP show significant negative effects on CO_2 emissions in lower and medium SDG9 regimes. This aligns with the environmental Kuznets curve concept discussed by Grossman and Krueger [6], suggesting that environmental degradation initially increases with economic development but then decreases after a certain level of economic growth is achieved. Interestingly, urbanization demonstrates a strong positive impact on reducing CO_2 emissions in the high SDG9 regime. This finding is supported by research from Glaeser and Kahn [5], who argue that cities in developed countries can be more environmentally friendly due to economies of scale and advanced technologies. It highlights that urban development, when supported by advanced green technologies, can drive significant reductions in CO_2 emissions.

26.5 Conclusion and Recommendation

This study investigates the impact of electric vehicle (EV) adoption on CO_2 emissions across 45 countries, using SDG9 (Industry, Innovation, and Infrastructure) as a threshold variable. By employing a double panel threshold regression, the study divides the sample into three distinct regimes based on SDG9 levels: low, medium, and high. The findings provide valuable insights into the relationship between EV use and environmental outcomes in different stages of industrial development. In countries with low SDG9 scores, increased EV adoption is associated with a slight but significant rise in CO_2 emissions. This suggests that in less developed industrial sectors, the environmental benefits of EVs are not fully realized, possibly due to reliance on carbon-intensive electricity generation or inefficiencies in early EV infrastructure. In medium-SDG9 countries, the effect of EV use on CO_2 emissions is negligible, indicating a transitional phase where the benefits of EVs have yet to materialize significantly. Conversely, in high-SDG9 countries, where green infrastructure and sustainable industry are more advanced, the adoption of EVs has a significant negative impact on CO2 emissions, demonstrating that these countries are better positioned to harness the environmental benefits of electric vehicles. Cleaner energy grids, mature EV technology, and supportive policies likely contribute to this positive outcome.

Numerous policy suggestions were drawn from the existing results. For countries with low and medium levels of green industrial development, governments should adopt a more comprehensive approach to pollution reduction, beyond simply promoting EV use. This includes enforcing stricter carbon emission standards across both the automotive and energy sectors, transitioning towards cleaner energy sources for EV charging, and focusing on extending battery lifespan. Importantly, these countries should prioritize the development of EV infrastructure—such as charging stations and supply chains—before expecting meaningful environmental benefits from EV adoption. In countries with high levels of green industrial development, policymakers should further accelerate EV adoption by offering financial incentives, expanding the charging network, and phasing out support for conventional vehicles in favor of electric ones. These governments should also invest in renewable energy to ensure that the rise in EV use contributes to actual reductions in carbon emissions. Moreover, innovation in environmentally sustainable battery technology, alongside the development of efficient recycling and disposal methods, will be critical for ensuring the long-term environmental benefits of electric vehicles.

Acknowledgments We are grateful for financial support from the Centre of Excellence in Econometrics, Faculty of Economics, Chiang Mai University.

References

1. Albrechtowicz P (2023) Electric vehicle impact on the environment in terms of the electric energy source—case study. Energy Rep 9:3813–3821
2. Arrhenius S (1896) XXXI. On the influence of carbonic acid in the air upon the temperature of the ground. The London, Edinburgh, and Dublin Philosophical Magazine and Journal of Science 41(251):237–276
3. Brutschin E, Pianta S, Tavoni M, Riahi K, Bosetti V, Marangoni G, Van Ruijven BJ (2021) A multidimensional feasibility evaluation of low-carbon scenarios. Environ Res Lett 16(6):064069
4. Estevão J, Lopes JD, Penela D (2023) SDG9 and the competitiveness: employing mixed methods to understand how countries can use science to compete. Technol Forecast Soc Chang 187:122178
5. Glaeser EL, Kahn ME (2010) The greenness of cities: carbon dioxide emissions and urban development. J Urban Econ 67(3):404–418
6. Grossman GM, Krueger AB (1995) Economic growth and the environment. Q J Econ 110(2):353–377
7. Hansen BE (1999) Threshold effects in non-dynamic panels: estimation, testing, and inference. J Econ 93(2):345–368
8. Hawkins TR, Singh B, Majeau-Bettez G, Strømman AH (2013) Comparative environmental life cycle assessment of conventional and electric vehicles. J Ind Ecol 17(1):53–64
9. Holland SP, Mansur ET, Muller NZ, Yates AJ (2016) Are there environmental benefits from driving electric vehicles? The importance of local factors. Am Econ Rev 106(12):3700–3729
10. Jochem P, Plötz P, Ng WS, Rothengatter W (2018) The contribution of electric vehicles to environmental challenges in transport. Editorial 64:1
11. Koengkan M, Fuinhas JA, Teixeira M, Kazemzadeh E, Auza A, Dehdar F, Osmani F (2022) The capacity of battery-electric and plug-in hybrid electric vehicles to mitigate CO2 emissions: macroeconomic evidence from european union countries. World Electric Vehicle J 13(4):58
12. Moro A, Lonza L (2018) Electricity carbon intensity in European member states: impacts on GHG emissions of electric vehicles. Transp Res Part D: Transp Environ 64:5–14
13. Onn CC, Mohd NS, Yuen CW, Loo SC, Koting S, Abd Rashid AF et al (2018) Greenhouse gas emissions associated with electric vehicle charging: the impact of electricity generation mix in a developing country. Transp Res Part D: Transp Environ 64:15–22
14. Qiao Q, Zhao F, Liu Z, Jiang S, Hao H (2017) Comparative study on life cycle CO2 emissions from the production of electric and conventional vehicles in China. Energy Procedia 105:3584–3595
15. Sun D, Kyere F, Sampene AK, Asante D, Kumah NYG (2023) An investigation on the role of electric vehicles in alleviating environmental pollution: evidence from five leading economies. Environ Sci Pollut Res 30(7):18244–18259
16. Xia X, Li P (2022) A review of the life cycle assessment of electric vehicles: considering the influence of batteries. Sci Total Environ 814:152870
17. Xue Y, Nie T, Cui Y, Liu X, Chen J, Wu X et al (2023) Prediction of air pollution reduction benefits and atmospheric environmental quality improvement effects from electric vehicle deployment in Beijing, China. Int J Environ Sci Technol 20(10):10973–10982
18. Yang Z, Mock P, German J, Bandivadekar A, Lah O (2018) On a pathway to de-carbonization–a comparison of new passenger car CO2 emission standards and taxation measures in the G20 countries. Transp Res Part D: Transp Environ 64:53–69
19. Zhao X, Hu H, Yuan H, Chu X (2023) How does adoption of electric vehicles reduce carbon emissions? Evidence from China. Heliyon 9(9):e20296
20. Hao, H., Geng, Y., & Sarkis, J. (2016). Carbon footprint of global passenger cars: Scenarios through 2050. Energy, 101, 121–131.

21. Andersson, Ö., & Börjesson, P. (2021). The greenhouse gas emissions of an electrified vehicle combined with renewable fuels: Life cycle assessment and policy implications. Applied Energy, 289, 116621.
22. Li, W., Long, R., Zhang, L., Cheng, X., He, Z., & Chen, F. (2021). How the uptake of electric vehicles in China leads to emissions transfer: An Analysis from the perspective of inter-provincial electricity trading. Sustainable Production and Consumption, 28, 1006–1017.

Chapter 27
A Holistic Approach to Energy Consumption in Kuwait Academic Buildings: Energy Literacy, Behavior Change, and Technological Participation

Majdi M. Alomari, Hania EL-Kanj, Ayse Topal, and Nafesah I. Alshdaifat

Abstract This research describes a process of implementing energy efficiency and sustainability at the Australian University (AU) in Kuwait. The outlined steps and their respective outcomes are based on prior studies that examined key aspects of energy efficiency, including energy literacy assessments, behavior change interventions, energy-efficient technologies, and educational strategies. This concluding study integrates these steps into a cohesive and comprehensive process aimed at achieving sustainable energy practices at AU. The process's objective is to provide a strategic and actionable roadmap that combines educational, behavioral, and technological interventions, while placing an emphasis on the interconnected nature of each phase. This holistic approach offers valuable insights for academic institutions striving to improve energy efficiency and promote sustainability.

Keywords Academic Buildings · Energy Consumption · Energy Efficiency · Energy Literacy · Sustainability

27.1 Introduction

In the recent past, buildings in Kuwait have garnered significant attention due to their high energy consumption costs [1]. The improvement of building energy efficiency is influenced both directly and indirectly by occupational behavior. However,

M. M. Alomari (✉) · H. EL-Kanj
Australian University (AU), Mishref, Kuwait
e-mail: m.alomari@au.edu.kw

A. Topal
Nigde Omer Halisdemir University, Nigde, Turkey

N. I. Alshdaifat
JoVision, Hamburg, Germany

L. Chen (ed.), *Clean Energy Research*, Green Energy and Technology,
https://doi.org/10.1007/978-3-031-99312-1_27

occupant behavior is often treated as secondary or less significant compared to energy-saving technological solutions. Notably, energy behavior feedback interventions can result in an initial improvement of up to 15% in energy savings [2]. Behavior changes complement energy literacy in reducing energy consumption [3]. Behavioral strategies such as feedback mechanisms and incentives for sustainable actions have proven to significantly influence energy-saving behaviors [4, 5]. In Kuwait, cultural factors like collective values and community engagement further enhance the impact of such targeted interventions.

Energy consumption in academic buildings is a pressing issue, especially in regions like Kuwait, where extreme climate conditions and significant energy demand converge. A comprehensive approach to addressing this challenge involves focusing on three key areas: energy literacy, behavior change, and technological participation. Research underscores the importance of energy literacy, awareness and understanding of energy usage and its environmental impact in fostering sustainable practices. For example, DeWaters and Powers [6] and Brounen et al. [7] found that higher energy literacy correlates with reduced energy consumption behaviors. Similarly, there are several studies about energy conservation, sustainability, and energy awareness in the Gulf Cooperation Council (GCC) countries [8–11]. However, efforts to integrate energy literacy programs remain limited, highlighting a gap in leveraging education to curtail energy-intensive practices in Kuwait's academic institutions.

Drawing from these insights, this study explores the interconnections among energy literacy, energy behavior, and technology adoption among students and staff at the Australian University (AU) in Kuwait. It outlines a structured process aimed at enhancing energy efficiency in AU's academic buildings through a series of steps, each with specific expected outcomes. To achieve this, the study focuses on the following sub-objectives:

- Sub-objective 1: Evaluate the energy literacy levels of students, teaching staff, and non-teaching staff by identifying their awareness, knowledge, and energy-related behaviors, highlighting strengths and gaps.
- Sub-objective 2: Explore and analyze energy-efficient technologies that align with the structural and operational characteristics of academic buildings in Kuwait, particularly at AU.
- Sub-objective 3: Conduct a systematic review of empirical evidence to identify the most effective interventions for fostering sustainable energy behaviors within academic environments.
- Sub-objective 4: Assess the impact of integrating theoretical education with practical activities, such as energy audits, to reduce energy consumption and cultivate sustainable practices.
- Sub-objective 5: Collect and analyze feedback from building users about the performance of current energy systems, focusing on behavior, system acceptance, and areas for enhancement.

These objectives aim to deliver actionable insights and a strategic roadmap. By integrating educational initiatives, behavioral interventions, and technological

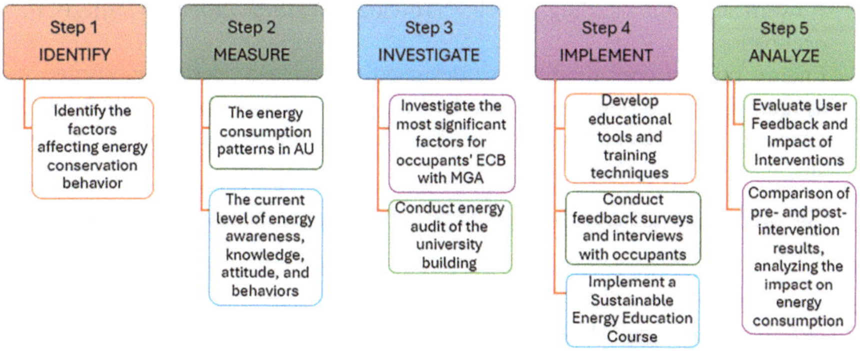

Fig. 27.1 Flowchart of study

advancements, the study seeks to advance energy efficiency and sustainability across academic institutions in Kuwait. Figure 27.1 shows the flowchart of this study.

27.2 Methodology

This methodology is designed to provide a clear and actionable roadmap for enhancing energy efficiency at AU through a combination of education, behavioral change, and technological participation. Each step is interconnected to ensure a comprehensive understanding of the factors influencing energy conservation and the development of practical solutions for sustainable energy practices.

Specific methods will be implemented to address the goals and objectives:

1. Measure energy literacy: Two survey questionnaires were conducted on students to measure opinions, beliefs, attitudes, and behaviors based on the theory of planned behavior (TPB) at the Australian University of Kuwait. The first questionnaire was administered during the initial trimester of the school year to assess the students' level of energy literacy (awareness, knowledge, and energy consumption behavior. The second questionnaire, which will provide all predictor variables, will be administered during the final trimester after the students are involved in the energy educational program.
2. An energy auditing procedure is used to establish energy conservation flows, identify the most effective energy efficiency options, and consider building occupants' efficiency improvements data collection at campus buildings. Measures of building to inform the best possible future efficient technologies options based on electricity use in buildings and building characteristics.
3. Multi-group analysis (MGA) was applied using Partial Least Squares Structural Equations Modelling (PLS-SEM) to investigate the factors for occupants' energy conservation behavior to detect any significant differences among the groups the

Kuwaiti academic building, which was divided into three groups in terms of their occupation.

4. Energy literacy education course designed to increase energy knowledge and awareness and encourage students to create positive energy behavior. The energy course emphasizes not only content knowledge but also has a positive effect on students' energy behaviors.

The methods to achieve these goals and objectives are shown in Fig. 27.2.

27.3 Results and Discussions

The primary aim of this study was to analyze the links among energy efficiency technologies, energy literacy, and behavior changes in university buildings, with a focus on understanding how energy literacy impacts behavior and how interventions can be designed to improve energy conservation. The study followed a structured procedure to achieve these objectives, beginning with surveys and ending with an energy literacy course. The steps and methodology adopted for this study were essential for identifying the factors that influence energy-saving behavior in academic environments.

Step 1: Influence of Environmental Knowledge on Energy-Related Occupant Behavior

The first survey aimed to identify the most effective behavior change interventions for improving energy use in academic buildings [12, 13]. It examined the cognitive (knowledge, understanding, skills), affective (attitudes, sensitivity), and behavioral (intention, participation, action) areas affecting energy-related understandings. The results indicated that attitude was the most significant factor influencing energy-related behavior, as it strongly impacts intention. The study found that when university users exhibit more positive attitudes towards energy conservation, they are more likely to engage in energy-saving behaviors.

The findings suggest that interventions should focus on enhancing positive attitudes, using both written and verbal communication tools to spread the message of energy conservation.

Step 2: Energy Literacy Assessment

The second step involved assessing the energy literacy levels of the occupants at the Australian University of Kuwait (AU) using an energy literacy survey [14, 15]. Data from the survey and subsequent interviews revealed key reasons behind students' low energy literacy. One major finding was that students' lack of performance in energy literacy was linked to a lack of sufficient knowledge and awareness about energy conservation. In contrast, faculty members showed more engagement and better understanding of energy-related issues. This difference highlighted the importance of understanding the factors that influence energy behavior at different levels of university staff and students. The analysis also emphasized that energy

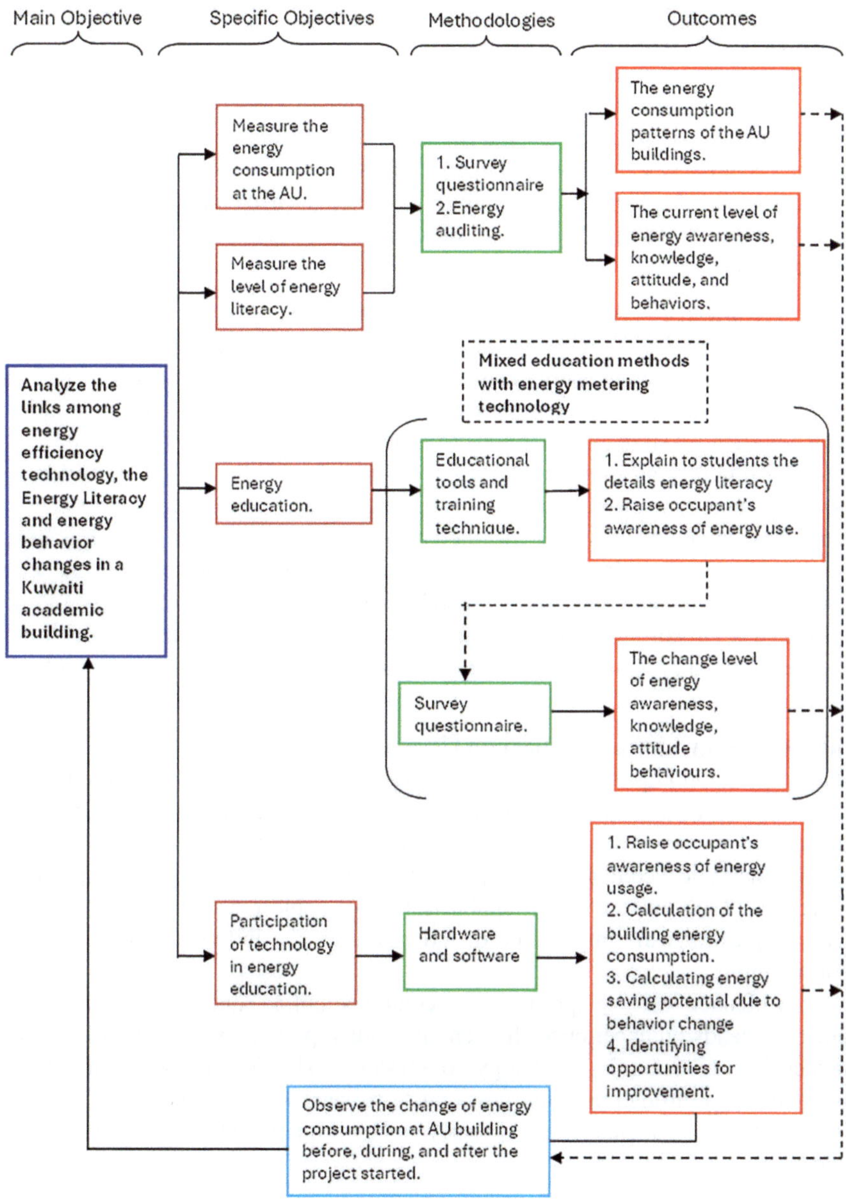

Fig. 27.2 The framework of research methodology

conservation behaviors are critical for transforming academic buildings into more sustainable campuses.

The survey also revealed that both students and faculty perceived energy conservation as an important issue, with a strong belief that engaging in energy-saving practices could significantly benefit their academic institutions. The survey results helped establish baseline levels of energy literacy and pointed out areas for further educational development. Additionally, it identified the need for interventions aimed at improving the energy knowledge of students, as well as the need for faculty-specific strategies.

Step 3: Multi-group Analysis (MGA) and Energy Auditing

In the third step, multi-group analysis (MGA) was used to determine the most significant factors affecting energy conservation behaviors among students at AU [16]. The analysis identified several key factors, such as attitudes and subjective norms, which influenced students' intentions to engage in energy-saving behaviors.

Following this, an energy audit was conducted in collaboration with students to assess the energy consumption patterns at AU. The audit revealed the most energy-consuming areas and provided a detailed understanding of the existing energy usage across different buildings. This information was crucial for selecting effective energy-saving technologies and for establishing a comprehensive energy literacy plan. The results of the audit were used to design targeted interventions, including the development of an educational program focusing on reducing energy use through behavioral and technological solutions.

Step 4: Energy Literacy Education Course

Based on the findings from the previous steps, an energy literacy education course [17] was developed to increase awareness and knowledge about energy conservation among university occupants. The course was designed to provide students with not just theoretical content but also practical methods for engaging in energy-saving behaviors. The course content emphasized sustainable practices, energy-efficient technologies, and the importance of individual responsibility in reducing energy consumption.

This blended learning approach [18] combined online educational modules with practical hands-on experiences like energy audits, providing students with a comprehensive understanding of energy conservation. The effectiveness of the course was measured through pre- and post-course surveys, which indicated a significant improvement in students' energy literacy and their intention to adopt energy-saving behaviors.

Key Action Steps and Outcomes

The study followed a series of key actions to achieve its objectives. These actions and their corresponding outcomes are summarized in Table 27.1:

Table 27.1 The modules of the energy literacy course

Key action	Step 1 Outcomes
Literature review of TPB	An extended theory of planned behavior (TPB) model was developed.
Developing a survey on energy conservation behavior	Research data was acquired through the survey.
Conducting survey	Data was collected to analyze the links between energy behavior and attitudes.
PLS-SEM analysis	The TPB model was tested using partial least squares structural equation modeling (PLS-SEM). Key variables such as energy conservation behavior, intention, subjective norm, attitude, and perceived behavioral control were assessed.
Key action	**Step 2 outcomes**
Measuring energy consumption patterns at AU	Total energy use and the most commonly used energy sources at AU were identified.
Assessing current energy literacy	Levels of energy literacy for students, teaching, and non-teaching staff were assessed.
Key action	**Step 3 outcomes**
Investigating energy conservation behavior with MGA	Significant factors influencing energy behavior among AU occupants were identified.
Conducting energy audit	The most energy-intensive areas and potential energy-efficient solutions were determined.
Key action	**Step 4 outcomes**
Developing educational tools	The optimal education method for an energy literacy course was determined.
Implementing sustainable energy education course	A sustainable energy education course was implemented online, aiming to increase energy literacy and knowledge about sustainability.

27.4 Conclusions

This study demonstrates the interconnections among energy literacy, behavioral change, and technological advancements in promoting energy efficiency and sustainability in academic buildings. The research provides a clear and structured procedure that can be applied to enhance energy practices in university settings. The findings underscore the importance of fostering positive attitudes towards energy conservation and emphasize the need for comprehensive educational programs that combine theoretical knowledge with practical application. The successful integration of these steps at AU provides a model for other universities seeking to enhance their sustainability efforts and reduce energy consumption.

Acknowledgments This research has been funded by the Australian University (AU), Kuwait. The authors would like to express their sincere gratitude to AU for providing the resources, facilities, and support necessary to carry out this research.

References

1. Khan MI, Bicer Y, Asif M, Al-Ansari TA, Khan M, Kurniawan TA, Al-Ghamdi SG (2024) The GCC's path to a sustainable future: navigating the barriers to the adoption of energy efficiency measures in the built environment. Energy Convers Manage X 23:100636
2. Hu S, Yan D, Azar E, Guo F (2020) A systematic review of occupant behavior in building energy policy. Build Environ 175:106807
3. Bosnjak M, Ajzen I, Schmidt P (2020) The theory of planned behavior: selected recent advances and applications. Eur J Psychol 16(3):352–356
4. Azizi ZM, Azizi NSM, Abidin NZ, Mannakkara S (2019) Making sense of energy-saving behaviour: a theoretical framework on strategies for behaviour change intervention. Procedia Comput Sci 158:725–734
5. Wolske KS, Gillingham KT, Schultz PW (2020) Peer influence on household energy behaviours. Nat Energy 5(3):202–212
6. DeWaters JE, Powers SE (2011) Energy literacy of secondary students in New York state (USA): a measure of knowledge, affect, and behavior. Energy Policy 39(3):1699–1710
7. Brounen D, Kok N, Quigley JM (2013) Energy literacy, awareness, and conservation behavior of residential households. Energy Econ 38:42–50
8. AlKhars M, Miah F, Qudrat-Ullah H, Kayal A (2020) A systematic review of the relationship between energy consumption and economic growth in GCC countries. Sustainability 12(9):3845
9. Al-Maamary HM, Kazem HA, Chaichan MT (2016) Changing the energy profile of the GCC states: a review. Int J Appl Eng Res 11(3):1980–1988
10. Al Shidi H, Sulaiman H, Amoatey P (2016) Shifting to renewable energy to mitigate carbon emissions: initiatives by the states of Gulf Cooperation Council. Low Carbon Econ 7(3):123–136
11. Elbannan M, Maarouf I, Abdo M (2022) Recent review of energy conservation plans and targets in the Gulf Cooperation Council region: barriers and challenges. WIT Trans Ecol Environ 260:299–311
12. Alomari MM, El-Kanj H, Topal A (2021) Analysis of energy conservation behavior at the Kuwaiti academic buildings. Int J Energy Econ Policy 11(1):219–232
13. Alomari MM, Hania EK, Topal A, Alshdaifat NI (2023) Exploring the impact of the COVID-19 pandemic on energy literacy and conservation behavior in academic buildings of Kuwait. Heliyon 9(11):e21474
14. Alomari MM, El-Kanj H, Topal A, Alshdaifat NI (2024) Improving energy literacy among Kuwaiti university students with a tailored course on sustainable energy practices. Glob J Eng Educ 26(1):6–12
15. El-Kanj H, Alomari MM, Topal A, Alshdaifat NI (2022) Assessment of energy literacy among university occupation: the case of Kuwait. Int J Appl Eng Res 7(1):1–13
16. Alomari MM, El-Kanj H, Topal A, Alshdaifat NI (2022) Energy conservation behavior of university occupants in Kuwait: a multigroup analysis. Sustain Energy Technol Assess 52:102198
17. Alomari MM, El-Kanj H, Topal A, Alshdaifat NI (2025) Designing an effective energy literacy course: a multi-phase approach tailored to Kuwaiti students' needs. World Trans Eng Technol Educ 23(1)
18. Alomari MM, El-Kanj H, Alshdaifat NI, Topal A (2020) A framework for the impact of human factors on the effectiveness of learning management systems. IEEE Access 8:23542–23558

Chapter 28
Hydrogen Fuel Cell Vehicles: Policy Review, Consumer Adoption, and Strategic Implications for Hong Kong's Path to Carbon Neutrality

Z. Zeng, K. C. Chan, S. C. Fu, and Christopher Y. H. Chao

Abstract The global shift towards green energy and zero-emission transportation is a key priority, driven by international climate agreements and national sustainability goals. Hydrogen fuel cell vehicles (HFCVs) offer a promising solution for sustainable transport, with zero emissions, efficient energy use, and performance comparable to traditional internal combustion vehicles. This paper reviews the development, adoption, and challenges of HFCVs, focusing on policy frameworks, technological advancements, and consumer attitudes in key hydrogen-promoting regions such as China, Japan, South Korea, the United States, and the European Union. This study uses a systematic literature review approach to examine factors influencing consumer adoption, vehicle characteristics, and infrastructure requirements. It also explores strategic recommendations for promoting HFCVs in Hong Kong, aligning with the region's carbon neutrality goals for 2050. The findings emphasise the need for integrated policy support, infrastructure development, and public awareness to commercialise hydrogen fuel cell technologies successfully.

Keywords Hydrogen Fuel Cell Vehicles (HFCVs) · Carbon neutrality · Zero-emission transportation

Z. Zeng · K. C. Chan (✉) · S. C. Fu
Department of Building Environment and Energy Engineering, The Hong Kong Polytechnic University, Hong Kong, China
e-mail: oscar-kc.chan@polyu.edu.hk

C. Y. H. Chao
Department of Building Environment and Energy Engineering, The Hong Kong Polytechnic University, Hong Kong, China

Department of Mechanical Engineering, The Hong Kong Polytechnic University, Hong Kong, China

L. Chen (ed.), *Clean Energy Research*, Green Energy and Technology,
https://doi.org/10.1007/978-3-031-99312-1_28

28.1 Introduction

Since the turn of the century, corporations and individuals worldwide have grown increasingly aware of the environment's critical role in humanity's future. Companies in various countries have gradually shifted from profit-oriented to sustainability-oriented operations, influencing global development trends. At the 21st Conference of the Parties (COP21) in Paris in 2015, 196 United Nations member states reached a landmark agreement on climate change. The primary aim was to limit the increase in global average temperature to well below 2 °C above pre-industrial levels while striving to further limit the temperature increase to 1.5 °C. The agreement also focuses on achieving a global peak in greenhouse gas emissions as soon as possible to attain climate neutrality by the mid-twenty-first century. Achieving these targets would significantly reduce the risks and impacts of climate change. Globally, there is a focused development on green and sustainable industries, with sustainable energy technologies being a primary area of interest.

Energy consumption is typically associated with carbon emissions. Thus, low-carbon or zero-carbon alternative energy sources are required to meet environmental goals. The concept of hydrogen as an energy source dates back to the early nineteenth century, with the first hydrogen fuel cell vehicle developed in 1830. It was not until the 1960s that hydrogen fuel cells found practical applications.

Hydrogen fuel cell vehicles (HFCVs), a type of zero-emission vehicle, offer unique advantages over other zero-emission vehicles. HFCVs can match the performance of internal combustion engine vehicles in power and are more efficient and quicker to refuel than electric vehicles (EVs), so it is suitable for medium- and heavy-duty applications, including buses, trucks, and other commercial transport vehicles. Crucially, HFCV produces no carbon emissions during operation. When hydrogen is used as a fuel in fuel cells, it is catalysed at the anode into protons and electrons; the electrons travel through an external circuit generating electricity, while the protons move through the electrolyte to the cathode, where they combine with oxygen and electrons to produce water and heat [1]. Similar to EVs, this process does not generate any nitrogen oxides, hydrocarbons or other analogous pollutants.

Compared to batteries, hydrogen fuel cells have a higher energy density, which allows HFCVs to achieve longer driving ranges with relatively smaller sizes and weights. EVs are often limited by their battery capacity and charging infrastructure, while HFCVs can refuel quickly at hydrogen stations, offering greater convenience compared to the long charging times required for EVs. Hydrogen's high energy density also means that the vehicle can provide more power with less weight, making it suitable for long-distance travel and heavy-duty transportation. It can be refuelled in approximately 5 to 10 minutes at hydrogen refuelling stations, a time frame that is comparable to the refuelling process for conventional gasoline vehicles. This rapid refuelling capability makes HFCVs particularly suitable for high-demand sectors such as public transport and logistics, where quick turnaround times are essential.

From a supply chain perspective, emissions from HFCVs occur during both upstream and midstream processes such as hydrogen production and transportation. To achieve efficient utilisation of hydrogen energy, optimising the production chain is a critical step [2]. David et al. developed a supply chain optimisation model for the transportation sector, considering hydrogen purity and purification requirements [3]. AH Azadnia conducted a comprehensive risk assessment of the green hydrogen supply chain, offering insights for relevant policy and regulatory development [4]. Existing research on hydrogen supply chain network design (HSCND) published in academic journals has been examined, revealing and categorising shortcomings such as handling uncertainty and raw material issues, as well as the lack of intertemporal integration planning [5]. In China, hydropower and nuclear power demonstrate the lowest life cycle primary energy consumption. At the same time, hydrogen production from natural gas, coal and grid electricity lacks advantages, showing similar results in life cycle greenhouse gas emissions [6]. Regarding the industrial chain for HFCVs, global HFCV supply chains face various supply risks: the United States in vehicle components, China across the entire supply chain, and Japan in platinum sourcing. A universally present risk in HFCV production worldwide is the gas diffusion membrane [7].

Currently, "grey hydrogen" constitutes 80–85% of all hydrogen production, typically produced through steam-methane reforming and electrolysis (splitting water with electricity) [8]. If carbon capture and storage are used to prevent emissions during the steam reforming (grey hydrogen) or gasification (brown hydrogen) processes, this type of hydrogen is termed "blue hydrogen." Blue hydrogen manufacturing is seen as an optimal transitional method to the ultimate goal of widespread "green hydrogen" production and application, which would realise truly zero-emission transportation. However, the high cost of green hydrogen production and the need for efficient, large-scale hydrogen storage and transport solutions currently limit its competitiveness. Addressing these challenges is essential for hydrogen to become a viable long-term solution for reducing carbon emissions.

To expedite the application of hydrogen energy all over the world, trials are being undertaken based on their specific circumstances. The European Union has implemented several policies, including the Emissions Trading System and the European Green Deal. These policies aim to reduce emissions through increased energy efficiency, promoting renewable energy sources, and reforming transportation and building standards [9]. The United States has established federal emission standards for passenger cars and light trucks, implemented CO_2 emission standards for aircraft [10], and released climate adaptation plans for over 20 federal agencies to enhance the resilience of federal facilities, employees, resources and operations to the impacts of climate change [11]. Japan announced its ambition to achieve carbon neutrality by 2050, unveiling a "Green Growth Strategy" that identifies 14 sectors with growth potential and provides action plans for these sectors [12]. Furthermore, Japan introduced a "Zero Carbon Emission Roadmap" to achieve this goal through green innovation and investment [13]. Canada's "2030 Emission Reduction Plan" aims to reduce greenhouse gas emissions by 40–45% from 2005 levels by 2030, paving the way for net-zero emissions by 2050 [14].

According to predictions by the China Hydrogen Energy Alliance, by 2050, China could have over three million HFCVs with the potential of establishing around 10,000 hydrogen refuelling stations. Concurrently, the Hong Kong government set multiple sectoral targets in the "Hong Kong 2050 Climate Action Plan" to achieve carbon neutrality by 2050 [15]. In fact, transportation emissions account for about 18% of Hong Kong's total emissions. The action plan includes the "Clean Air Plan," which will implement relevant strategies to promote new energy transportation and accelerate the low-carbon transition, aiming for the electrification of the transportation system by 2035 [16]. Measures will also be taken as part of the electric vehicle roadmap to achieve zero emissions by 2050 [17] and the 2024 Policy Address. In June 2024, the Hong Kong government announced the latest "Hong Kong Hydrogen Energy Development Strategy," aiming to drive the development of the hydrogen industry by refining regulations, setting standards, aligning with the market, and advancing cautiously through four strategic areas, in response to climate change challenges. This helps propel Hong Kong towards carbon neutrality, develop new productive forces, and maintain its international competitiveness.

Therefore, Hong Kong plans to adopt hydrogen energy as one of the solutions. If the goal is to broadly introduce hydrogen and fuel cell technology into Hong Kong's energy system, tailor-made research on the factors influencing people's purchase of HFCVs will play a crucial role in policy formulation in this field. Globally, there is a positive attitude towards HFCVs, and many scholars have studied the policies, market development, and consumer behaviour related to hydrogen energy and HFCVs in various regions. This paper summarises the content of relevant literature and offers suggestions tailored to Hong Kong's situation, which can provide a reference for policy formulation.

The structure of this paper is as follows: Section 28.1 provides an overview of the current context, a brief introduction to hydrogen energy and fuel cells, and outlines the theoretical foundation. Section 28.2 describes the research methodology, including the criteria for searching and filtering articles. Section 28.3 synthesises different studies on factors influencing the purchase of HFCVs, summarising the policies of pioneering regions in HFCV applications. Section 28.4 discusses the current situation and goals in Hong Kong and offers related recommendations, discussing the limitations of the article. Section 28.5 summarises the results achieved in this article.

28.2 Method

To comprehensively gather and analyse global research on HFCVs and related hydrogen policies, applications and factors influencing purchase decisions, this study employed a systematic literature review methodology. Therefore, the literature collection focused on academic research papers, policy documents and industry research reports. To ensure the timeliness and breadth of the literature, databases such as Google Scholar, Web of Science and Science Direct were utilised to collect documents about HFCVs and hydrogen energy. The search keywords included

'HFCV policy review', 'HFCV adoption', and 'hydrogen fuel cell vehicles market share', using Boolean logic ("Hydrogen fuel Cell Vehicles"or HFCV or "Fuel Cell Vehicles" or FCV in subject terms) AND ("Policy Review" or "Policy" or "Adoption" or "Market Share" in title/abstract) AND ("China"or "German" or "Japan" or "Korea" or "US" in title/abstract).

To ensure the credibility of the policy content, the official websites of various governments and international organisations served as primary sources for policy documents. These included the National Development and Reform Commission of China (NDRC), the Ministry of Economy, Trade, and Industry of Japan (METI), Department of Energy of the United States (DOE), the European Commission (EC), and the Ministry of Trade, Industry, and Energy of South Korea (MOTIE). Also, hydrogen development reports published by international organisations such as the International Energy Agency (IEA) and the International Renewable Energy Agency (IRENA) were also utilised.

To capture the latest trends in global hydrogen and HFCV development, the temporal scope of the literature and policy documents was limited to 2015–2024. During this period, several countries launched hydrogen development strategies, and there were significant advancements in the application and technological breakthroughs of HFCVs. The detailed inclusion/exclusion criteria were shown in Table 28.1.

The initial search by Google Scholar yielded over 16,500 articles, and the Web of Science search results had 80 articles. A selection process was conducted focusing on the academic literature that addresses methods for promoting HFCVs and factors influencing their adoption, studies on HFCV markets, and research on hydrogen energy industry development and technological innovation. The geographical scope of the research includes major hydrogen-promoting countries and regions such as China, Japan, South Korea, the United States and the European Union, incorporating their relevant academic contributions into the review. Literature solely focused on hydrogen production or other areas (such as chemical and non-transport sectors) that do not pertain to HFCVs; articles lacking empirical data support or with insufficient policy interpretation were excluded. Outdated literature outside the specified time range was generally only included in the analysis if it was

Table 28.1 Inclusion/exclusion criteria

Inclusion criteria	Exclusion criteria
1. Focusing on approaches to the promotion of HFCVs and the factors affecting the promotion of HFCVs	1. Involving only hydrogen production or other areas (e.g., chemical, non-transportation)
2. Study on the HFCV market and the development and technological innovation of the hydrogen energy industry	2. Not supported by empirical data or inadequate policy interpretation
3. The study area covers the major hydrogen-promoting countries and regions	3. Outdated non-classical literature outside the timeframe
4. Empirical studies	4. Without peer-reviewed
5. English	5. Other languages

considered classic. Through this filtering process, the most representative 50 articles were selected from approximately 16,500 related documents, covering the three main research directions of hydrogen policy, market promotion and technological development.

In sum, these articles contain both consumer-side interviews and surveys to study consumer attitudes and perceptions of HFCVs, as well as market- and policy-side planning to discuss how to popularise HFCVs faster and more comprehensively from an industrial perspective.

28.3 Results

Global Policy on Hydrogen Energy and Hydrogen Fuel Cell Vehicles

The global market for HFCVs is witnessing significant growth, driven by concerted efforts from governments and automotive manufacturers to advance this clean energy technology. In 2020, South Korea emerged as the global leader in HFCV sales, recording 10,906 units sold, marking an increase of 114.6% from the previous year, with an ambitious target to reach 1.8 million HFCVs by 2030. Japan, a global frontrunner in hydrogen and fuel cell technology, boasts the highest number of hydrogen refuelling stations and aims to achieve a fleet of 800,000 HFCVs by 2030. China‘s rapid development in the HFCV sector led to sales of 7352 units in 2020, making it the third highest globally, to reach one million HFCVs by 2030 [13, 18, 19].

The United States and Germany are also actively promoting the development of HFCV technology. The United States has shown notable ownership rates of HFCVs but needs to catch up in the establishment of hydrogen refuelling infrastructure. On the other hand, Germany has made progress in building hydrogen stations, though its HFCV sales figures remain modest. Globally, the HFCV market is transitioning from its nascent phase to a growth phase, with expectations for continued expansion [20]. According to a study by Polaris Market Research, the global HFCV market was valued at approximately $151 million in 2022, with projections to reach about $6.288 billion by 2032 at a compound annual growth rate of approximately 45.2% [21]. This underscores the immense growth potential of the HFCV market.

Despite the relatively smaller number of HFCVs compared to battery EVs and plug-in hybrid EVs, the introduction of HFCVs is just at the beginning. In contrast, the initial commercial production of battery EVs commenced several years ago, and it took over a decade to reach a global vehicle stock in the hundreds of thousands to millions [22]. This historical perspective highlights the rapid adoption trajectory expected for HFCVs in the broader transition towards sustainable transportation technologies. Many countries and regions have enacted relevant laws and regulations to promote the application of HFCVs and facilitate the progress of hydrogen energy penetration. Some important related laws/policies/strategies implemented worldwide are listed in Table 28.2.

To enhance the adoption of HFCVs, the South Korean government has adopted a multifaceted approach. This includes financial incentives to reduce purchase costs, investments in research and development to foster innovation, and the expansion of charging infrastructure to improve convenience. Additionally, the government has engaged in extensive public education to increase awareness and implemented strict emission standards to limit the sale of conventional vehicles. It has also set

Table 28.2 Important related laws/policies/strategies worldwide

Country/region	Related laws/policies/strategies	Details
China	Hydrogen Energy Industry Development Plan (2021–2035) [18]	Aims for approximately 50,000 fuel cell vehicles by 2025 and the construction of several hydrogen refuelling stations. Targets 100,000–200,000 tons of renewable energy-based hydrogen production annually.
Japan	Basic Hydrogen Strategy [24]	Increase the production capacity of clean hydrogen, including green hydrogen produced through renewable energy sources such as wind and solar power, and blue hydrogen produced from fossil fuels combined with carbon capture and storage (CCS) technology. Develop hydrogen infrastructure, including a network of hydrogen refuelling stations, to support the growth of hydrogen fuel cell vehicles (HFCVs) and other hydrogen applications.
South Korea	Hydrogen Economy Roadmap [19]	The south Korean government aims to produce 6.2 million HFCVs and build 1200 hydrogen refuelling stations by 2040. Hydrogen is a key focus of national development, promoting applications in the automotive and power sectors.
United States	US National Clean Hydrogen Strategy and Roadmap [25]	Supports research and development of hydrogen fuel cell technologies, providing tax incentives and grants. California plans to have 50,000 HFCVs on the road and 200 hydrogen refuelling stations by 2030.
Germany	National Hydrogen Strategy [26]	By 2045, the demand for hydrogen and its derivatives in Germany could increase to between 360 and 500 TWh of hydrogen and approximately 200 TWh of hydrogen derivatives. Germany plans to increase its domestic electrolysis capacity from 5 gigawatts (GW) in 2020 to 5 to 10 GW by 2030. The German federal government and the states (Länder) plan to make several billion euros available from federal and state funds to promote hydrogen generation, build the necessary infrastructure, and enable its use.
European Union	European Hydrogen Strategy [27]	The EU's hydrogen strategy aims to install at least 6 GW of renewable hydrogen electrolysers by 2024, scaling to 40 GW by 2030. It projects the production of one million tonnes of renewable hydrogen by 2024 and anticipates investments of €24–42 billion in electrolysers and €65 billion for hydrogen infrastructure by 2030. By 2050, clean hydrogen is expected to meet 24% of the world's energy demand, potentially reaching €630 billion annual sales.

penetration targets for HFCVs and mandated their procurement in the public sector to demonstrate their environmental benefits. Despite these efforts, the market acceptance of HFCVs has been lower than expected, highlighting the need for more robust policy frameworks and market strategies [23].

In Germany, the promotion of HFCVs is supported through the National Innovation Program for Hydrogen and Fuel Cell Technology (NIP), which provides fiscal incentives such as exemptions from vehicle road taxes and reductions in sales profit taxes. Furthermore, non-financial incentives include special authorisations for low-emission vehicles. The construction of hydrogen stations is orchestrated by the National Organisation for Hydrogen and Fuel Cell Technology (NOW), complemented by public awareness campaigns to foster recognition of sustainable transport options. The government also supports the market development of HFCVs through low-interest loans to reduce purchasing costs [28].

The Japanese government, aiming to reduce dependence on nuclear and fossil fuels, has launched a strategic initiative since April 2014 to establish a "hydrogen society." This initiative is articulated through the "Hydrogen and Fuel Cell Strategy Roadmap," which delineates three implementation phases. The government subsidises the cost of HFCVs, making vehicles like Toyota's "Mirai" more affordable. Additionally, Japan is committed to expanding its hydrogen fuel infrastructure and promoting its hydrogen technology both domestically and internationally [29].

The United States Department of Energy (DOE), through its H2@Scale initiative, actively advances the research, development, and commercialisation of HFCVs and related technologies. It has established clear cost targets, such as reducing the cost of fuel cells to $30/kW and hydrogen to $4/kg, to enhance competitiveness with existing technologies. Additionally, DOE's Energy Materials Network (EMN) accelerates the discovery of advanced materials through a multi-laboratory research alliance, providing a solid R&D foundation and policy support for the promotion of HFCVs [30].

The experiences of the United States, Germany and South Korea, with their proactive policies and regulations, have successfully driven the development of the HFCV industry, offering valuable lessons for other nations. The United States has established a supportive framework for HFCV technology through the Energy Policy Act and the national hydrogen economy roadmap. At the same time, federal and state government cooperation, such as the Freedom CAR program, has facilitated the transition to a hydrogen economy. Germany's National Innovation Program for Hydrogen and Fuel Cell Technology and the Electric Mobility Act provide policy and financial incentives that have propelled the HFCV market. South Korea, through laws such as the Alternative Energy Technology Development and Promotion Act and the Hydrogen Economy Roadmap 2040, has delineated a development blueprint for the hydrogen industry ecosystem, with the Hydrogen Economy Promotion and Hydrogen Safety Management Act providing legal support and safety standards for the hydrogen industry. These countries demonstrate that a clear policy framework, interdepartmental cooperation, financial incentives, infrastructure development, and the establishment of safety and standards are key factors in driving the development of the HFCV industry [31].

For China, the strategic importance of the hydrogen and fuel cell industries is articulated in national-level plans such as the "Energy Technology Revolution Innovation Action Plan (2016–2030)" and "Made in China 2025," which provide financial subsidies and tax incentives. According to the "Notice on Promoting the Demonstration Application of Fuel Cell Vehicles," a "reward instead of subsidy" approach is implemented to incentivise cities participating in demonstration projects [32]. The "Hydrogen Energy Industry Standard System Construction Guide (2023 Edition)," issued by the National Standardisation Management Committee, aims to regulate and guide the standardised development of the hydrogen energy industry [33].

The "Medium and Long-Term Plan for Hydrogen Energy Industry Development (2021-2035)" released by the National Development and Reform Commission and the National Energy Administration identifies hydrogen as a crucial component of the national energy system. It outlines the development of the hydrogen industry by 2035 [31]. The "Medium and Long-Term Development Plan for the Automotive Industry" highlights the roadmap for fuel cell technology development and emphasises fuel cell power systems as a key area for breakthroughs in the research and industrialisation of new energy technologies. The "New Energy Vehicle Industry Development Plan (2021–2035)" proposes the development of advanced modular power battery and fuel cell system technologies and supports regions that are equipped to conduct commercial demonstration operations of fuel cell vehicles [34].

The "Hydrogen Fuel Cell Vehicle Safety Guide (2019 Edition)" focuses on addressing safety issues in design and application, providing guidance and references for fuel cell vehicle developers, manufacturers, service personnel, and consumers [35]. Internationally, China has led the revision of the United Nations technical regulations on the safety of HFCVs, enhancing the safety standards for these vehicles globally. This leadership underscores China's commitment to advancing the safety and technology of HFCVs both domestically and worldwide [36].

Hong Kong, with its unique conditions, cannot solely rely on local resources to develop and utilise hydrogen energy. However, the government exhibits a strong willingness to promote hydrogen energy. In pursuit of carbon neutrality, the government has outlined some context-specific strategies in "The Strategy of Hydrogen Development in Hong Kong," referred to as the "Innovation-driven development and self-reliance, safety first and adoption of clean and low-carbon energy, market-led with government guidance, prudent application led by demonstration" approaches. Instead of directly applying the relevant regulations and policies of other countries, Hong Kong has improved the existing Gas Safety Ordinance (GSO) (Cap. 51) on these bases, formulated standards for the hydrogen energy technology industry chain, complemented the development of the market, and progressively promoted standardised management by actively carrying out application demonstrations.

Drawing on emission reduction policies from the maritime industry, a suite of policy measures can be developed to promote HFCVs. Implementing carbon and fossil fuel taxes will increase the costs of traditional vehicles, thereby creating market competitiveness for HFCVs. Establishing standards for zero-emission vehicles and progressively phasing out fossil fuels will highlight the environmental benefits of HFCVs. Government investment in hydrogen infrastructure and financial support for research and development will help reduce the production costs of HFCVs. International cooperation plays a crucial role in sharing technology and coordinating standards, accelerating the development of hydrogen fuel infrastructure across nations. By setting clear transition targets and roadmaps, a structured shift towards clean transportation can be ensured. These measures will collectively foster the adoption of HFCVs and drive the green transformation of the transportation sector [37].

Factors Influencing Consumers to Purchase HFCVs

Individual Level Individual-level factors are crucial determinants in the decision-making process for purchasing behaviours, encompassing an individual's educational background, economic status, social position, and family situation. Research indicated that among current users of HFCVs, particularly in the Los Angeles metropolitan area in California, USA and in Aichi Prefecture, Japan, there is a prevalence of older individuals, with interviewed American vehicle owners predominantly over the age of 45 [38], and more than 80% of Japanese owners exceeding 50 years of age [39]. This demographic trend may be attributed to the correlation between age and increased personal wealth accumulation, enhanced social status and expanded cognitive awareness due to life experiences, making these individuals more inclined to invest in high-cost but environmentally friendly products.

Regarding household income, most HFCV owners report earnings above the national average [38, 39], suggesting that the higher initial cost of HFCVs directly influences the purchasing decisions among varying income levels, with higher income serving as a facilitative factor. Additionally, the level of education, familial circumstances, and the number of vehicles owned can significantly impact purchasing decisions. Educational attainment affects individual awareness of environmentally friendly products, influencing purchasing choices [40]. Survey results reveal that over half of the interviewed vehicle owners possess advanced degrees, supporting the correlation between educational level and the propensity to purchase.

Driving habits and personal needs also play a substantial role; for instance, consumers requiring a vehicle for long-distance, zero-emission travel or those with mountainous commuting routes may find HFCVs particularly suitable. This factor suggests that certain considerations discussed may have a more profound impact on the decision-making processes of younger generations. In summary, age, income and awareness influence purchasing behaviours.

Vehicle-Specific Factors In addition to individual factors, another crucial determinant influencing the selection of various vehicles is the performance characteristics inherent to the vehicle itself. Many prospective transport buyers consider the intrinsic features and efficacy of the vehicle. Currently, the development of HFCVs is still in its nascent stages, yet the technology meets operational requirements. Moreover, HFCVs combine the rapid refuelling capabilities of internal combustion engine vehicles with the low emissions of EVs, offering a unique and irreplaceable competitive advantage. The range capability is another advantage of HFCVs over battery EVs.

However, the relatively high cost of HFCVs may exacerbate negative purchasing intentions of potential users in the initial stages of market introduction. As highlighted in some studies, safety concerns also impact purchasing decisions [41]. The effect of price on purchasing intentions varies across different studies, but it is evident that price is one of the key attributes considered by consumers. As the market matures, the influence of vehicle price is likely to diminish [42, 43], and people might even be willing to pay a premium for HFCVs if the operating costs and refuelling convenience are comparable to those of internal combustion engine vehicles [44].

Regarding safety issues, there is no direct evidence to suggest that they are a definitive factor disrupting purchasing intentions, but they are undoubtedly a consideration. If regulatory bodies possess a deep understanding and preparedness for HFCV safety issues, this could enhance public confidence in the safety of HFCVs, thereby impacting purchasing decisions. Furthermore, if safety concerns are adequately addressed and recognised by the public, it could reduce negative perceptions of HFCVs, thereby enhancing their market acceptance and sales.

Infrastructure-Level Factors The convenience of infrastructure development is a primary concern for many consumers [45]. The expansion of the HFCV market necessitates not only the completion of infrastructure but also widespread societal acceptance. This section will also focus on the production and transportation of hydrogen, which are critical aspects to consider when discussing hydrogen stations. According to interviews with current HFCV owners, the extent and convenience of the hydrogen refuelling station network are decisive factors for their continued use of HFCVs. In regions where hydrogen stations are not readily accessible, individuals often need to plan refuelling times and routes in advance. They may require a secondary means of transportation to address potential issues such as the unavailability of HFCVs or operational halts at hydrogen stations [38]. Respondents from Japan and the United States indicate that the scarcity of hydrogen stations and the lack of refuelling infrastructure are major barriers to purchasing HFCVs [39, 41]. Therefore, the strategic placement of hydrogen stations will be a crucial element in the widespread adoption of HFCVs in the future.

Regarding social acceptance, Liu Jie et al. suggest that HFCVs need to leverage "word-of-mouth" marketing to enhance market recognition due to their credibility, cost-effectiveness, and broad impact [40]. Particularly in the early stages of industrialisation, consumers require improved awareness of HFCVs through product experience, energy requirements, and safety assurances to boost societal confidence and promote consumption. This approach not only facilitates the integration of

HFCVs into everyday life but also underpins the foundational acceptance necessary for their broader acceptance and use.

As Elon Musk has pointed out, the storage and transportation of hydrogen pose significant challenges due to high costs and safety risks. Efficient and safe storage and transportation of hydrogen are critical for its subsequent applications. Currently, there are four primary methods of hydrogen storage: compressed hydrogen, liquid hydrogen, metal hydrides (hydrogen storage alloys), and carbon nanotube storage. Compressed and liquid hydrogen are relatively mature technologies. However, compressed hydrogen, stored in large pressurised tanks, presents potential hazards, and liquid hydrogen poses challenges for long-term storage due to its natural evaporation if not used. Hydrogen storage alloys offer safety and convenience but have limited storage capacity per unit volume and are relatively costly. Demonstrations of these technologies are ongoing. Carbon nanotube storage remains in the early stages of development and is far from commercialisation. In summary, the usage of compressed and liquid hydrogen is limited, and development efforts are focused on enhancing the storage density of hydrogen storage alloys. The advancement of carbon nanotube technology may provide an excellent future material for hydrogen storage.

There are two common methods for hydrogen transportation. One method involves hydrogen transport trucks that pressurise or liquefy hydrogen and transport it to the destination using specialised tank trailers. The second method utilises hydrogen supply pipelines, injecting hydrogen directly into pipelines made of special materials or existing natural gas pipelines. Currently, most plans are based on mixed natural gas pipelines. Due to hydrogen's excellent combustion-enhancing properties, injecting hydrogen up to 10% of the total volume into the pipeline does not compromise the safety of the pipeline and can improve the performance of natural gas. Europe has already conducted pipeline demonstration projects. However, the hydrogen must undergo a pressurisation process, which is very costly and energy intensive. Additionally, the vehicles, pipelines, and containers used for transporting hydrogen can cause "hydrogen embrittlement," necessitating strict regular maintenance. The high cost of pressurisation remains one of the main reasons for reducing the competitiveness of the hydrogen market.

Market and Adoption Factors The marketing strategies for HFCVs are a pivotal factor at the macro level. Research has elucidated strategic recommendations for fostering the acceptance of early adopters of fuel cell vehicles (FCVs). These recommendations include the pre-emptive establishment of hydrogen refuelling infrastructure, promoting the sustainable production of hydrogen, developing hybrid hydrogen energy models with home-charging capabilities, implementing government subsidies to reduce initial costs, enhancing education about hydrogen safety, and emphasising the longer driving range advantages of HFCVs [41]. These strategies directly address major consumer concerns and provide an empirical basis for the marketing of HFCVs.

Scholars have proposed a series of market strategies for HFCVs to facilitate their commercialisation in China. These strategies span national strategy formulation, technological innovation, industrial chain construction, public education, business model innovation, market segmentation, and international cooperation. They emphasise enhancing technological maturity through demonstration projects, reducing market risks, and foreseeing a roadmap for the large-scale marketisation of HFCVs [46].

In Japan, the HFCV market and application strategies reflect a collaborative effort between the government and corporations to facilitate the commercialisation of HFCVs through a series of measures. Government subsidy policies have reduced consumer purchase costs, while there is a proactive construction of Hydrogen Refuelling Stations (HRS) to support infrastructure development. Regarding technological research and development, Japan continues to invest in enhancing fuel cell performance and reducing costs, with targets to install 1.4 million and 5.3 million residential fuel cell units by 2020 and 2030, respectively, and to establish a corresponding HRS network. Furthermore, the Japanese government is adjusting regulations to reduce HRS costs and accelerating technological innovation from research to market through a tripartite cooperation framework. The Hydrogen Society Strategic Roadmap sets three stages of development goals, from market expansion to technological maturity and the construction of a comprehensive supply chain. These integrated strategies demonstrate Japan's resolute determination and comprehensive planning in promoting HFCVs and their technological development [29].

California has driven the development of the HFCV market through a series of strategies, including zero-emission vehicle regulations, construction of hydrogen fuel supply station infrastructure, fiscal and non-fiscal incentives, and public-private partnerships. The strategic focus covers the light passenger car market, gradually expanding to buses and heavy trucks, utilising economies of scale to reduce costs. Concurrently, California emphasises the development of renewable hydrogen, promoting its application in HFCVs through legislative and market incentive measures. Despite challenges such as cost, technology, infrastructure, and public awareness, California's experience provides a useful reference for the global development of the HFCV market [47].

28.4 Discussion

As a global financial and trade hub, Hong Kong has adopted several innovative approaches to promote HFCVs. Hong Kong possesses unique advantages in promoting HFCVs. With zero tariffs on imports and exports and its position at the forefront of technological innovation, these advantages can be leveraged to maximise the efficiency of HFCV deployment. Hong Kong has prioritised pilot projects to assess the safety performance of HFCVs, gather local data and experience, and subsequently develop a set of standards tailored to Hong Kong's specific context. Additionally, pilot projects for hydrogen-powered public transportation, including hydrogen fuel cell buses and taxis, are currently underway to evaluate the feasibility of hydrogen as a solution for providing efficient, low-carbon public transportation services.

All the summaries in this paper are intended to provide theoretical support for the future application of HFCVs and policy formulation in Hong Kong, so the feasibility of HFCV promotion in Hong Kong will be discussed next.

To achieve carbon neutrality by 2050, the Hong Kong Special Administrative Region government announced a plan in the "2022 Policy Address" to cease the registration of new gasoline and hybrid private vehicles by 2035 or earlier. According to the "2023–24 Budget," the government has allocated HK$200 million to trial hydrogen fuel cell electric double-decker buses and heavy vehicles, accelerating the low-carbon transition of new energy transportation. HFCVs, favoured for their rapid refuelling and long-distance capabilities, are becoming the preferred choice for heavy transport, aiding the green transformation of the public transit system.

The rollout of hydrogen fuel cell buses faces technical and safety challenges. However, the government has established an interdepartmental committee to address these issues while exploring regulations aligned with green technology advancements. By implementing the "Hong Kong 2050 Climate Action Plan" and the "Hong Kong Electric Vehicle Roadmap," the government commits to increasing the use of new energy vehicles and plans to establish a fully electrified transport system by 2035, moving towards the 2050 zero emissions goal. These steps demonstrate Hong Kong's proactive efforts in promoting clean transportation and achieving environmental goals, potentially leading the region in sustainable transit solutions.

To comprehensively develop hydrogen energy, Hong Kong established a cross-departmental Hydrogen Energy Working Group in 2022, pooling expertise from various sectors to promote the application of hydrogen energy and subsequently approved more than ten pilot projects, gradually extending hydrogen energy promotion across various industries. In the absence of formally established standards, provisional standards were introduced, including "Practice Guidelines for Hydrogen Fuel Vehicles and Repair Workshops," "Hydrogen Refuelling Station Practice Guidelines," and "Guidelines for Quantitative Risk Assessment of Hydrogen Energy Systems." Given Hong Kong's unique geographical environment, the development and utilisation of hydrogen energy require multi-party collaboration, and the uncertainty surrounding green hydrogen remains significant. Consequently, Hong Kong's current strategy for hydrogen energy development is structured around four main areas: enhancing regulations, establishing standards, aligning with the market, and cautious advancement. The priorities for the development of the hydrogen energy industry are defined as setting standards, regional cooperation, technological demonstration, and investment promotion.

In the market survey featured in the report "Market Study on the Use of Hydrogen in Green Transportation in Hong Kong," key data reveals a positive outlook from Hong Kong's industrial sector on the development potential of HFCVs. While 40.9% of respondents see challenges in locally producing enough hydrogen, 62.5% are optimistic about hydrogen imports from Mainland China, and 42.0% consider imports from countries like Australia and Chile as viable options. Regarding safety, 68.2% of respondents do not view HFCVs as more hazardous in operation compared to traditional fuel vehicles, indicating a trust in the safety of hydrogen use. Economic factors are critical, with 56.8% of respondents identifying the price of hydrogen as a significant factor affecting the acceptance of HFCVs, and 54.5% are open to a cost difference of up to 20% compared to gasoline vehicles. Furthermore, 87.5% of respondents agree that hydrogen will be an important energy source for future pollution reduction and decarbonisation efforts, and 73% emphasise the government's pivotal role in fostering the hydrogen industry. Although 51.1% of respondents still need to be ready to engage in hydrogen-related businesses, these data points highlight the potential and expectations within the hydrogen sector in Hong Kong [48].

The market survey insights underscore the Hong Kong industrial sector's optimistic outlook on the potential for HFCVs, with a significant proportion of respondents recognising the importance of hydrogen as a key energy source for future decarbonisation efforts. While challenges such as local production limitations and economic factors like hydrogen pricing are acknowledged, there is a strong endorsement for the role of government in facilitating the hydrogen industry's growth through policy support and infrastructure development. Despite the noted obstacles, the data indicates a growing interest in the hydrogen sector, highlighting the need for collaborative efforts to address these challenges and capitalise on the opportunities hydrogen technology presents.

Policy Implications

Combining relevant literature and the Hong Kong market, promoting HFCVs can be strategically segmented into three phases, ensuring systematic and phased implementation of hydrogen energy development.

Phase One—Introduction and Awareness In this initial phase, public knowledge of HFCVs needs to be improved, necessitating education and awareness campaigns to enhance understanding of HFCVs and hydrogen energy among citizens or potential users. It is advisable to conduct related exhibitions and daily promotional activities, as well as utilise advertising and news articles for public education. Concurrently, organising online and offline promotional events will increase public participation. This stage should involve environmental advocates and public figures in HFCV test-driving events to collect user feedback, which should be used to improve relevant HFCV features and achieve early product iterations. Also, it is necessary to establish related academic programs within various universities in Hong Kong to cultivate specialised talent capable of addressing potential technical and commercial challenges in the future. Additionally, relevant departments should plan and construct hydrogen refuelling stations to meet future city-wide refuelling needs.

Phase Two—Growth Period As the concept of hydrogen environmental protection becomes widely accepted and HFCV technology matures, this stage becomes optimal for promoting HFCVs. The government should initiate fiscal incentives such as vehicle purchase subsidies, tax reliefs, and increasing taxes and operational costs for high-pollution vehicles while offering reductions for hydrogen refuelling and parking fees to enhance HFCVs'market competitiveness and attract more citizens to purchase. Early adopters should also receive certain incentives. Moreover, government departments should update their fleets to use HFCVs as official vehicles, serving as role models. Infrastructure development should also continue to improve, ensuring that hydrogen stations are competitive in convenience compared with traditional gas stations and electric vehicle charging stations. Concurrently, Hong Kong can capitalise on its financial prominence and market potential to draw international and domestic enterprises, institutions, and skilled personnel to foster the growth of the hydrogen economy.

Phase Three—Technological Optimisation and Market Deepening This phase should focus on optimising hydrogen production, transportation and storage technologies. Specifically, transitioning from traditional "grey hydrogen" production to environmentally friendly "green hydrogen" is the key to achieving zero emissions by 2050. Continuing the promotion and refinement of HFCV technology can ensure that HFCV is more advantageous in terms of performance and cost. Additionally, enhancing domestic and international cooperation and exchanges can improve the international competitiveness of hydrogen technologies and gradually expand the market to achieve large-scale commercialisation of HFCVs.

The implementation of these strategies requires the collective effort of the government, businesses, and the public. Through this series of comprehensive measures, Hong Kong can make significant progress in hydrogen energy promotion and provide valuable experiences for global hydrogen applications.

Limitations

The limitations of this paper are as follows: Firstly, due to time constraints, it was not possible to collect and summarise all the literature and policies related to hydrogen energy and HFCVs. Instead, only representative articles were reviewed. The interdependencies among various influencing factors mean that they do not act in isolation to affect decisions; rather, the cumulative or combined effect of different factors may exert varying degrees of influence. Future studies may need to construct models to determine the weight of each factor and thus quantify their impact. Additionally, the number of HFCV owners interviewed needs to be expanded to ensure the generalisability of the data.

The second limitation concerns the applicability of the recommendations proposed, which are based on the specific contexts of China and Hong Kong. Whether these recommendations can be adopted by most countries remains a subject for discussion. Future efforts should focus on identifying countries worldwide where the promotion of HFCVs is feasible and propose tailored solutions appropriate to their specific conditions.

The third point is that this paper does not have a detailed economic model, which is a crucial aspect of discussing economic benefits. Subsequent researchers might focus on establishing such models and exploring practical methods for implementing relevant policies. This approach will enhance understanding of the economic implications of advancing hydrogen energy technologies and help refine strategies for their promotion.

28.5 Conclusion

This article comprehensively examines the various factors influencing the purchase of HFCVs. Currently, the transition to green transportation has emerged as a global trend. The Hong Kong government, aiming to achieve zero emissions in its transportation system by 2050, is committed to promoting the widespread adoption of HFCVs. However, introducing new zero-emission vehicles necessitates consideration of multifaceted factors. The article categorises the factors affecting consumer purchases into six areas and analyses them sequentially. Individually, the main factors are the level of environmental awareness and income; sufficient understanding of environmental issues combined with a relatively good income can foster acceptance of HFCVs. Regarding the vehicle itself, it must meet criteria such as safety, exceptional range, comparable drivability to ICEVs and EVs, and reasonable fuel costs.

From a societal and infrastructure perspective, the layout and construction of hydrogen refuelling stations significantly influence the frequency of HFCV usage. Many current HFCV drivers' concerns are centred on the convenience of hydrogen stations. If refuelling stations are not accessible near homes or workplaces, additional planning for refuelling routes is necessary, and alternative vehicles may be required as backups. Broad social acceptance of HFCVs can greatly encourage purchase motivations. In fact, to address the deployment issues of hydrogen stations, the storage and transportation of hydrogen also need to be standardised, as this represents a technical challenge.

Furthermore, support from governments and relevant institutions is crucial. The article summarises related policies from countries such as Germany, Japan, the USA, and China, providing references for future policy development in Hong Kong. The marketing of HFCVs is also vital, as well as gathering relevant market policies from China, Japan, and the USA to identify market strategies that can be emulated. Insights are also provided for the transformation of logistics and freight transportation.

Acknowledgement This research project (Project Number: 2023.A6.226.23B) is funded by the Public Policy Research Funding Scheme of the Government of the Hong Kong Special Administrative Region.

References

1. Hydrogen and Fuel Cell Technologies Office—Hydrogen Delivery. https://www.energy.gov/eere/fuelcells/hydrogen-delivery. Last accessed 6 Nov 2024
2. Zhao F, Mu Z, Hao H, Liu Z, He X, Victor Przesmitzki S, Ahmad Amer A (2020) Hydrogen fuel cell vehicle development in China: an industry chain perspective. Energ Technol 8(11):2000179
3. Wickham D, Hawkes A, Jalil-Vega F (2022) Hydrogen supply chain optimisation for the transport sector–focus on hydrogen purity and purification requirements. Appl Energy 305:117740
4. Azadnia AH, McDaid C, Andwari AM, Hosseini SE (2023) Green hydrogen supply chain risk analysis: a European hard-to-abate sectors perspective. Renew Sust Energ Rev 182:113371
5. Li L, Manier H, Manier MA (2019) Hydrogen supply chain network design: an optimization-oriented review. Renew Sust Energ Rev 103:342–360
6. Ren L, Zhou S, Ou X (2020) Life-cycle energy consumption and greenhouse-gas emissions of hydrogen supply chains for fuel-cell vehicles in China. Energy 209:118482
7. Xun D, Sun X, Geng J, Liu Z, Zhao F, Hao H (2021) Mapping global fuel cell vehicle industry chain and assessing potential supply risks. Int J Hydrog Energy 46(29):15097–15109
8. US Energy Information Administration—Hydrogen production, https://www.eia.gov/energyexplained/hydrogen/production-of-hydrogen.php. Last accessed 6 Nov 2024
9. European Commission—The European Green Deal https://commission.europa.eu/strategy-and-policy/priorities-2019-2024/european-green-deal_en. Last accessed 6 Nov 2024
10. United States Environmental Protection Agency- Climate Change Regulatory Actions and Initiatives, https://www.epa.gov/climate-change/climate-change-regulatory-actions-and-initiatives, last accessed 2024/11/6
11. The White House- FACT SHEET: Biden-Harris Administration Releases Agency Climate Adaptation Plans, Demonstrates Leadership in Building Climate Resilience. https://www.whitehouse.gov/briefing-room/statements-releases/2024/06/20/fact-sheet-biden-harris-administration-releases-agency-climate-adaptation-plans-demonstrates-leadership-in-building-climate-resilience/. Last accessed 6 Nov 2024
12. Ministry of Economy, Trade and Industry- Green Growth Strategy Through Achieving Carbon Neutrality in 2050. https://www.meti.go.jp/english/policy/energy_environment/global_warming/ggs2050/index.html. Last accessed 6 Nov 2024
13. Ministry of Economy, Trade and Industry—Japan's Roadmap to "Beyond-Zero" Carbon. https://www.meti.go.jp/english/policy/energy_environment/global_warming/roadmap/. Last accessed 6 Nov 2024
14. Government of Canada- Canada's climate plans and targets. https://www.canada.ca/en/services/environment/weather/climatechange/climate-plan/climate-plan-overview.html. Last accessed 6 Nov 2024
15. HKSAR Government, Hong Kong's Climate Action Plan 2050 (2021). https://cnsd.gov.hk/wp-content/uploads/pdf/CAP2050_booklet_en.pdf. Last accessed 8 Nov 2024
16. HKSAR Government, Clean air plan for Hong Kong 2035 (2021). https://www.eeb.gov.hk/sites/default/files/pdf/Clean_Air_Plan_2035_eng.pdf. Last accessed 8 Nov 2024
17. Environment and Ecology Bureau, HKSAR (2021) Hong Kong Roadmap on Popularisation of Electric Vehicles. https://www.eeb.gov.hk/sites/default/files/pdf/EV_roadmap_eng.pdf. Last accessed 8 Nov 2024
18. National Development and Reform Commission- Medium and Long-term Plan for the Development of the Hydrogen Energy Industry (2021–2035). https://www.ndrc.gov.cn/xxgk/zcfb/ghwb/202203/t20220323_1320038.html. Last accessed 7 Nov 2024
19. International Energy Agency-Korea Hydrogen Economy Roadmap 2040. https://www.iea.org/policies/6566-korea-hydrogen-economy-roadmap-2040. Last accessed 7 Nov 2024
20. Allied Market Research- Hydrogen Fuel Cell Vehicle Market Size, Share, Competitive Landscape and Trend Analysis Report, by Vehicle Type, by Technology, by Range: Global

Opportunity Analysis and Industry Forecast, 2023–2032. https://www.alliedmarketresearch.com/hydrogen-fuel-cell-vehicle-market. Last accessed 7 Nov 2024
21. Globe Newswire- Global Hydrogen Fuel Cell Vehicles (HFCV) Market Share Envisaged to Reach USD 62.88 Billion by 2032, at 45.2% CAGR: Polaris Market Research. https://www.globenewswire.com/en/news-release/2023/03/13/2625817/0/en/Global-Hydrogen-Fuel-Cell-Vehicles-HFCV-Market-Share-Envisaged-to-Reach-USD-62-88-Billion-by-2032-at-45-2-CAGR-Polaris-Market-Research.html. Last accessed 7 Nov 2024
22. Üçok MD (2023) Prospects for hydrogen fuel cell vehicles to decarbonize road transport. Discov Sustain 4(1):42
23. Moon S, Kim K, Seung H, Kim J (2022) Strategic analysis on effects of technologies, government policies, and consumer perceptions on diffusion of hydrogen fuel cell vehicles. Energy Econ 115:106382
24. Agency for Natural Resources and Energy- The 6th Strategic Energy Plan. https://www.enecho.meti.go.jp/en/category/others/basic_plan/. Last accessed 7 Nov 2024
25. US Department of Energy, US National Clean Hydrogen Strategy and Roadmap. https://www.hydrogen.energy.gov/library/roadmaps-vision/clean-hydrogen-strategy-roadmap. Last accessed 7 Nov 2024
26. The Federal Government of German-The National Hydrogen Strategy. https://www.bmwk.de/Redaktion/EN/Hydrogen/Dossiers/national-hydrogen-strategy.html. last accessed 7 Nov 2024
27. https://eur-lex.europa.eu/legal-content/EN/TXT/?uri=CELEX:52020DC0301. Last accessed 7 Nov 2024
28. Wang X, Wang S, Liu W, Wu MT (2019) Research on incentive policies for hydrogen energy in key application fields in Germany and their implications. Modern Business 23:44–45
29. Behling N, Williams MC, Managi S (2015) Fuel cells and the hydrogen revolution: analysis of a strategic plan in Japan. Econ Analysis Policy 48:204–221
30. Miller EL, Thompson ST, Randolph K, Hulvey Z, Rustagi N, Satyapal S (2020) US Department of Energy hydrogen and fuel cell technologies perspectives. MRS Bull 45(1):57–64
31. Azni MA, Md Khalid R (2021) Hydrogen fuel cell legal framework in the United States, Germany, and South Korea—a model for a regulation in Malaysia. Sustainability 13(4):2214
32. Report of Market Prospective and Investment Strategy Planning on China Fuel Cell Industry(2024–2029) https://bg.qianzhan.com/report/detail/f8ac09ee48294313.html. Last accessed 7 Nov 2024
33. Hydrogen Energy Industry Standard System Construction Guide (2023 Edition). https://www.gov.cn/zhengce/zhengceku/202308/content_6897986.html. Last accessed 24 Aug 2024
34. New Energy Vehicle Industry Development Plan (2021–2035) http://heic.org.cn/newshow.asp?id=604. Last accessed 24 Aug 2024
35. Hydrogen Fuel Cell Vehicle Safety Guide (2019 Edition). https://www.kelionline.com/categories/detail/12623. Last accessed 7 Nov 2024
36. Ministry of Industry and Information Technology of the People's Republic of China. https://wap.miit.gov.cn/jgsj/zbys/gzdt/art/2023/art_53cd034d5d59469cbc1e03d6a6a217d0.html. Last accessed 7 Nov 2024
37. Latapí M, Davíðsdóttir B, Cook D, Jóhannsdóttir L, Radoszynski AM, Karlsson K (2024) Hydrogen fuel cells in shipping: a policy case study of Denmark, Norway, and Sweden. Mar Policy 163:106109
38. Lopez Jaramillo O, Stotts R, Kelley S, Kuby M (2019) Content analysis of interviews with hydrogen fuel cell vehicle drivers in Los Angeles. Transp Res Rec 2673(9):377–388
39. Khan U, Yamamoto T, Sato H (2021) Understanding attitudes of hydrogen fuel-cell vehicle adopters in Japan. Int J Hydrog Energy 46(60):30698–30717
40. Liu J, Chen T, Hu B (2023) Consumer acceptance under hydrogen energy promotion policy: evidence from Yangtze River Delta. Int J Hydrog Energy 48(30):11104–11112
41. Hardman S, Shiu E, Steinberger-Wilckens R, Turrentine T (2017) Barriers to the adoption of fuel cell vehicles: a qualitative investigation into early adopters attitudes. Transp Res A Policy Pract 95:166–182

42. Li W, Long R, Chen H, Chen F, Zheng X, He Z, Zhang L (2020) Willingness to pay for hydrogen fuel cell electric vehicles in China: a choice experiment analysis. Int J Hydrog Energy 45(59):34346–34353
43. Greene DL, Lin Z, Dong J (2013) Analyzing the sensitivity of hydrogen vehicle sales to consumers' preferences. Int J Hydrog Energy 38(36):15857–15867
44. Lipman TE, Elke M, Lidicker J (2018) Hydrogen fuel cell electric vehicle performance and user-response assessment: results of an extended driver study. Int J Hydrog Energy 43(27):12442–12454
45. Hardman S, Tal G (2018) Who are the early adopters of fuel cell vehicles? Int J Hydrog Energy 43(37):17857–17866
46. Han W, Zhang G, Xiao J, Bénard P, Chahine R (2014) Demonstrations and marketing strategies of hydrogen fuel cell vehicles in China. Int J Hydrog Energy 39(25):13859–13872
47. Trencher G (2020) Strategies to accelerate the production and diffusion of fuel cell electric vehicles: experiences from California. Energy Rep 6:2503–2519
48. Hong Kong Hydrogen Economy Study and Report. https://campaigns.hkpc.org/hubfs/APAS/Report%20-%20Market%20Study%20on%20the%20Use%20of%20Hydrogen%20in%20Green%20Transportation%20in%20Hong%20Kong_Complete.pdf. Last accessed 7 Nov 2024

Chapter 29
The Impact of Climate, Economic and Geopolitical Shocks on Energy Prices and Risk/Return Coefficient Behavior in Mid- and Long-Term: With Technical Analysis and Structural Break Test

Aynur Pala

Abstract This paper examines the behavioral changes of the time series of fossil fuel energy prices and their variation coefficients in response to global economic, geopolitical, and climatic shocks. For this purpose, we used sub-contents of the IMF Energy Price Indices: petroleum, propane, natural gas and coal price indices for the period of 1992–2023. Technical trend analysis and multiple structural break tests developed by Bai and Perron (2003) methodologies were used together to define the most important break date among all of them. Finally, multiple shocks experienced together with economic and climate shocks create long-term breakouts, while single-cause shocks appear to create medium-term fluctuations. The most important break date was defined as 2013 and pre/post-sample was compared with the risk/return approach. The results show that while the average risk-return ratio of all fossil fuel prices is 3 in pre-break sub sample, the coefficients increased nine and six times in crude oil and propane prices, respectively; remained in one and half and one times in natural gas and coal prices in post-sub sample. This means that crude oil and propane, both petroleum derivatives, were riskier than coal and natural gas, which are also used in electricity production. It shows that coal and natural gas are more preferable in investments than petroleum derivatives due to their use in electricity generation, which is clean energy, and the opportunities offered by carbon markets.

Keywords Fossil Fuel Commodity Price · Structural Breaks

A. Pala (✉)
Istanbul Okan University, Tuzla, Istanbul, Türkiye
e-mail: aynur.pala@okan.edu.tr

L. Chen (ed.), *Clean Energy Research*, Green Energy and Technology,
https://doi.org/10.1007/978-3-031-99312-1_29

29.1 Introduction

Commodities represent natural resources among the factors of production and, therefore, form the initial step in the economic cycle. This structure positions commodity prices as a leading indicator in tracking the direction of the global economy. Commodities are classified into three categories: energy, industrial inputs, and food. Energy encompasses fossil fuels, including crude oil, natural gas, propane, and coal.

Fossil fuels—including petroleum, natural gas, and coal—remain essential energy sources across multiple sectors due to their versatility and energy density. These resources are widely utilized for heating, industrial production processes, transportation, and electricity generation. Petroleum, for instance, serves as a key energy source for transportation and heating, and also plays a critical role as a raw material in the manufacture of plastics and chemicals such as propane. It contributes approximately 3% to global electricity generation. Natural gas is predominantly used for heating and electricity generation, accounting for about 23% of electricity production in 2023, and also serves as a feedstock for chemicals like propane. Coal continues to be a significant contributor, particularly in electricity generation—making up 36% of global production—as well as in steel manufacturing (via coke) and industrial heating. Derivatives like propane, produced from both oil and natural gas, are important for residential and industrial heating, as well as in manufacturing processes. Sixty-one percent of the world's electrical production is made up of coal and natural gas. Coal and natural gas prices diverge from petroleum and propane because they are used for electrical production. The renewable energy transformation trend goes on, but in the mid-term, electrical production based on coal and natural gas is still important. The carbon market effect on coal and natural gas power production continues due to the cost level of renewable energy generation. Propane is made up of petroleum and natural gas, and its price is affected by petroleum and natural gas.

Since the early 1990s, energy markets have undergone profound transformations, influenced by a confluence of economic, geopolitical, and environmental factors. These shocks induce structural breaks in the movements of commodity price series and their interrelationships in the medium and long term. Understanding and managing these impacts is crucial for researchers, policymakers, corporate managers, and investors. A key characteristic of energy commodity price series during this period has been the presence of structural breaks, or sudden, significant shifts in price trends, which reflect changes in underlying market dynamics. These breaks are often associated with major economic shocks, geopolitical events, and the increasing prominence of climate change and sustainability considerations.

The 1990s marked a pivotal decade for energy markets, with increasing global interdependence, the deregulation of energy markets, and the integration of emerging economies. Early in the decade, the collapse of the Soviet Union and the Gulf War (1990–1991) led to significant disruptions in global oil supply, which contributed to the first major structural break in oil price trends. As global markets became more interconnected, the 1997 Asian financial crisis highlighted the sensitivity of energy prices to shifts in global demand, while the rise of oil-importing economies, particularly China, began to have lasting effects on long-term price trends.

Geopolitical events such as the September 11 attacks (2001) and the Iraq War (2003) introduced further volatility into energy markets, with oil prices exhibiting erratic fluctuations. During this period, energy prices often followed upward trajectories driven by supply disruptions, speculative behavior, and concerns over future energy security. These events also underscored the growing role of geopolitical risk in determining energy price stability.

In the early 2000s, rising concerns over climate change and environmental sustainability began to exert pressure on energy markets, setting the stage for a new set of structural breaks. The 2008 global financial crisis, which resulted in sharp declines in global oil demand, caused another major structural break in energy prices. Despite the subsequent recovery in oil prices, the ongoing emphasis on sustainability, carbon reduction, and the global transition to renewable energy created a new set of challenges for the energy market, culminating in another break in 2013 that marked the end of the long-term upward trend in energy prices that had been characteristic of the preceding decades.

This shift was further exacerbated by the global push towards low-carbon technologies and the introduction of climate-related policies, such as the Kyoto Protocol and the Paris Agreement. These environmental and regulatory factors led to significant changes in the energy market's structure, with clean energy technologies beginning to gain a foothold in the global energy supply. The post-2013 period witnessed a weakening of the long-standing price trends, as the influence of climate change, environmental policies, and renewable energy transitioned from being peripheral concerns to central drivers of energy price movements.

The COVID-19 pandemic of 2020 marked yet another key event in energy markets, triggering a sharp decline in energy demand due to global lockdowns. However, the recovery in energy prices during the post-pandemic period has been tempered by the intensification of climate-related pressures and ongoing geopolitical instability, such as the 2022 Russia-Ukraine war. These developments suggest that energy prices are no longer subject solely to traditional economic and supply-demand dynamics but are increasingly influenced by broader, more complex forces, including climate change, sustainability agendas, and geopolitical risk.

This introduction provides a broad overview of the structural breaks in energy commodity prices since 1990, a period marked by heightened volatility and a profound transformation in how the energy market functions. Understanding these structural breaks is crucial for policymakers, investors, and scholars, as they provide insights into the resilience of energy price trends, the interplay of various factors influencing energy markets, and the shifting dynamics of global energy supply and demand. This paper will further explore the nature and causes of these structural breaks, focusing on the impact of economic, geopolitical, and climate-related shocks on energy commodity prices.

29.2 Literature Review

Energy prices, particularly those of fossil fuels such as crude oil, natural gas, coal, and propane, have been the subject of extensive academic research. Given the fundamental role of energy markets in global economic stability, understanding structural breaks in energy price series is crucial for forecasting, policy-making, and risk management. Structural breaks refer to significant and often permanent shifts in the underlying data-generating process of a time series, which can alter trends, volatility, and interrelationships with other economic variables. This literature review aims to synthesize key studies on the identification, causes, and economic implications of structural breaks in energy price series.

The early literature on energy price volatility and structural breaks focused primarily on the oil market, driven by the 1970s oil crisis. However, as energy markets became more integrated and global in the 1990s, researchers began to examine the behavior of energy prices with an emphasis on structural shifts. Pindyck [1] was among the first to highlight the role of structural shifts in oil prices, focusing on the unpredictable behavior of oil prices in the post-1970s period. His work identified that the oil price series exhibited non-stationarity with significant structural breaks, particularly during periods of geopolitical tensions. Hamilton [2] expanded on this concept by examining the long-term relationship between oil prices and economic recessions. He demonstrated that sharp increases in oil prices, driven by supply-side shocks, could lead to structural breaks in the macroeconomy, influencing both the pricing of oil and its interrelationships with other economic variables.

As the availability of high-frequency data improved and econometric methods advanced, the 2000s saw a proliferation of studies applying formal tests for structural breaks. The development of tests for detecting multiple breaks in time series data allowed researchers to more accurately capture regime shifts in energy prices. Bai and Perron [3] developed the most widely used framework for detecting multiple structural breaks in time series data. Their method allows for the identification of multiple breaks in both the intercept and slope of a series, which is particularly useful for modeling energy prices. Using this method, several studies examined energy price data from the 1990s and early 2000s, finding evidence of multiple structural breaks, particularly in response to oil price shocks and the global integration of energy markets.

Zivot and Andrews [4] applied Bai and Perron's methodology to oil price data, identifying structural breaks during the 1973 oil embargo and the 1990–1991 Gulf War. Their results suggested that oil prices exhibit nonlinearities and structural shifts that make simple linear models inadequate for forecasting future price movements. Ciferri and Delli Gatti [5] applied Bai and Perron's methodology to investigate the impact of geopolitical shocks, particularly the Gulf War, on oil price dynamics. They found that the oil price series underwent a major structural break in the early 1990s, after which oil prices followed a more volatile and unpredictable pattern. Blanchard and Gali [6] examine the impact of the 2000s global economic slowdown and its effects on energy markets. They find that during the global downturn, oil prices were significantly affected by both demand-side reductions and financial market disturbances, which contributed to short-term volatility in energy prices.

Kilian and Hicks [7] document the relationship between the global recession in the early 2000s and oil price fluctuations. The study highlights that the economic crisis led to a weakening of demand, which in turn resulted in lower oil prices during this period. Fattouh [8] discusses the aftermath of the 9/11 attacks and their influence on oil prices. The study shows how the attacks led to increased geopolitical uncertainty and fear of supply disruptions, which caused an upward spike in oil prices in the short term. While this shock was temporary, it contributed to the long-term volatility in energy prices during the early 2000s.

By the 2010s, the literature on energy price structural breaks had shifted toward examining the effects of a wider range of geopolitical, economic, and climate-related shocks. Researchers increasingly sought to understand how these shocks led to regime shifts in energy prices, particularly within the context of global oil price cycles. Hamilton [9] provides an extensive analysis of the 2008 mortgage crisis and its impact on oil prices. He identifies that the financial crisis led to a substantial reduction in oil demand, causing a significant decline in oil prices from mid-2008 to early 2009. His study shows how this crisis triggered a structural break in oil price trends.

Ritter and Zheng [10] also discuss the oil price dynamics in the aftermath of the financial crisis, noting that while demand for oil fell sharply in 2008, the subsequent recovery in oil prices was driven by supply-side constraints and speculative activities in energy markets. Piraud [11] explores the impact of the European sovereign debt crisis, particularly on energy prices. The study finds that oil prices were volatile during the 2010–2012 period, driven by uncertainties in global financial markets and concerns about economic recovery in Europe, which had a ripple effect on global energy markets.

Jammoul and Kassim [12] examined structural breaks in crude oil prices from 1986 to 2011 using Bai and Perron's multiple break methodology. They found that geopolitical events, such as the 2003 Iraq War and the 2007–2008 financial crisis, caused significant shifts in oil price trends. The study concluded that these structural shifts could not be explained solely by supply and demand fundamentals but were also influenced by broader geopolitical dynamics and financial market speculation. Kilian [13] focused on the role of demand shocks in driving oil prices, identifying a structural break around the 2003 Iraq invasion. He also showed that, beyond supply disruptions, global demand increases, particularly from emerging economies like China, have contributed to structural shifts in energy markets. Kilian and Murphy [14] examine how geopolitical shocks like the Arab Spring affected energy prices, particularly oil and natural gas. The study identifies the disruption of oil production in countries like Libya as a significant contributor to the price increases observed in 2011–2012. They also emphasize how geopolitical instability can lead to permanent shifts in oil price levels, particularly in a supply-constrained market.

Alhajji and Hu [15] examined the impact of the 2008 global financial crisis on oil prices, utilizing structural break tests to show that the crisis led to a change in the oil price transmission mechanism. Their results suggested that post-crisis oil price dynamics were influenced by new market conditions, including changes in investor behavior, geopolitical uncertainty, and regulatory shifts in energy markets.

In the late 2010s and early 2020s, the growing emphasis on climate change and sustainability led to new studies examining how these factors could trigger structural breaks in energy price series. Research began to consider how the global push towards renewable energy and the implementation of climate policies, such as carbon pricing and emissions reductions, could reshape energy price dynamics. Sadorsky [16] explored the effects of climate-related policies on energy price series, applying structural break tests to examine the period from 2000 to 2018. The study found evidence of a break in the mid-2010s, driven by the growing prominence of renewable energy sources and international climate agreements like the Paris Agreement. This structural change indicated that energy prices would likely face downward pressure as global investment shifted towards clean energy technologies. Arezki et al. [17] investigated the role of climate change in triggering structural shifts in energy markets, especially oil prices. Their findings suggested that climate-related policies, such as the Paris Agreement, and growing concerns over fossil fuel dependence, contributed to the structural break observed in energy price movements around 2015. They emphasized that energy prices would likely face long-term instability due to the increasing integration of sustainability factors into energy markets. Cansado-Bravo and Rodríguez-Monroy [18] examined the influences of structural breaks on the long-term properties of Brent crude oil, gasoil, low-sulfur fuel oil, natural gas, and coal over the period 2002–2018. They observed that only fuel oil prices show evidence of mean-reverting behavior. It highlights nonrenewable energy trends and the importance of the policy mix in the transition to a low-carbon energy system. They said that changes in slope during the declining trend since 2011 that continued into 2014–2015 are also detected.

The COVID-19 pandemic and subsequent economic disruptions created a unique environment for analyzing energy price structural breaks. The sharp declines in demand during the global lockdowns and the subsequent recovery raised questions about the resilience of energy price trends. In addition, geopolitical tensions such as the 2022 Russia-Ukraine war further complicated energy price dynamics. Bose and Ozturk [19] applied structural break tests to investigate the effects of the COVID-19 pandemic on oil and natural gas prices. Their results indicated a significant structural break in 2020, with a sharp fall in prices followed by a rapid recovery as economies began reopening. The study argued that the pandemic had accelerated changes in energy demand patterns and had triggered a shift toward more flexible, market-based pricing mechanisms.

Sadorsky [20] discusses the break in the upward trend of fossil fuel prices around 2013. His findings suggest that the rise of renewable energy technologies, along with increasing climate-related pressures, contributed to a structural break in energy price trends, marking the end of the long-term upward trajectory seen in the 1990s and 2000s. Boucher et al. [21] also emphasize that post-2013, the energy market began to stabilize at a lower price level, reflecting the transition towards more sustainable energy sources. They suggest that while geopolitical events like the

Russia-Ukraine war continue to drive short-term price fluctuations, the underlying structural shift towards clean energy and sustainability is reshaping the global energy market.

29.3 Data and Methodology

We used sub-contents of the IMF Energy Price Indices: petroleum, propane, natural gas and coal price indices for the period of 1992–2023. Technical trend analysis and multiple structural break tests developed by Bai and Perron [3] methodologies were used to define the most important break date among all detected breaks. The variation coefficient of indices helps us to compare instruments between sub-break samples and control/management of portfolio allocation.

Data

The IMF Commodity Price series and indexes are the most popular monitoring tools. The energy price indexes cover petroleum, natural gas, coal and propane price indexes. The variable codes and definitions are presented in Table 29.1.

Methodology

We define the most important break dates on fossil fuel energy prices using technical trend analysis and the Bai and Perron structural break test together. Technical trend analysis and Bai and Perron's [3] structural break test help us to gain a detailed look at the structural change in the energy price series.

This methodology of detecting multiple breakpoints was developed in the late 1990s [22, 23]. However, efficient and robust algorithms for dating multiple structural changes were developed only in the early 2000s [3, 24]. As a result, the broad practical application of the methodology for dating multiple structural changes has

Table 29.1 Variable code and definition

Code	Variable definition	Source
POILAPSP	Crude oil (petroleum), price index, 2016 = 100, simple average of three spot prices; Dated Brent, West Texas Intermediate, and the Dubai Fateh	IMF
PNGAS	Natural gas price index, 2016 = 100, includes European, Japanese, and American Natural Gas Price Indices	IMF
PCOAL	Coal price index, 2016 = 100, includes Australian and South African Coal	IMF
PPROPANE	North American Spot LPG Propane Price/Mont Belvieu LST	IMF

started relatively recently. We generate a sub-sample by defining a break date. After defining break, it will be compared with the variation coefficients of fossil fuels variation coefficient among pre-break and post-break sub-samples.

Trendlines are one of the most important tools used in technical analysis. "When the price series breaks the trend line upwards or downwards, it indicates the end of the current trend and suggests the possibility of a new trend or a consolidation period. According to technical analysis principles, such breakouts often signal a change in market dynamics, which could lead to a shift in trend or a period of consolidation." (Murphy, J. J. (1999), [25]).

We create three comparative time series graphs with technical trend analysis for oil-gas, oil-coal and oil-propane. We also try to catch the most important break date in these series using a long-term trend line break. The series was transformed by a logarithmic scale to eliminate inflation.

The most important break date will be defined and pre-/post-sample will be compared with the variation coefficient (CV), namely, the risk/return approach. It helps us to compare instruments between sub-break samples and the control/management of portfolio allocation.

The variation coefficient is used in finance to measure risk. According to Markowitz [26], the CV helps compare the risk and return profiles of different assets for portfolio optimization. CV is a valuable metric in assessing investment stability and volatility in diverse markets.

The variation coefficient (CV) is calculated as the ratio of the standard deviation to the mean, as a percentage:

$$\mathrm{CV} = \left(\frac{\sigma}{\mu} \cdot 100 \right)$$

where

σ is the standard deviation
μ is the mean of the dataset

An asset with a higher CV may be considered riskier than an asset with a lower CV, which helps investors assess the relative risk and consistency of different investments.

29.4 Results

Structural Break Test

We applied the Bai and Perron procedure of detecting the breakpoints for $m = 0, \ldots, 4$, and found the evidence supporting the presence of structural breaks in the residuals from the equilibrium relationship. Bai and Perron [3] advocate that BIC is suitable for a model selection criterion. RSS and BIC can be used as model selection criteria. Table 29.2 reports the breakpoints for m-segmented models as well as the associated total residual sum of squares (RSS) and the Bayesian information criterion (BIC). The results of the breakpoint detection process can be shown as follows.

Table 29.2 reports the breakpoints for m-segmented models as well as the associated total residual sum of squares (RSS) and the Bayesian information criterion (BIC). Break test results show that the most important breakpoint (when $m = 1 \ldots 9$) is detected in the years of the 1997 Asian Crisis, 2000 Global Economic Crisis, 2002/2003 Twin Towers Terror Attack/Gulf War, 2006 Kyoto Protocol, 2008 Mortgage Crisis, 2012 Arab Spring and 2020 pandemic. Commodity series break points agree very well with the results of technical trend analysis. During the full period, all of the series have followed long-term uptrends from 1992 to 2013. There are many mid-term downward correction trends in that long-term upward trend channel, such as 1997–1999, 2002–2003, 2008, and 2012. In the technical trend analysis results, we found that all series have one most important break in the long term. The long-term trend line of all energy price series has broken down at (coal 2012, propane/petroleum 2013, and natural gas 2014), after the 2012 Arab

Table 29.2 Results of detecting multiple breakpoints

M	Breakpoints								RSS	BIC
POILAPSP										
0									26361	312.6
1				2004					8500	283.3
2				2004			2013		6633	282.3
3				2004		2009	2013		3785	271.3
4				2004		2009	2013	2020	2794	268.5*
PNGAS										
0									345953	395
1								2020	172650	379.7
2		2002						2020	115667	373.8
3			2003				2014	2020	93827	374
4	1999			2004			2014	2020	83641	377.3*
PCOAL										
0									216621	380
1								2020	102205	362.9
2					2006			2020	56236	362.6
3			2003		2006			2020	54382	356.6
4			2003		2006		2012	2020	52814	350.7*
PPROPANE										
0									3.9	30.7
1			2003						2.3	20.9
2			2003				2013		1.5	13.2
3	1999			2004			2013	2020	1.3	15.931*
4	1999			2004			2013	2020	1.2	20.3

m denotes the number of breakpoints. *BIC* denotes the Bayesian information criterion, and *RSS* denotes the total residual sum of squares

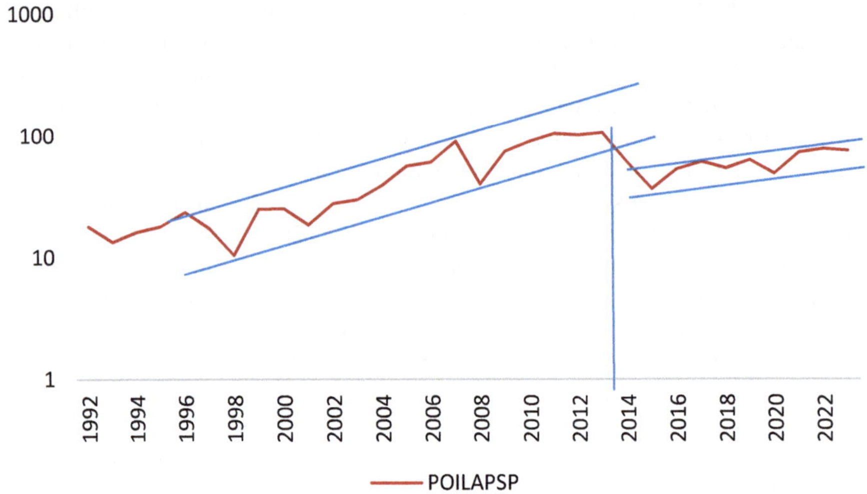

Fig. 29.1 Crude oil price index, IMF

Spring. The new trend channel with a mild upward slope following the break is illustrated in Fig. 29.1.

We can look at structural break/s in the fossil fuel price series with technical trend analyses. The long-term trend break point is consistent with the structural break test.

Technical Trend Analysis

We create three comparative time series graphs with technical trend analysis for oil-gas, oil-coal and oil-propane. We also try to catch the most important break date in these series using a long-term trend line break. The series was transformed by a logarithmic scale to eliminate inflation.

The crude oil price index, presented in Fig. 29.1. The log-scaled Crude Oil Price Index experienced a breakdown in its long-term upward trend in 2013. During the preceding uptrend channel, crude oil prices were significantly impacted by several global economic and geopolitical events, including the 1997 Asian financial crisis, the 2000 global economic downturn, the 2006 Kyoto Agreement, the 2008 mortgage crisis, and the 2012 Arab Spring. In each of these instances, crude oil prices faced mid-term corrections that temporarily reversed the broader upward movement, but ultimately returned to their upward trajectory after the crises subsided. However, in 2013, the long-term uptrend channel was definitively broken, marking a significant shift in the price dynamics.

Following the 2006 Kyoto Protocol—considered one of the most impactful climate agreements—the period from 2007 to 2013 saw crude oil prices failing to reach, let alone exceed, the 2007 all-time high. This period coincided with the growing global focus on climate change, which, along with increasing sustainability efforts by countries and international institutions, began to exert downward pressure on oil prices. The rise of biofuels produced from agricultural products further diminished the upward pressure on oil prices that had been associated with geopolitical events, including the 2012 Arab Spring.

The COVID-19 pandemic in 2020 added another layer of disruption to the energy markets, serving as a stark reminder of the profound effects of global warming. The pandemic, in particular, exposed the vulnerability of fossil fuel markets, especially oil, which has long been one of the primary contributors to global warming. This unprecedented event led to a dramatic shift in oil price behavior, signaling that the dynamics of the energy market had been fundamentally altered.

To fully understand the broader implications of these shifts, it is also important to examine the evolutionary trends in the price series of other fossil fuels, such as natural gas, coal, and propane, over the same period. These commodities, while sharing many of the same economic and geopolitical influences, have exhibited distinct price behaviors and patterns in response to global events and environmental pressures.

The natural gas price index comparatively with oil price index, presented in Fig. 29.2. During the 2004–2013 period, global production of natural gas increased, driven not only by the shale boom in North America but also by new discoveries in other regions (e.g., Russia, the Middle East, and Australia). Major exporters like Russia, Qatar, and Algeria expanded their production capacity, contributing to an oversupply in the global market, which put downward pressure on prices. While global production of natural gas surged, demand growth was not as robust, particularly in the aftermath of the 2008 global financial crisis. Economic downturns in Europe and Japan, along with slower-than-expected growth in emerging markets, led to weaker-than-expected demand growth. In Europe, for instance, natural gas demand stagnated due to the economic recession and the rising adoption of renewable energy sources.

The coal price index, presented in Fig. 29.3. A significant factor influencing the coal price break in 2011 was the rising global supply of natural gas, particularly from the United States due to the shale gas boom. Shale gas production significantly lowered natural gas prices, making it a more competitive and cleaner alternative to coal, especially for power generation. In countries like the U.S. and Europe, natural gas began to replace coal in electricity generation, reducing demand for coal and contributing to a downward shift in coal prices.

In 2011, there were significant supply-side disruptions in coal markets, particularly from key producing countries like Australia and Indonesia. For instance, heavy floods in Queensland, Australia, in late 2010 and early 2011 caused severe disruptions to coal mining operations and exports. While this initially led to price spikes, the subsequent restoration of supply led to a sharp correction in 2011, contributing to the break in coal prices.

The Fukushima nuclear disaster in Japan in March 2011 had a significant impact on energy markets, including coal. In the aftermath of the disaster, Japan, which had

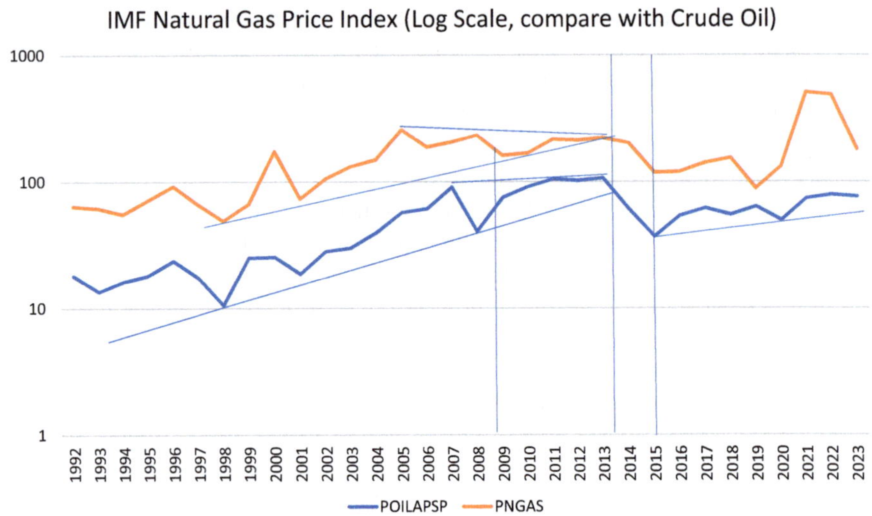

Fig. 29.2 Natural gas price index, IMF

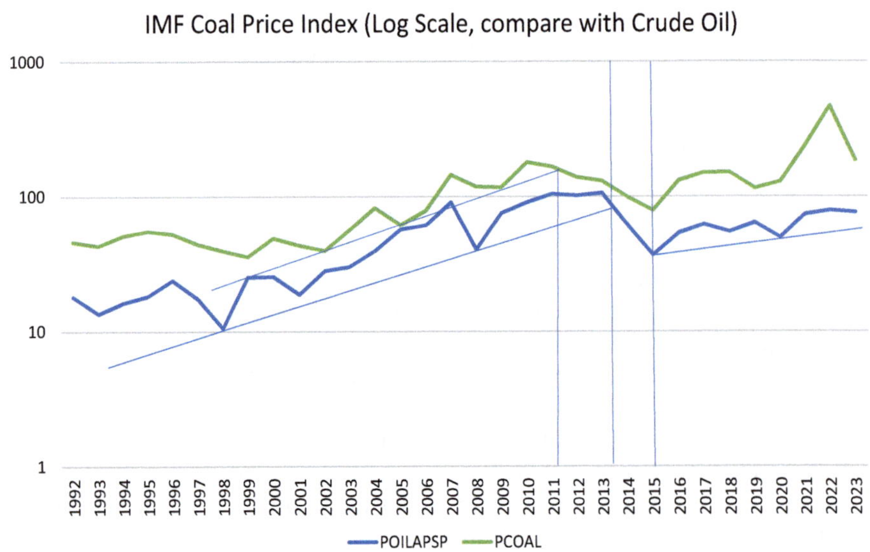

Fig. 29.3 Coal price index, IMF

been a major consumer of coal for electricity generation, began to increase its reliance on other energy sources, including liquefied natural gas (LNG) and renewable energy. This shift in demand affected coal prices globally, contributing to the structural break in 2011.

The propane price index is presented in Fig. 29.4. Propane is produced from both crude oil and natural gas, so its price is affected by the same global climate change

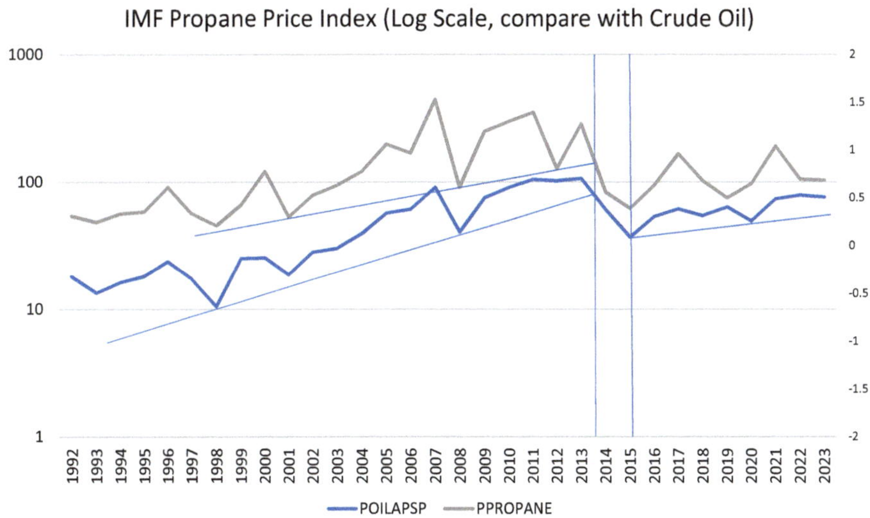

Fig. 29.4 Propane price index, IMF

pressures affecting fossil fuel markets. Propane prices experienced a major trend reversal in 2013.

In the technical trend analysis, energy price indexes showed a long-term uptrend during the period of 1992–2013 (11 years). The long-term trend line was broken down in 2013 due to climate change pressure during 2006–2013. Post break time, the energy price has a lower average return according to pre-break time.

During the long-term up-trend, it is observed that some mid-term escalations are due to economic, geopolitical and climatic reasons; the 1997 Asian crisis, the 2000 global economic crisis, 2002 Twin Tower Terrorist Attack, 2003 US-Iraq War, rising climate change problem and action 2006 Kyoto Protocol, 2008 mortgage crises, 2012 Arab Spring. It is observed that from 2006 to 2013, all series did not test new high values due to climate change and economic turmoil.

In the post-2013 break, fossil fuel price series started a new uptrend having a lower mean and standard deviation compared to the pre-break. Inside the new uptrend there are some shocks based on pandemic/climatic and geopolitical risk.

Figure 29.5 compares the results of technical trend analysis with the structural break dates identified by the Bai and Perron [3] methodology. The results of the technical trend analysis and structural break test show that the most important break is in 2013. Risk/return ratio, namely variation coefficient (VC), changes by pre-/post-2013 sustainability break. The variation coefficient of crude oil price is 2.7 at pre-break 2013 and post-break 23.3. Crude oil price variation coefficient rises 7–8 times according to the previous break time value. The propane price variation ratio is 2.7 to 16.1, 5 times higher than before the break. This value shows that crude oil and propane have been the riskier asset among other energy assets, corresponding to the period before the 2013 break.

Technical Trend Break	BP(03) Breaks	Petroleum	Propane	Coal	N.Gas	Shocks
mid-term fluctuation	1999		x		x	Economic
mid-term fluctuation	2003/2004	x (2004)	x (2004)	x (2003)	x (2004)	Jeopolitic (Gulf War)
mid-term fluctuation	2006			x		Climatic (Kyoto)
mid-term fluctuation	2009	x				Economic (2008 mortgage)
Long-term trend breakdown	**2012/2013/2014**	x (2013)	x (2013)	x (2012)	x (2014)	**Jeopolitic (Arab Spring) / Economic / Climatic (increasing pressure)**
mid-term fluctuation	2020	x	x	x	x	Climatic

Fig. 29.5 Compare with the technical trend and Bai and Perron [3] breaks. Note: *x* indicates the common break dates identified by both technical analysis and the Bai-Perron [3] test

Table 29.3 Risk, return, and variation coefficient for the full sample and pre-/post-2013 sub-samples calculated by the 5YMA[a] series

Sample	Full (1995–2023)			Pre (1995–2013)			Post (2014–2023)			CVpost-CVpre	CV pre/post
	$\overline{X}$	s	CV	$\overline{X}$	s	CV pre	$\overline{X}$	s	CV post	Diff.	% growth
POILAPSP	11.4	40.1	3.5	16.2	43.5	2.7	1.4	31.6	23.3	20.6	771
PNGAS	15.7	65.6	4.2	13.4	44.9	3.3	20.4	99.1	4.9	1.5	45.5
PCOAL	10.4	37.3	3.6	8.3	29.1	3.5	14.6	52.1	3.6	0.1	1.9
PPROPANE	1.9	49.3	25.4	19.3	51.6	2.7	2.8	44.5	16.1	13.4	502.2

[a]*5YMA* five-year moving average; $\overline{X}$ mean; *s* standart deviation; *CV* coefficient of variation

Finally, multiple shocks experienced together with geopolitical, economic, and climate shocks create a long-term break (2013), while single-cause shocks appear to create medium-term fluctuations. The most important break date was defined as 2013 and pre-/post-sample were compared with the risk/return approach. These findings are supported by Boucher et al. [21]. Boucher et al. [21] also emphasize that post-2013, the energy market began to stabilize at a lower price level, reflecting the transition towards more sustainable energy sources.

Table 29.3 presents the risk, return, and variation coefficient for the full sample, as well as for the pre- and post-2013 sub-samples, based on the 5-year moving average (5YMA) series. Coal has become a relatively advantageous investment compared to other fossil fuel energy commodities. Propane price variation indicates that they will not be preferred in portfolios with extremely high variance and low return (a 5 times variation coefficient compared to the pre-2013 break. Energy prices, particularly those of crude oil and natural gas, have a direct impact on the price of propane. Propane is a crude oil and natural gas derivative.

Figure 29.6 shows the five-year moving average of the standard deviation of fossil energy price indexes, revealing long-term volatility patterns in the market. Specifically, after the pandemic, coal and natural gas volatility reached a new high. The global green recovery is facing a significant threat due to the escalating consumption of coal and the announcement of new coal development plans by several leading nations. (power plant expansion, and climate change policy).

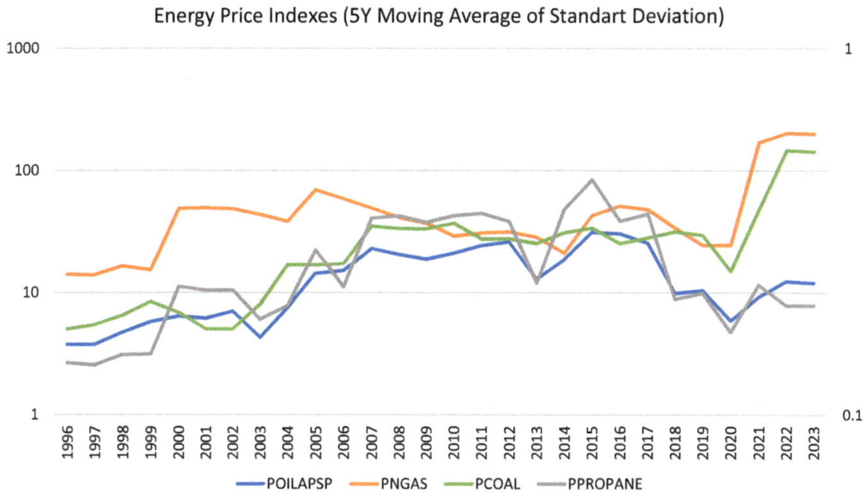

Fig. 29.6 Five-year moving average of the standard deviation of fossil energy price indexes, IMF

29.5 Conclusion

Energy (fossil fuel) price has been affected by global economic, geopolitical and climatic shocks. Specifically, it is important to understand the effect of global warming and the sustainability trend on energy price for researchers, politicians, energy commodity investors, traders, importers and exporters. This paper aims to examine energy price time series behavior using technical trend analysis and structural break test together for a smart and clear look. The data period covers 1992 to 2023. We used the IMF energy price index sub-blocks, petroleum, natural gas, propane and coal prices. To detect structural break/s in series, technical trend analysis and structural break test were applied based on Bai and Perron [3, 23] and Zeileis et al. [24]. The changes and structural breaks caused by the economic, geopolitical, and climate shocks experienced over the past 30 years have been confirmed through trend analysis and the structural break tests developed by Bai and Perron. There is one long-term break in 2013 based on climate change effect (since 2006) and five mid-term breaks (downside correction) in the long-term uptrend channel at 1999, 2003, 2008, 2013 based on economic/geopolitical shocks. Findings show that energy price behavior faces the hardest evolution step.

Since the 1990s, the effects of economic, geopolitical, and climate shocks on energy prices (fossil fuels) have been examined. It has been observed that economic and geopolitical issues have contributed to the fluctuations within the long-term upward trends in energy price series. The economic shocks causing these medium-term fluctuations include the 1997 Asian financial crisis, the 2000 global economic crisis, the 2008 mortgage crisis, and the 2010 European debt crisis. Geopolitical shocks include events such as the 2002 Twin Towers terrorist attacks, the 2003 Iraq War, and the 2012 Arab Spring. Furthermore, with the rise of global warming, climate change, and the resulting droughts and floods, the signing of the Kyoto Protocol in 2006 and the subsequent shift towards reducing carbon emissions and fostering sustainability have placed greater pressure on energy prices, surpassing the effects of economic crises and geopolitical conflicts.

Ultimately, the year 2013 marked a significant turning point when the long-term upward trend in the prices of fossil fuels such as crude oil, natural gas, propane, and coal, which had been in place since the 1990s, was broken. This break in 2013, contrary to historical trends, represented an important signal that global warming, climate change, global green transformation, and sustainability had gained a stronger influence on the global energy market than events such as the 2012 Arab Spring. Our findings were supported by Sadorsky [20] and Boucher et al. [21].

Since 2015, a new but weaker trend has been observed, with energy prices attempting to stabilize on a lower trajectory. The emergence of this weak upward trend in energy prices has continued to be under pressure since the 2020 pandemic, with the intensifying effects of climate change reminding the world more forcefully of its significance. Climate pressures, such as the 2022 Russia-Ukraine war and the 2024 Israel-Palestine conflict, continue to hinder strong upward responses in energy prices.

In conclusion, energy commodities derived from fossil fuels are undergoing a challenging evolutionary phase in the face of the growing impacts of climate change and the increasing emphasis on sustainability in recent years.

Our finding is that the long-term uptrend broke down in 2013, which is supported by Cansado-Bravo and Rodríguez-Monroy [18]. And we can see the start of the new uptrend after the 2013 break (mid-term declining period during 2011–2015). They found similar results, but it does not provide a broad look. According to our study, energy prices have a long-term uptrend from 1992 to 2013. Energy prices face some mid-term decline based on economic, geopolitical and climatic shocks within the long-term up-trend. Specifically, post-Kyoto (2006) sub-samples, fossil fuel energy prices series, especially crude oil and propane, face the pressure of sustainability/green pressure. All series could not reach their new highest level since 2006 due to climatic shock, food crises and Russian-Ukraine invasion and finally the long-term uptrend line of indexes was broken down in 2013. Generally, energy price pushback from the long-term trend line is based on geopolitical shocks like the 2002 Twin Tower terrorist attack, although in the 2012 Arab Spring, energy prices could not rise due to climate change/sustainability pressure. We can see from the energy prices graphs that the long-term trend has shifted to a lower mean and slope under the pressure of climatic problems and sustainability. The variation coefficient of crude oil price is 2.7 at pre-break in 2013 and post-break 20.6. Crude oil price variation coefficient rises 7–8 times according to the previous break time value. This value shows that crude oil has been the riskier asset among other energy assets.

References

1. Pindyck RS (1991) The dynamics of commodity markets: the case of oil prices. Energy J 12(2):1–26. https://doi.org/10.5547/ISSN0195-6574-EJ-Vol12-No2-1
2. Hamilton JD (1996) This is what happened to the oil price-macroeconomy relationship. J Monet Econ 38(2):215–220. https://doi.org/10.1016/0304-3932(96)01243-5
3. Bai J, Perron P (2003) Computation and analysis of multiple structural change models. J Appl Economet 18(1):1–22. https://doi.org/10.1002/jae.726
4. Zivot E, Andrews DWK (2002) Further evidence on the great crash, the oil price shock, and the unit root hypothesis. J Business Econ Stat 20(1):25–44. https://doi.org/10.1198/073500102288618668
5. Ciferri D, Delli Gatti D (2004) Geopolitical shocks and the volatility of oil prices. Energy Econ 26(4):807–821. https://doi.org/10.1016/j.eneco.2003.10.001
6. Blanchard O, Gali J (2007) The macroeconomic effects of oil price shocks: a dynamic general equilibrium approach. J Monet Econ 54(4):1233–1257. https://doi.org/10.1016/j.jmoneco.2006.12.008
7. Kilian L, Hicks B (2007) The 2000–2002 oil price shocks and their impact on the U.S. economy. Energy J 28(4):1–34. https://doi.org/10.5547/ISSN0195-6574-EJ-Vol28-No4-1
8. Fattouh B (2007) Oil prices and the geopolitical landscape: 9/11 and beyond, Oxford Institute for Energy Studies. Retrieved from https://www.oxfordenergy.org
9. Hamilton JD (2009) Causes and consequences of the oil shock of 2007–2008. Brook Pap Econ Act 2009(1):215–261. https://doi.org/10.1353/eca.0.0026
10. Ritter M, Zheng F (2010) Global financial crisis and oil price dynamics: insights and challenges. Energy Policy 38(11):7554–7562. https://doi.org/10.1016/j.enpol.2010.07.054
11. Piraud C (2012) Energy markets and the European debt crisis: oil price volatility and economic uncertainty. Energy Econ 34(6):1674–1683. https://doi.org/10.1016/j.eneco.2012.06.012
12. Jammoul N, Kassim K (2013) Structural breaks in crude oil prices: the impact of geopolitical events and financial crises. Energy Econ 40:448–457. https://doi.org/10.1016/j.eneco.2013.09.003
13. Kilian L (2009) The impact of the 2008 oil shock on the global economy. National Bureau of Economic Research (NBER) Working Paper No. 15006. https://doi.org/10.3386/w15006
14. Kilian L, Murphy DP (2014) The role of geopolitics in the 2011 oil price spike. Energy J 35(3):1–27. https://doi.org/10.5547/01956574.35.3.1
15. Alhajji AF, Hu B (2015) The impact of the global financial crisis on oil prices and the transmission mechanism. Energy Econ 50:197–205. https://doi.org/10.1016/j.eneco.2015.06.001
16. Sadorsky P (2019) Climate change and energy prices: structural breaks and long-term trends. Energy Econ 83:1–15. https://doi.org/10.1016/j.eneco.2019.07.008
17. Arezki R, Bruckner M, Dube G (2020) Climate change, oil prices, and energy transitions: a structural break analysis. Energy Policy 141:111453. https://doi.org/10.1016/j.enpol.2020.111453
18. Cansado-Bravo P, Rodríguez-Monroy C (2020) The effects of structural breaks on energy resources in the long run. Evidence from the last two oil price crashes before COVID-19. Designs 4(4):49. https://doi.org/10.3390/designs4040049
19. Bose S, Ozturk I (2022) Impact of COVID-19 pandemic on oil and natural gas prices: structural break analysis and future predictions. Energy 238:121597. https://doi.org/10.1016/j.energy.2021.121597
20. Sadorsky P (2020) Energy price trends in the wake of climate change policies and renewable energy technologies. Energy Policy 137:111099. https://doi.org/10.1016/j.enpol.2019.111099
21. Boucher A, Zahariev K, Xu X (2023) Global energy price dynamics: the role of geopolitical shocks, climate change, and sustainability post-2013. Energy Econ 104:105636. https://doi.org/10.1016/j.eneco.2023.105636
22. Bai J (1997) Estimating multiple breaks one at a time. Econ Theory 13(3):315–352. https://doi.org/10.1017/S0266466600005831

23. Bai J, Perron P (1998) Estimating and testing linear models with multiple structural changes. Econometrica 66(1):47. https://doi.org/10.2307/2998540
24. Zeileis A, Leisch F, Hornik K, Kleiber C (2003) Testing and dating of structural changes in practice. Comput Stat Data Anal 44(1–2):109–123. https://doi.org/10.1016/S0167-9473(03)00030-6
25. Murphy JJ (1999) Technical analysis of the financial markets. Prentice Hall, New York
26. Markowitz H (1952) Portfolio selection. J Finance 7(1):77. https://doi.org/10.2307/2975974

Index

L. Chen (ed.), *Clean Energy Research*, Green Energy and Technology,
https://doi.org/10.1007/978-3-031-99312-1

MIX
Papier aus verantwortungsvollen Quellen
Paper from responsible sources
FSC® C105338

If you have any concerns about our products,
you can contact us on
ProductSafety@springernature.com

In case Publisher is established outside the EU,
the EU authorized representative is:
Springer Nature Customer Service Center GmbH
Europaplatz 3, 69115 Heidelberg, Germany

Printed by Libri Plureos GmbH
in Hamburg, Germany